Clinical Pharmacology of Cerebral Ischemia

Contemporary Neuroscience

Clinical Pharmacology of Cerebral Ischemia, edited by **Gert J. Ter Horst**
and **Jakob Korf,** 1997

Molecular Mechanisms of Dementia, edited by **Wilma Wasco** and
Rudolph E. Tanzi, 1997

Neurotransmitter Transporters: Structure, Function, and Regulation,
edited by **Maarten E. A. Reith,** 1997

Motor Activity and Movement Disorders: Research Issues and Applications,
edited by **Paul R. Sanberg, Klaus-Peter Ossenkopp,** and **Martin
Kavaliers,** 1996

Neurotherapeutics: Emerging Strategies, edited by **Linda M. Pullan**
and **Jitendra Patel,** 1996

*Neuron–Glia Interrelations During Phylogeny: II. Plasticity
and Regeneration,* edited by **Antonia Vernadakis**
and **Betty I. Roots,** 1995

*Neuron–Glia Interrelations During Phylogeny: I. Phylogeny
and Ontogeny of Glial Cells,* edited by **Antonia Vernadakis**
and **Betty I. Roots,** 1995

The Biology of Neuropeptide Y and Related Peptides, edited
by **William F. Colmers** and **Claes Wahlestedt,** 1993

Psychoactive Drugs: Tolerance and Sensitization, edited
by **A. J. Goudie** and **M. W. Emmett-Oglesby,** 1989

Experimental Psychopharmacology, edited by **Andrew J. Greenshaw**
and **Colin T. Dourish,** 1987

Developmental Neurobiology of the Autonomic Nervous System, edited
by **Phyllis M. Gootman,** 1986

The Auditory Midbrain, edited by **Lindsay Aitkin,** 1985

Neurobiology of the Trace Elements, edited by **Ivor E. Dreosti**
and **Richard M. Smith**

Vol. 1: *Trace Element Neurobiology and Deficiencies,* 1983
Vol. 2: *Neurotoxicology and Neuropharmacology,* 1983

Clinical Pharmacology of Cerebral Ischemia

Edited by

Gert J. Ter Horst
Jakob Korf

University of Groningen, The Netherlands

Humana Press ✳ Totowa, New Jersey

© 1997 Humana Press Inc.
999 Riverview Drive, Suite 208
Totowa, New Jersey 07512

For additional copies, pricing for bulk purchases, and/or information about other Humana titles, contact Humana at the above address or at any of the following numbers: Tel.: 201-256-1699; Fax: 201-256-8341; E-mail: humana@interramp.com

This publication is printed on acid-free paper. ∞
ANSI Z39.48-1984 (American Standards Institute) Permanence of Paper for Printed Library Materials.

Cover illustration: From Fig. 3 in Chapter 2, "Imaging of Stroke," by Henk Stevens, Hugo M. L. Jansen, and Jacques DeReuck, and Jakob Korf.

Cover design by Patricia F. Cleary.

Photocopy Authorization Policy:

Printed in the United States of America. 10 9 8 7 6 5 4 3 2 1

Library of Congress Cataloging-in-Publication Data

Clinical pharmacology of cerebral ischemia / edited by Gert J.
    Ter Horst, Jakob Korf.
        p.   cm.
      Includes index.
      ISBN 0-89603-378-3 (alk. paper)
      1. Cerebral ischemia—Chemotherapy. 2. Cerebral cortex—Effect of
drugs on. I. Ter Horst, Gert J.. II. Korf, J.
      [DNLM: 1. Cerebral Ischemia—drug therapy. 2. Cerebral Ischemia—
physiopathology. WL 355 C6412 1997]
RC388.5.C58 1997
616.8'1—dc20
DNLM/DLC
for Library of Congress   96-43894
                                                                          CIP

Preface

Stroke is one of the main causes of death, occupying a position immediately following cardiovascular disease and cancer. However, many patients struck down by stroke survive and experience the consequences of the insult for many years, often at the emotional, motor, and intellectual levels. This group makes a large demand on both available therapeutic care and medical treatment. Thus far, neither warning signals nor treatments have been found that might lead to the prevention of the occurrence or a reduction of the clinical symptoms of a stroke, although much clinical and preclinical research has already been directed to the development of neuroprotective strategies. The main drawback in all clinical studies is the therapeutic window, which has been found to be very small. Most clinical trials were thus initiated far beyond the point of irreversibility. Preclinical stroke research has produced insights into the mechanisms of cell death and repair, and tested numerous pharmacological and other treatments that were often shown to be effective in a variety of animal stroke models. In the clinical setting, however, these treatments resulted in few, if any, demonstrable neuroprotective effects because they were initiated late.

The aim of *Clinical Pharmacology of Cerebral Ischemia* is to evaluate our current knowledge regarding aspects of neurodegeneration and protection in stroke. In the last decades, neurodegeneration after a stroke was considered to be a cause of necrosis. Nowadays, however, apoptosis or programmed cell death is believed to be an additional cell death mechanism after stroke, a process that involves activation of cell death genes and protein synthesis. Programmed cell death phenomena most likely play a critical role in the penumbra, a zone several centimeters wide that surrounds the area of acute damage, and also in distant areas exhibiting delayed neuronal death. Thus the penumbra and delayed neuronal death are the current targets in stroke treatment studies. In the penumbra, the preservation of neuronal integrity and function may be feasible. This should reduce the number of disabilities and the requests for care, and enhance the quality of life of stroke patients. To achieve neuroprotection one must have a thorough knowledge of fundamental processes that govern neuronal functioning and survival after cerebral ischemic events. In addition to the clinical symptoms, the severity and the development of stroke should be visualized and quantified with neuroimaging technologies. Various aspects of neuronal survival and death are discussed in *Clinical Pharmacology of Cerebral Ischemia* and wherever possible related to pharmacological interventions. The book includes chapters on calcium homeostasis, glutamate toxicity, the role of free radicals, glycine and hormones, and gene activation. There are chapters dedicated to neuroimaging of stroke, clinical trials, and the role of cerebral immune activation. For a long while, the brain has been considered an immune privileged site, but recent studies show that immune activation may be part of the series of events that take place after cerebral ischemic events. Thus, we have dedicated a chapter to animal stroke models for preclinical studies.

Clinical Pharmacology of Cerebral Ischemia is a detailed, informative volume intended to serve as a standard reference work of value not only to established neurologists and research scientists, but also to beginning investigators and biomedical students interested in stroke research and treatment.

Gert J. Ter Horst
Jakob Korf

Contents

Contributors

GREGORY W. ALBERS • *Stanford Stroke Center, Stanford University Medical Center, Stanford, CA*

STEVEN P. BUTCHER • *Department of Pharmacology, Fujisawa Institute of Neuroscience, University of Edinburgh, Scotland, UK*

GINEKE I. DE JONG • *Department of Animal Physiology, Groningen Center for Behavioral and Cognitive Neurosciences, University of Groningen, The Netherlands*

JACQUES H. A. DE KEYSER • *Department of Neurology, Groningen Center for Behavioral and Cognitive Neurosciences, University of Groningen, The Netherlands*

JACQUES DeREUCK • *Department of Neurology, University Hospital of Ghent, Belgium*

ELKAN GAMZU • *Cambridge NeuroScience, Cambridge, MA*

KONSTANTIN-ALEXANDER HOSSMAN • *Department of Experimental Neurology, Max-Planck Institute for Neurological Research, Cologne, Germany*

HUGO M. L. JANSEN • *Department of Neurology, University Hospital of Ghent, Belgium*

KENSUKE KAWAI • *Department of Neurosurgery, Teikyo University School of Medicine, Tokyo, Japan*

JOHN S. KELLY • *Department of Pharmacology, Fujisawa Institute of Neuroscience, University of Edinburgh, Scotland, UK*

JAKOB KORF • *Department of Biological Psychiatry, Groningen Center for Behavioral and Cognitive Neurosciences, University of Groningen, The Netherlands*

KENNEDY R. LEES • *University Department of Medicine and Therapeutics, Western Infirmary, Glasgow, Scotland, UK*

PAUL G. M. LUITEN • *Departments of Animal Physiology and Biological Psychiatry, Groningen Center for Behavioral and Cognitive Neurosciences, University of Groningen, The Netherlands*

TOMOHIRO MATSUYAMA • *Fifth Department of Internal Medicine, Hyogo College of Medicine, Nishinomiya, Japan*

LAURA J. McINTOSH • *Department of Biological Sciences, Stanford University, Stanford, CA*

KEITH W. MUIR • *Department of Neurology, Institute of Neurological Sciences, Southern General Hospital, Glasgow, Scotland, UK*

CSABA NYAKAS • *Haynal University of Health Sciences, Budapest, Hungary*

TIHOMIR P. OBRENOVITCH • *Department of Neurochemistry, Institute of Neurology, London, UK*

ANTONIO POSTIGO • *Department of Biological Psychiatry, Groningen Center for Behavioral and Cognitive Neurosciences, University of Groningen, The Netherlands*

ROBERT M. SAPOLSKY • *Department of Biological Sciences, Stanford University, Stanford, CA*

JOHN SHARKEY • *Department of Pharmacology, Fujisawa Institute of Neuroscience, University of Edinburgh, Scotland, UK*

HENK STEVENS • *Department of Biological Psychiatry, Groningen Center for Behavioral and Cognitive Neurosciences, University of Groningen, The Netherlands*

BAUKE STUIVER • *Department of Animal Physiology, Groningen Center for Behavioral and Cognitive Neurosciences, University of Groningen, The Netherlands*

KIYOSHI TAKAGI • *Department of Neurosurgery, Teikyo University School of Medicine, Tokyo, Japan*

AKIRA TAMURA • *Department of Neurosurgery, Teikyo University School of Medicine, Tokyo, Japan*

GERT J. TER HORST • *Department of Biological Psychiatry, Groningen Center for Behavioral and Cognitive Neurosciences, University of Groningen, The Netherlands*

DAVID C. TONG • *Stanford Stroke Center, Stanford University Medical Center, Stanford, CA*

CHRISTOPH WIESSNER • *Department of Experimental Neurology, Max-Planck Institute for Neurological Research, Cologne, Germany*

MIDORI A. YENARI • *Stanford Stroke Center, Stanford University Medical Center, Stanford, CA*

1

Stroke

Prevalence and Mechanisms of Cell Death

Gert J. Ter Horst and Antonio Postigo

1. STROKE

Cerebrovascular accident (CVA) is a clinical definition used to describe symptoms of an acute neurological disorder caused by disturbance of the cerebral blood supply. Intracerebral and subarachnoid hemorrhages account for approx 20% of CVAs and 80% are of the ischemic type. Stroke defines all conditions in which the duration of the CVA symptoms exceed 24 h. Ischemic CVAs exhibiting short duration of neurologic dysfunction usually not exceeding 10–15 min are transient ischemic attacks (TIAs). A TIA may be a warning sign of an impending stroke. Adequate supply of oxygen and glucose are necessary for the proper functioning of the brain. Minor changes in the cerebral oxygen and glucose supply may invoke damage that is irreversible because the brain has very limited repair capabilities. The most important disadvantage to neural repair is the inability of the neurons to divide. This implies that all neuronal loss caused by CVA is irreversible and that it permanently affects the functioning of the brain. Prevention of a first stroke or recurrent strokes through reduction of known risk factors is the most effective strategy for controlling this devastating disease, but for the near future, such total elimination of stroke is unlikely. Therefore, treatment strategies to limit ischemic brain injury must be identified and developed.

1.1. Incidence

The incidence of stroke lies below that of cancer and myocardial infarction at approximately 1 per 1000 people *(1–4)*. However, it is age and sex dependent *(5)*. In the age group of 80 yr, the incidence of stroke will reach values of 20 per 1000 people *(5,6)*. The incidence rate is slightly higher among males in all age groups; mean annual first-ever stroke incidence rates age-adjusted to the world population are 1.32 in males and 0.77 per 1000 people in females *(5,7)*. Almost 20% of the people that suffer a stroke die in the first week, 33% do not survive the first year *(3,5)*. These mortality rates are high, but two-thirds of the stroke patients survive and will experience the permanent physical and mental disabilities caused by the neuropathology for many years. Not only the patient, but also the family and the society are affected because very often the victim is stripped of mobility, language, and intellect. Very simple everyday activities like feeding and toiletting become the responsibility of the spouse,

From: Clinical Pharmacology of Cerebral Ischemia *Edited by: G. J. Ter Horst and J. Korf Humana Press Inc., Totowa, NJ*

1

family members, or other care providers. In the Netherlands, approx 70% of all first-ever stroke patients are admitted to the hospital. After the first month, 8% of the stroke survivors are still hospitalized, 75% have returned home, and 19% are discharged to a nursing home or rehabilitation center. The average duration of stay of stroke patients in nursing homes or rehabilitation centers is approx 470 d, and after 6 mo, 81% of all stroke survivors in the Netherlands live at home *(5)*. The annual cost of the acute and chronic care of stroke patients is approx $30 billion in the United States *(1)*, $1 billion in the Netherlands *(5)*, and $154 million in New Zealand *(8)*. Annual per person costs in the Netherlands are estimated at $42,930 for women and $37,630 for men, of which 47% is related to treatment of comorbid diseases, including stroke-related cardiovascular diseases.

1.2. Etiology

Hemorrhagic strokes may be situated intra- or extracerebrally. Usually, the definition refers to a condition in which bleeding has occurred in the basal ganglia and the capsula. Hypertension, hematomas, or tumors are possible causes of intracerebral hemorrhage. Causative factors for a subarachnoid or subdural hemorrhage, respectively are a basal cerebral artery aneurysm rupture and cranial trauma *(6)*.

The causes of ischemic stroke are numerous and include large artery atherosclerosis, small vessel occlusion, embolisms *(9)*, and thrombosis. Also, several risk factors have been identified, like the use of alcohol and oral contraceptives, migraine *(10)*, diabetes mellitus, systemic lupus erythematosis *(11)*, cardiac valvular disease *(12–14)*, age, female sex *(15)*, smoking and hypertension *(16)*, vasospasm, a previous stroke or transient ischemic attack *(17,18)*, and infectious diseases *(13,14,19)*. Hypertension, age, TIAs, and acute alcohol intoxication are associated with the highest relative stroke risk *(20)*. Increased systolic blood pressure (>120 mmHg) is more strongly related to ischemic stroke than increased diastolic blood pressure *(21,22)*. Increasing age is a second important risk factor. At the age of 65, the incidence rate in the Netherlands is 0.003% in women and 0.005% in men, and then it steadily increases to reach values of 2.2 and 1.8% in the over-85-yr-old group *(5)*. The slightly higher incidence in women is most likely associated with the higher number of women older than 65 years in the general population. Also, TIAs occur more frequent in women than in men, with estimated prevalence rates in the Netherlands of 5.6 and 3.4 per 1000, respectively *(23)*. In the United States, regional differences were found in TIA prevalence, with estimates varying from 0.2 per 1000 in the Lehigh Valley (PA) area *(24)* to 2.4 per 1000 in the Rochester, NY area *(25)*. In approx 50% of the reported TIA cases, it is followed by a CVA within a period of 5 yr *(26)*.

In a recent study, it was shown that infection was significantly more common among stroke patients than control subjects within 4 wk and within 1 wk before cerebral ischemia or examination *(13)*. In particular, respiratory tract, gastrointestinal, and urinary tract infections were all more frequent among patients than among control subjects. Bacterial infections significantly increased the risk of cerebrovascular ischemia. Although viral infections were also more common in patients than controls, but in the statistical analysis they were not identified as a significant risk factor. Recent infection was also clinically associated with more severe stroke, which may be related to hyperthermia, a well known deleterious factor in ischemic stroke. Grau and coworkers

(14) have shown that cortical infarcts of the middle cerebral artery (MCA) predominate in patients reporting a recent infection. Such infarcts appear to be owing mostly to embolic mechanisms, which suggests a relationship between inflammatory mechanisms, regional cerebral endothelial damage *(27)*, and stroke. This aspect, as well as the role of the immune system in progression of the cerebral damage after ischemia will be discussed in Chapter 10.

1.3. Pathogenesis

Cerebral ischemia results from decreased or interrupted blood supply, which leads to reduced availability of glucose and oxygen in the territory of the affected vascular bed(s). This will cause a cellular energy crisis. As a final strategy to prevent death, the cells in the ischemic area initiate anaerobic glycolysis. However, the energy obtained with anaerobic glycolysis can not totally compensate for the energy shortage. It provides a very small fraction of the amount of energy needed for the neuronal survival. Moreover, it leads to production of lactate. Shortage of energy interrupts the activity of the cellular ion pumps and, therefore, the intracellular calcium and extracellular potassium concentrations increase, within 1–2 min after the onset of the ischemia. Thereafter, the extracellular concentrations of neurotransmitters increase, in particular of glutamate and dopamine, and edema occurs. To some extent, this early damage is reversible. Continuation of the ischemic condition, however, rapidly leads to extensive irreversible damage resulting from a so-called "ischemic cascade" (*see* Section 2.).

The mechanisms described above occur not only in the ischemic core but also threaten the areas around the infarct that are bombarded with waste products from the ischemic cascade and confronted with a decreased blood flow *(1)*. This leads to formation of a so-called "ischemic penumbra;" an area with reduced neuronal functioning around the infarct. The reduction or prevention of the cell death in the ischemic penumbra is the main target of pharmacological intervention studies. Recovery of the neuronal functioning in the ischemic penumbra may account for spontaneous improvements of symptoms after stroke.

1.4. Symptoms

Well known clinical symptoms following stroke are paralysis, unilateral disturbance of autonomic functioning *(28–30)*, aphasia *(31)*, dizziness or vertigo *(32)*, and impairment of vision and balance. Acute cardiovascular side effects are the most important reason for hospitalization of stroke patients. Right hemisphere stroke has been associated with occurrence of supraventricular tachycardia and left hemisphere stroke with ventricular arrhythmias *(33)*. Cerebral infarction may reduce cardiac autonomic activity because it affects the suprasegmental stimulation of the primary autonomic nuclei in the brainstem and thoracic spinal cord *(34–36)*. Cortical sites participating in cardiovascular regulation, some receiving blood supply from the middle cerebral artery, have been localized in the cingulate, medial prefrontal, and insular cortices *(36–38)*.

2. PATHOPHYSIOLOGY OF CELL INJURY

Several mechanisms have been proposed to explain the pathophysiology of ischemic cerebrovascular disease, including increased excitotoxicity, calcium overload, free radical formation, immune activation, inhibition of protein synthesis, and alterations in gene expression (for review, *see* ref. *1* and Chapter 8). There are two fundamental types of cell death, namely necrosis and apoptosis *(39)*. Necrosis, or degenerative cell death, is essentially accidental in its occurrence and is the outcome of severe injurious changes in the environment of the cells. Apoptosis, or programmed cell death, is considered an active process of gene-directed self destruction with a biologically meaningful, homeostatic function *(39,40)*. In the central nervous system after ischemia immediate, early, and delayed neuronal death is recognized. Immediate neuronal damage most likely is of the necrotic type, whereas apoptosis may be the main cause of the delayed neuronal death. Necrosis often involves cell swelling, whereas apoptosis is characterized by nuclear and cytoplasmic condensation *(39–41)*. In this section, we review mechanisms that may be responsible for ischemic, necrotic cell death and/or the initiation of the intracellular mechanisms leading to apoptosis. Our current knowledge about the sequence of intracellular events associated with neuronal apoptosis also will be discussed.

2.1. Ischemic Cascade

The ischemic cascade is a series of events triggered by a reduced cerebral glucose and oxygen supply that are responsible for the cell injury. Various aspects of the ischemic cascade will be reviewed in depth in the following chapters of the book.

Generally, the evolution from reversible to irreversible damage in necrosis involves progressive derangements in energy and substrate metabolism. The energy needs of the brain are supplied by aerobic metabolism of glucose and oxygen in the mitochondrial respiratory chain in which adenosine diphosphate (ADP) is phosphorylated to adenosine triphosphate (ATP). Most of the ATP thus generated is used for maintenance of the intracellular homeostasis and ATP-driven membrane ion pumps for the stabilization of the transmembrane concentration gradients of sodium, potassium, and calcium, in particular. Neuronal impulse conduction and synaptic functions rely on these concentration gradients. During ischemia, a very small amount of ATP can be produced by anaerobic glycolysis. However, the glycogen stores in the brain are small and are depleted within minutes after the onset of ischemia. Then, uncompensated leakage of ions across the cell membrane occurs that results in cell swelling (edema) and a persistent membrane depolarization that is accompanied by release of the neurotransmitters glutamate and dopamine. Derangements of the calcium homeostasis in particular are considered important for the progression of the cell injury. Glycogen is metabolized to lactic acid, which causes progressive intracellular acidosis and the subsequent release of the organic bound iron. Free iron may catabolize the formation of free radicals—highly reactive oxygen species with unpaired electrons in the outer orbit. Eventually, these processes result in necrosis and fragmentation of the cell membranes or apoptosis and disintegration of the cell. The neuronal debris is removed by microglial cells, the immunocompetent cells of the brain.

2.1.1. Edema

Edema is a condition of excess accumulation of fluids. Ischemic brain edema has a vasogenic and a cytotoxic component *(42)*. Vasogenic edema is the most common form of brain edema and is attributed to increased permeability of the brain capillary endothelial cells. Intracellular swelling of neurons, astrocytes, and endothelial cells is called cytotoxic edema. Cytotoxic edema is associated with a reduction of the extracellular space and caused by failure of the ATP-dependent sodium–potassium pump. Failure of this ATP-driven pump allows the influx of Na^+, which is accompanied by the influx of Cl^- and water, thereby causing cell swelling *(1,39)*.

Vasogenic edema is associated with impaired reperfusion of the ischemic tissue *(43,44)*, which has been attributed to a variety of factors, such as increased blood viscosity, endothelial and perivascular glial swelling *(45)*, endothelial blebs protruding into the capillary lumen *(46)*, platelet aggregation *(47,48)*, and vessel obstruction by leukocyte attachment to the endothelial surface *(49,50)*. Alterations of platelet function or vascular tone and permeability are ascribed to liberation of unesterified fatty acid metabolites from the ischemic tissue, in particular to the metabolites of arachidonic acid, i.e., the prostaglandins and thromboxanes *(51)*. Trials in experimental animals in which thromboxane formation was suppressed after ischemia with cyclooxygenase or thromboxane synthase blockers showed a reduction of brain edema *(52)*, enhanced reperfusion *(52–54)*, and reduced neuronal death *(55)*. Leukocyte attachment to the vascular surface involves altered gene regulation in the endothelial cells and expression of cell surface glycoproteins of the selectin and immunoglobulin families that enable the docking of the leukocytes. The docking is regulated by leukocyte cell-surface molecules of the integrin family *(56,57)*. Endothelial expression of the surface glycoproteins, intercellular adhesion molecule-1 (ICAM-1), and E- and P-selectin, facilitate leukocyte adhesion to the endothelium and are the ligands for the integrins on the surface of the leukocytes. The expression of endothelial E-selectin and ICAM-1 mRNA is up regulated in the affected vascular beds between 6 h and 5 d after ischemia and shows peak values after 12 h *(56)*. Administration of anti-ICAM-1 antibodies immediately upon reperfusion of the ischemic tissue gave a significant reduction of the volume of the lesion after ischemia in experimental animals *(57,60)*. Expression of ICAM-1 and E-selectin is found exclusively in vascular endothelial cells after stimulation with proinflammatory cytokines, in particular tumor necrosis factor-alpha (TNFα) and interleukin-1 *(58,59)*, which may also explain the increased stroke risk after bacterial infection *(13,14)*. Antiinflammatory agents like the glucocorticosteroids are successfully employed in the clinical treatment of vasogenic brain edema. These effects may be mediated by annexins, or lipocortins *(67)*; recently discovered inhibitors of phospholipase A2 *(61,65,66)*. Administration of the synthetic glucocorticosteroid methylprednisolone, has been shown to induce expression of annexins in the cerebral endothelial cells *(63,64)*. Annexins possess Ca^{2+} binding sites and may therefore contribute to stroke damage reduction not only by their potent antiinflammatory effects but also by improving the cellular calcium storage capacity *(62,68)*.

2.1.2. Calcium Overload

Due to tight regulatory controls involving voltage-gated and N-methyl-D-aspartate (NMDA)-linked, receptor-operated calcium channels *(74)*, endoplasmatic reticular and mitochondrial sequestration, Ca^{2+}-ATPase activity and Na^+/Ca^{2+} exchange, a 10,000-fold concentration gradient is maintained between the intra- and extracellular free Ca^{2+} concentration *(1,69,70,73)*. Calcium ions subserve a ubiquitous role in the organization of cell function as intracellular messengers and regulators of neurotransmitter secretion, electrical activity, cytoskeletal function, cell metabolism, and gene expression. Excessive influx of calcium into the cell under ischemic conditions is considered a major mechanism of cell injury and death *(69,72)*. Two phases of increases of the cytosolic free Ca^{2+} have been documented. Early Ca^{2+} influx across the plasma membrane that has been associated with altered function of the NMDA-linked calcium channels and other calcium transport systems, and the release of Ca^{2+} from intracellular stores such as the mitochondria and the endoplasmatic reticulum due to inhibition of the ATP-driven membrane ion pumps *(69,71,73,74)*. Also, the depletion of Mg^{2+} ions can promote Ca^{2+} accumulation *(75)*. The late phase of calcium influx is a consequence of nonspecific membrane leakage as part of the physical disruption of the plasma membrane during the process of cell swelling (*see* Section 2.1.1.).

The progressive increase of cytosolic free Ca^{2+} ions may have several deleterious effects, including activation of phospholipases and proteases, which promote the membrane damage, activation of ATP-ases, that sustains the cellular energy crisis, and mitochondrial accumulation of calcium. The accumulation of calcium in mitochondria strengthens the inhibition of the ATP production that was already instigated by the lack of glucose and oxygen. There is also some evidence for interrelationships of gene activation (*see* Chapter 8), in particular the expression of the immediate early genes c-fos, c-jun, c-myc, *Erg*-1, and jun-B, and changes in the intracellular and possibly intranuclear calcium concentrations *(76,77)*. For a review of the role of Ca^{2+} in the pathogenesis of stroke and efficacy of calcium uptake inhibitors in clinical trials and animal stroke models, we refer to Chapter 4.

2.1.3. Exicitotoxicity

The intracellular accumulation of Ca^{2+} during cerebral ischemia is attributed largely to release of the excitatory amino acid glutamate into the synaptic cleft and extracellular space from the presynaptic and astroglial storage pools. Moreover, owing to the breakdown of the ion gradients across neuronal and glial membranes in ischemia, the glutamate uptake mechanisms are inhibited that prevent the clearance of glutamate from the extracellular space. These findings have stimulated research of glutamate-induced cytotoxic mechanisms and neuron/glial interactions, and also the development of therapeutic strategies aimed at inhibition or prevention of glutamate-induced cell death.

2.1.3.1 GLUTAMATE

The neurotoxic action of the excitatory amino acid glutamate arises from its capacity to trigger a pathophysiological chain of events when it acts continuously and abusively on its receptors *(80,81)*. These receptors are divided on the basis of the amino acid sequence homology and pharmacological properties into the ionotropic and metabotropic subgroup. The ionotropic group is identified by their selective affinity for the agonists: NMDA, α-amino-3-hydroxy-5-methyl-4-isoxazole propionate (AMPA), and kainate. All members of the ionotropic glutamate receptor subgroup are receptor-channel complexes that regulate the transmembrane passage of Na^+, K^+, and Ca^{2+} ions. The NMDA channels possess Mg^{2+} and glycine binding sites. Extracellular Mg^{2+} blocks NMDA receptors in a voltage-dependent manner, it relieves the inhibitory effects of protons (pH) on this receptor *(78,79)*, and increases NMDA receptor affinity for glycine (for review, *see* Chapters 5 and 6). In a few animal studies, the therapeutic potential of 7-chlorokynurenic acid, a potent antagonist at the glycine-modulatory site on the NMDA receptor, in terms of neuroprotection following transient forebrain ischemia has been shown *(85,86)*. Moreover, this glycine antagonist exhibited free radical scavenger properties and inhibited lipid peroxidation *(86)*. The latter observations support the notion that a positive feedback may exist between the activation of glutamate receptors and free radical formation and that this interaction is responsible for the generation of ischemic brain damage *(82,87)* (*see* Chapter 7). Alternatively, neuroprotection for glutamate-induced cell death has been achieved in animal models with the selective NMDA-receptor antagonist MK-801, given either alone *(88)* or in combination with the Ca^{2+} entry blockers nicardipine *(89)* or nimodipine *(90)*, or the inhibitory neurotransmitter γ-aminobutyric acid (GABA) agonist muscimol *(91)*. The use of MK-801 in human stroke therapy, however, should be considered with caution. The MK-801 may cause side-effects such psychosis because it has been shown to interact with the phencyclidine (PCP, "angel dust") receptor; a drug known as an inducer of schizophreniform psychosis. The PCP receptor represents a site within the NMDA receptor channel complex *(92)*.

The members of the metabotropic glutamate receptor subgroup are G-protein coupled receptors that act through intracellular second messenger systems including inositol triphosphate, cyclic adenosine monophosphate, and calcium. The common feature of all glutamate receptors is that they elicit an increase in intracellular calcium. Studies regarding glutamate-induced cell death have focused for a long time solely on the NMDA-gated ion channel, but it is now recognized that all glutamate receptor types participate in derangement of calcium homeostasis after ischemia, albeit by distinct mechanistic routes.

The morphological characteristics of excitotoxic injury in vivo are consistent with a necrotic type of cell death. However, recent in vitro and in vivo studies have provided evidence that some neuronal populations may die via apoptosis *(82–84)*. Ankarcrona and coworkers *(84)* have reported that after exposure to glutamate, the mode of neuronal death may be determined by mitochondrial function; late onset formation of apoptotic nuclei and chromatin fragmentation in cerebellar granular cells after glutamate exposure was associated with the early recovery of mitochondrial membrane potentials and energy levels. Because programmed cell death involves protein synthesis it is energy dependent. The glutamate-induced programmed cell death may involve protein synthesis, that is regulated by expression of the immediate early genes c-fos, c-jun, and jun-B *(93,94)*.

2.1.3.2. NEURON/GLIA INTERACTIONS

Interactions between neurons and glial cells may be critical in determining the outcome of hypoxic–ischemic injury. Glial cells are the most numerous cell type in the central nervous system that fulfill roles like guidance of migrating cells, neurite outgrowth in development, the control of the ion composition of the extracellular space, and neurotransmitter uptake and inactivation in the adult brain. Moreover, glia may provide substrates for the neuronal energy metabolism *(95,96)*. The glial cell population is comprised of astrocytes, oligodendrocytes, and microglia cells. Astrocytes posses neurotransmitter receptors for example for glutamate *(97,98)* and serve a role in glutamate uptake to maintain low extracellular levels of this cytotoxic neurotransmitter *(1,99,100,102)*. After CVAs, however, a derangement of astrocyte functioning has been observed that may affect glutamate uptake mechanisms, as demonstrated in vitro

(99,101,102). In addition to the increased release of glutamate from the presynaptic storage vesicles, this inhibition of glutamate uptake may be one of the underlying causes for the increased extracellular levels of this excitatory amino acid (EAA) neurotransmitter after CVAs.

Derangement of neuron/glial interaction may contribute otherwise to damage development after CVAs. Recently, we have studied the astrocyte and microglial response after transient hypoxia/ischemia in rats (in preparation). Astrocytes disappeared immediately from the acutely infarcted areas, as could be revealed with glial fibrillary acidic protein (GFAP) immunoreactivity; a marker selective for astrocytes *(103)*. In the penumbra, however, the number of astrocytes increased during the following days, thus forming a glial scar *(104)*. At the same time, microglial activation and infiltration was observed in the core of the infarct. This microglial response was terminated several days later, simultaneously with a regrowth of astrocyte processes into the necrotic tissue from the glial scar. Astrocyte GFAP immunoreactivity was preserved much longer in areas that exhibited early and delayed neuronal damage, and only in areas that showed loss of GFAP immunoreactivity, microglial activation occurred. Microglial cells remove the neuronal debris, a process that involves formation of free radicals, proteases, and cytotoxic cytokines (Chapters 7 and 10) among other mechanisms, and they can secrete nitric oxide and EAAs *(105)*. These observations suggested that preservation of astrocyte functioning is critical for neuronal viability and survival after hypoxic-ischemic injury. Astrocytes have been shown to produce various neurotrophic factors; polypeptides that support the growth, differentiation, and survival of neurons *(106)*. It has become evident from in vitro and in vivo research that neuronal viability depends upon the collaboration of several growth factors from the neurotrophin (nerve growth factor [NGF], brain derived neurotrophic factor [BDNF], neurotrophin[NT]-3,4/5), the insulin-like, fibroblast or epidermal growth factor families *(107–110)*. The expression of neurotrophic factor mRNA has been shown to significantly increase in the affected brain areas within hours after ischemia *(111,112)*, and was preceded by induction of neuronal and glial c-fos, c-jun, jun-B, jun-D, *Krox 24*, *zif/268*, and *nur 77* mRNA expression *(111,113)*. This implies a possible role for these immediate early genes in initiation of neurotrophin gene expression in astrocytes after hypoxic-ischemic insults *(111)*. The effects of several neurotrophins on neuronal degeneration in tissue cultures of murine cortical cells have been examined. The neurotrophins BDNF, NT-3, and NT-4/5 attenuated the apoptotic cell death induced by the calcium channel antagonist nimodipine, but they potentiated the necrotic type of cell death induced by ischemia or NMDA *(114)*. These observations imply that effects of neurotrophins on neuronal viability may depend strongly upon hitherto unclarified microenvironmental conditions in the hypoxic-ischemic areas. This exemplifies that neuron/glia interaction in relation to release of neurotrophins needs to be investigated more thoroughly, in vitro in coculture systems and in vivo both in animal models and patients.

2.1.4. Reactive Oxygen Species

Reactive oxygen species or free radicals, like the superoxide ($°O_2-$) and peroxynitrite ($°NO_2-$) anions, hydrogen peroxide (H_2O_2), and the hydroxyl radical ($°OH-$), are defined as highly reactive molecules with an unpaired electron in the outer orbit. They are not biological curiosities but produced under aerobic conditions as byproducts of the biochemical machinery of the cell *(115)*. Microglia and macrophages employ free radicals for the removal of the cellular debris after brain injury *(105)*. In all eukaryotic cells, defense mechanisms have evolved that serve to protect the cell against the damaging effects of free radicals. The cellular defense involves both enzymatic and nonenzymatic free radical inactivation or scavenging. Superoxide dismutase, glutathione peroxidase, and catalase comprise the enzymatic defense, and the vitamins E (α-tocopherol) and C are the most well known members of the nonenzymatic defense system. Free radical-mediated cell damage involves, for example, peroxidation of proteins and membrane lipids, and DNA strand breaks.

Reperfusion after an hypoxic/ischemic insult is considered an important trigger for the generation of excess free radicals in the affected parts of the brain *(116–118)*. Also, the release of dopamine from presynaptic storage vesicles, which has been observed after hypoxic/ischemic injury *(119)*, may contribute to the generation of free radicals. Dopamine is metabolized by monoamine oxidase B (MAO-b) to hydrogen peroxide. This may be an alternative source of free radicals in dopaminergically innervated areas of the brain like the striatum. Moreover, iron and manganese ions accumulate in the ischemic brain that catalyze in the so-called Fenton reaction the conversion of hydrogen peroxide to the hydroxyl radical; the most noxious radical species known *(115,120)*. Other free radicals may derive from the activated microglial cells that accumulate in the injured areas *(105)*, and from the endothelial cells that produce nitric oxide; the endothelium derived relaxing factor *(121,122)*. Nitric oxide is converted to the peroxynitrite radical when it reacts with the superoxide anions *(123)*, which are abundantly produced in the hypoxic/ischemic tissue upon reperfusion.

The enzymatic free radical defense has evolved to a well-balanced system in which superoxide radicals are converted by superoxide dismutase (SOD) to hydrogen peroxide. Subsequently, the hydrogen peroxide is removed in the fore- and midbrain by glutathione peroxidase (GPx) *(124)* and in the hindbrain structures by catalase (Cat) *(125)*. Glutathione peroxidase catalyzes the conversion of glutathione (GSH) to glutathione disulfide (GSSG) that in turn is reduced by glutathione reductase *(126)*. The radical scavenging enzymes can inactivate most of the free radicals that are produced by the biochemical machinery of the cell under physiological conditions. However, the intracellular amounts of the scavenger enzymes are not sufficient to counteract the excess formation of reactive oxygen species after hypoxia/ischemia. This has been shown to induce in the affected areas of the brain a selective up-regulation of scavenger enzyme mRNAs and proteins *(127)*. For example, increased Cu/Zn-SOD and GPx mRNA expression has been found in the hippocampal CA1 area *(124,128)*, a region exhibiting delayed neuronal death after hypoxic/ischemic injury *(129)*. This illustrates that the gene expression in the CA1 pyramidal cells is affected by the insult, which is part of the mechanisms that lead to programmed cell death (Section 2.2.).

To reduce free radical-mediated hypoxic/ischemic injury, several pharmacotherapeutic approaches have been used to increase either the free radical scavenger enzyme content or the scavenging capacity of the brain. Polyethylene-glycol conjugated SOD (PEG-SOD) *(130,131)* and Cat *(131)*, Ebselen *(132)*, and phenyl-t-butyl-nitrone *(133)* are some examples of treatments that have effectively reduced the damage size in animal models for hypoxic/ischemic injury. Damage reduction also has been observed in transgenic mice that overexpress Cu/Zn-SOD, however, in animals that showed more then fivefold increased Cu/Zn-SOD expression, the hypoxic/ischemic injury increased *(134)*. Similar adverse effects have been reported after high-dosage iv and intracerebral donations of PEG-SOD *(130,131)*. These investigations illustrate the need for a balanced free radical scavenger enzyme activity cascade. Increased SOD activity without a concomitant increased GPx or Cat activity will lead to accumulation of hydrogen peroxide, which is a product for the Fenton reaction that generates the noxious hydroxyl radicals. L-Deprenyl, an inhibitor of MAO-b at high dosages, has been shown to reduce hypoxic/ischemic brain injury both prophylactically *(135)* and therapeutically *(136)*. This effect has been attributed not only to its MAO-b inhibiting effect but also to its capacity to increase simultaneously the SOD and Cat activity in the brain at low dosages *(137)*. Amounts and activity of GPx were not affected by the L-deprenyl treatment. These observations in animal models are promising but they also show the very small therapeutic window after hypoxic/ischemic injury.

2.1.5. Immune Activation

The brain has long been considered an immunologically privileged site because the blood-brain barrier was thought to be nonpermeable for mediators of inflammation and cells of the immune system. Recent investigations, however, have demonstrated that hypoxic/ischemic brain injury involves infiltration of leukocytes *(138,139)*, and possibly also T-lymphocytes *(140,141)*, macrophage/microglia activation *(139,140,142)*, induction of adhesion molecules on the cerebrovascular endothelium *(141,143)*, and expression of proinflammatory cytokines by brain microglia *(105,144–147)*, astrocytes *(145,148)*, and possibly some neurons *(148,149)*. The immunosuppressant drug FK506 has been shown to reduce hypoxic/ischemic injury in animal models *(150)*.

Research in the field of immune activation and stroke is mainly dedicated to mechanisms of neurodegeneration and protection that involve the mediators of inflammation, the cytokines. This rapidly expanding family of peptides includes the interleukins (IL), interferons, tumor necrosis factors (TNF), and the growth and cell stimulating factors. Effects mediated by interleukin-1 (IL-1) and TNFα are the best characterized cytokine responses to cerebral ischemia/reperfusion. The IL-1 mRNA and protein expression has been found to increase rapidly within the fields of damage after hypoxic/ischemic injury *(151,152)*. The IL-1 may induce beneficial effects by synthesis of NGF *(153)* which promotes neuronal survival, regeneration, and neurite outgrowth *(107,110)*. But, IL-1 expression may have adverse effects too. Interleukin-1 has been shown to stimulate astrocyte proliferation, brain edema, leukocyte infiltration, and endothelial expression of adhesion molecules *(154)*. It elicits release of arachidonic acid, nitric oxide, β-amyloid precursor protein, and corticotrophin releasing factor *(see* Chapter 9), all of which have been implicated in neurodegeneration, and stimulates the production of neurotoxic substances by glia (for review, *see* ref. *152)*. A significant reduction of infarct volume after MCA-occlusion in rats has been obtained with intracerebral administration of the recombinant IL-1 receptor antagonist *(152,155)*. These experiments illustrate not only the detrimental effects of IL-1 in the development of hypoxic/ischemic injury, but also they reveal possibilities for new therapeutic approaches in stroke patients.

The TNFα has been shown to generate in high doses adverse effects like selective cerebral microvascular injury *(27,156,157)*, increased expression of other proinflammatory cytokines like IL-1, IL-6, and interferon, and cytotoxic effects that may induce either necrotic or apoptotic cell death, depending on the state of the cell *(158)*. The TNFα-mediated apoptotic cell death may involve a 45 kD membrane-associated protein called APO-1/FAS; a member of the nerve growth factor and tumor necrosis factor receptor superfamily *(159,160)*. The FAS antigen is a well known apoptosis-associated cell surface molecule that has been found to be rapidly induced in the hippocampal CA1 area during the first 24 h of recirculation after ischemia *(161)*. In vitro, FAS expression has been induced in astrocytes and possibly microglia by interferon-γ and TNFα *(162)*. TNFα may exhibit beneficial effects at low doses, in regulating immune responses, growth and differentiation, and protecting cells against excitotoxic damage (148) and oxidative stress after hypoxia/ischemia. The latter neuroprotective mechanisms involve induction of gene and protein expression of for example heat shock protein *(72,163)*, which may play a role in development of ischemic tolerance *(164)*, and the free radical scavenger proteins, MnSOD and Cu/ZnSOD *(147)*. For a review of the role of immune activation in development of cerebral hypoxic/ischemic injury and possibilities for therapeutic interventions in patients, we refer to Chapter 10.

2.2. Molecular Mechanisms of Neuronal Apoptosis

In numerous fields of research, ranging from developmental neurobiology to cancer biology and immunology, cell death has drawn a lot of attention. It has become increasingly clear that cells do not always whither away when they are exposed to a lethal factor, but in most occasions die by a process called apoptosis. This term was originally coined for the type of cell death that shows a well-defined set of morphological events *(165)*. Sometimes, the term programmed cell death (PCD) is used to describe the forms of apoptosis that would require the activation of an intrinsic program, i.e., the synthesis of "killer proteins," for regulated suicide *(166)*. However, there also appear to be forms of apoptosis that rely on constitutively expressed proteins *(167)*. As proposed by Steller *(168)* these differences in observations may be unified by the hypothesis that apoptosis requires modulation of constitutively expressed proteins by inducable, newly expressed proteins in response to different stimuli. Therefore, as in most of the literature in this field, we use these terms interchangeably.

The so-called necrotic cell death is a pathological form of cell death that results from an acute injury to the cell. As a result of this injury, cells swell, lyse, and the expulse of cytoplasmatic material into the extracellular environment. This may lead to an immune response owing to the possible toxicity of the cellular content. Apoptosis is distinguishably different, since it results in the neat removal of the cell. First, after the death signal, the nucleus and cytoplasm condense and the membrane starts to bleb. The chromosomal DNA is then degraded into large (50–300 kb) and often into smaller (180–200 bases) fragments *(169)*. By this time, the membrane has completely lost its integrity and pieces of membrane engulf the remains of both cytoplasm and nucleus to form so-called apoptotic bodies. These bodies are phagocytosed and subsequently digested by macrophages and neighboring cells.

In the last 5 yr, many different molecules have been postulated to be involved in the apoptotic program. Immediate early genes such as c-fos *(170)* and c-jun *(171,172)* are in some way necessary for PCD, as well as cell cycle regulators like p53, c-myc, Rb-1, E1A, cyclin D1 and p34^{cdc2} (173–178). Furthermore, the identification of supposed killer and rescue proteins has progressed rapidly. However, in many cases the way in which these factors act in an apoptotic program, that is responsive to many factors in many different cells is still largely unknown. Even more, the same factor may make one cell commit suicide, whereas the neighboring cell(s) stay alive. Rubin et al. *(179)* made it clear which theoretical models are thinkable. Different elements may interact serially from inducing signal to death effector molecules, or in parallel. Alternatively, it may be that different inducers and their downstream molecules converge at the point of one integrating effector, which would be responsible for the marked morphological features. From this integrating effector onward it could become possible to establish the typical apoptotic phenomenon in different cell types under different circumstances. Together with Rubin et al. *(179)*, we argue that this parallel convergent model may be closest to the truth, although a common effector molecule of apoptosis has not yet been found.

We highlight some recent developments in the molecular biology of apoptosis, with emphasis on its occurrence in and relevance for neurobiology. In addition, present and future ways of intervening in apoptotic cell death shall be discussed.

2.2.1. Genes Involved in PCD in the Nematode Caenorhabditis elegans

Programmed cell death in *C. elegans* can be divided in four stages: decision to die, execution of death, engulfment of the cell by phagocytes, and subsequent degradation of the cell remains. Mutational analysis showed that 14 genes are involved in this process, three of which are implicated in the execution of death (168). The cell death defective-3 *(ced-3)* and *ced-4* genes are required for cell death, whereas the third gene, *ced-9*, protects cells from undergoing apoptosis. The protein encoded by the ced-4 gene is characterized and does not seem to be similar to any other known protein *(180)*. The ced-3 protein, however, appears to be a member of a cysteine protease family. Some of the known members of this family are the interleukin 1-β converting enzyme (ICE) *(181,182)*, nedd-2/Ich-1 *(183)*, and CPP32 *(184)*. In order to confirm the suspected role of these proteins in the apoptotic process, ced-3, ICE, and Ich-1 were overexpressed in several human and rat cell lines *(183,185)*. This resulted in the induction of cell death in these mammalian cells. However, it still is uncertain what the exact physiological role of ICE and ICE-like proteases in the apoptotic process is, and if they exert this role in every apoptotic cell death. That these proteases do have an important role seems to be supported by the finding that expression of the cowpox virus *crmA* gene, which is a potent inhibitor of ICE-like proteases, protects chicken dorsal root ganglion neurons from apoptosis due to nerve growth factor deprivation *(186)*. In our view, it is an important task for the future to unravel the mode of action of these proteases, as well as explore other ICE-homologs and their functions.

Besides some killer proteins of PCD, repressors of the process have also been identified. The *C. elegans ced-9* gene is largely homologous to the Bcl-2 family of cell death repressors *(187,188)*. The *bcl-2* gene was originally isolated from the t(14;18) translocation breakpoint in certain follicular lymphomas. When overexpressed in different types of cultured neurons, Bcl-2 represses the death owing to growth factor withdrawal *(189–192)*. The Bcl-2 overexpression in transgenic mice protected neurons from developmental cell death as well as from ischemia-induced cell death *(193)*. However, at the same time it became clear that there are also neurons that cannot be rescued by

Bcl-2 overexpression when deprived of growth factor *(192)*. Possibly, there are several survival proteins available to a cell that can either work together or act independently on the survival of the cell. A novel rescue protein that has recently been found is the baculovirus p35, that is structurally unrelated to Bcl-2. The p35 appears capable of suppressing differentially induced cell death in insect cells *(194)* in *C. elegans (195)*, and in a dopaminergic neuronal cell line *(196)*.

Adding to the complex picture of cell death regulation was the discovery of *bcl-2* gene family members (reviewed in ref. *197*). A protein called Bax, that is related to Bcl-2, heterodimerizes to Bcl-2 *(198)*. When critical residues in conserved domains of Bcl-2 were mutated, the heteromer could not be formed and cell death could no longer be prevented *(199)*. This led researchers to believe that Bax activity promotes cell death, unless it is bound to Bcl-2 *(199,200)*. It was therefore assumed that the Bcl-2-Bax pair constitutes a preset balance within cells in which the ratio between the two determines whether a cell accepts or denies a death order *(201)*. Further discoveries complexed, though not excluded this idea, since it was found that other Bcl-2 homologs protect against or induce cell death as well. These new Bcl-2-like proteins all have Bcl-2 binding capacities *(200)*, and therefore they may play a role in the proposed balance as well. The Bcl-x$_l$ (Bcl-x, long variant) protein protects against cell death in IL-3-dependent murine cells, probably by binding to Bax *(202)*. Bcl-x$_l$ knockout mice show a massive neuronal cell death during development *(203)*, thus confirming this protective role in vivo. Recently, it has been shown that the differential expression of members of the *bcl-2* gene family may be implied in the sensitivity to damage and the actual death following ischemia *(204,205)*. Bcl-2 is an integral membrane protein of the outer mitochondrion membrane, the nuclear envelope and the endoplasmatic reticulum *(206)* and several possible functions have been proposed. Some suggest a role in intracellular fluxes across membranes *(207)* or in rescuing from radical damage *(208,209)*, whereas others argue that Bcl-2 serves its function by interacting with R-ras, resulting in unknown intracellular signalling pathways *(210)*. Despite all this new knowledge, the exact physiological role of Bcl-2(-like) proteins remains unresolved.

2.2.2. Do Cell Death and Proliferation Share Common Features?

The first indications that the apoptotic cascade may involve proliferative characteristics came from observations that connected molecules like c-fos, c-myc, and hsp 70 contributed to cell death in the ventral prostrate epithelium following castration *(211)*. These were known to be involved in growth, so what could their function be during death? Many later studies confirmed the implication of some immediate early genes. During development c-fos-*lacZ* transgenic mice showed strong c-fos expression in areas in which massive cell death occurred *(170)*. Furthermore, a well defined apoptosis model was developed using NGF-deprived cultured sympathetic neurons from the superior cervical ganglion (SCG) *(166)*. Studies applying this model confirmed a role for c-jun, since its mRNA appeared to be induced in all dying neurons for an extended period of time, whereas c-fos only appeared transiently just before and during chromatin condensation *(171)*. Also, Ham and colleagues *(172)* showed that the level of c-jun protein significantly increases after NGF deprivation, whereas other Jun and Fos family members remain constant. Furthermore, overexpression of c-jun protein appeared to be an inducer of apoptosis in itself *(172)*. Lately, many investigations show an implication of early genes in the response to ischemia and status epilepticus-induced delayed neuronal death (reviewed in ref. *212*), but c-jun appears also to be involved in the response to axon damage in peripheral sensory neurons, which was interpreted as a regenerative action *(213)*. However, again supporting a role in a damaging response instead of a physiological rescue attempt is the association of c-jun protein induction with β-amyloid-induced apoptosis in hippocampal neurons *(214)*.

The c-myc proto-oncogen, which was shown to be involved in cell proliferation in many studies, has recently also been demonstrated to induce apoptosis in rat fibroblasts *(177)*. This again prompted the idea of an interaction of the cell cycle machinery and the suicide program.

Another finding by Freeman and coworkers *(175)* gave further rise to this notion. They adopted reverse transcriptase-polymerase chain reaction (RT-PCR) in the SCG neuronal apoptosis model mentioned above, in order to determine whether certain mRNAs of cell cycle regulators are induced after NGF deprivation. In this experiment, they observed that cyclin D1 mRNA was induced at the time when neurons become committed to die. All the other transcripts they studied decreased, so this seemed to support the hypothesis that, perhaps somewhere downstream c-jun activation, the selective induction of cyclin D1 without the normal molecular environment needed for G_1 to S phase transition, would lead to an abortive re-entry of the cell cycle. This re-entry, either owing to cyclin D1 induction or some other mechanism, would then cause apoptosis in postmitotic neurons *(179)*. However, in the initial study by Freeman et al., no protein levels were measured, nor was cyclin-dependent kinase (cdk) activity determined. Furthermore, it was not shown whether cyclin D1 induction actually causes apoptosis. Recently, Wiessner *(215)* reported an increase in cyclin D1 mRNA in the striatum and the hippocampus after transient global ischemia. This signal, however, seemed to correlate with activated microglial cells. In our laboratory, we studied the cyclin D1 protein levels in adrenalectomy induced apoptosis in the dentate gyrus of the hippocampus and found induction of this protein only in cells showing a microglia-like morphology in the hilus, but not in the granule cells of the dentate gyrus (Postigo, Van der Werf, Krugers; in preparation). In this latter model, an induction of another cell cycle-related molecule, namely p53 was reported *(216)*. The functions that p53 appears to have in inducing both growth arrest and apoptosis are still largely unknown and they may not be coupled *(217)*. Via DNA damage, p53 can however induce p21 *(218,219)*, a cdk inhibitor of the cyclin E-cdk2 complex, which is involved in the G_1-S transition (for review, *see* ref. *220*).

Lastly, Shi et al. *(176)* reported that the premature activation of the serine-threonine kinase p34^{cdc2} in YAC-1 lymphoma cells leads to an apoptotic-like cell death, that could be inhibited by addition of excess kinase peptide substrate.

2.2.3. Putting the Pieces Together

So far, we focused on major areas of research on apoptosis. Unfortunately, in most models it is unknown what is cause and what is consequence in the chain of events. For instance, suppose that in a number of apoptotic cell deaths the induction of the c-jun protein is essential for the onset of apoptosis. What would lie upstream of this event? Perhaps a signal transduction cascade ending at Jun kinases (JNK) *(221)* specifically inducing c-jun? And how can it be that different stimuli such as NGF deprivation, corticostrone withdrawal, and β-amyloid addition to the cellular environment, to name a few, lead to the same induction of c-jun? Would these very different stimuli to a cell follow different or similar paths to the nucleus? May there be even more signalling pathways from membrane to nucleus to achieve the turning on of modulatory elements of the program? Interesting new developments in this field are for instance that p21Ras proteins appear to be necessary and sufficient to rescue SCG neurons from death by disabled intracellular Trk receptor signalling *(222)*. Also, in PC-12 cells it was observed that the dynamic balance between growth-factor activated extracellular-regulated kinase (ERK) and stress-activated JNK-p38 pathways is important in determining the apoptotic response *(223)*. In general, this kind of investigation will give rise to a more thorough understanding of the early factors determining life and death of a cell.

In the direction of killer and rescue genes, there is also a lot of uncertainty about the exact modes of action of known molecules. Hopefully a lot of working mechanisms will be clarified, so that their place on the apoptotic map can be determined. Little is known about the actual executors of the apoptotic morphology, so research will probably focus in the coming years on new proteases to fulfill the hypothesized common effector role. Novel information may also come from the study of the fruit fly *Drosophila melanogaster*, in which most programmed cell deaths seem to be regulated by one common mechanism. This would involve the *reaper* gene *(224)* that is thought to act as a regulatory protein upstream of cell death effectors *(168)*.

Coming back to the role of cell cycle regulators, are they really the factor that cannot be circumvented when a cell has started his way through the apoptotic cascade? Because despite the indications concerning the involvement of cell cycle genes in apoptosis that were described above, there still is little evidence of a functional necessity of these genes during cell death in vivo.

Some interesting experiments put a new light on a hypothesis by Heintz *(225)*. He argued that a cell can respond to oncogene activation by either proliferation or death, depending on its differentiation state. Thus, in the case of mature, postmitotic neurons, any proliferative trigger would result in an appropriate suicide induction. Feddersen et al. performed experiments in which the SV40 T-antigen was put under the control of a Purkinje cell specific promoter, thus causing apoptosis *(226)*. This death was, however, prevented when the retinobalstoma (Rb) protein binding site was deleted from the T-antigen. This apparent central role for this cell cycle molecule confirms findings in the pRb knockout mice in which both ectopic mitosis and formidable apoptosis seems to take place in the developing brain *(227)*. This again supports the vision that, in this case, loss of Rb inhibition on the cell cycle leads to either proliferation or programmed cell death depending on the state of differentiation of the cell.

In two mouse mutants *Lurcher* and *staggered* cerebellar granule cell death owing to loss of target innervation is preceded by cyclin D and proliferating cell nuclear antigen (PCNA) induction. Also, cells start to incorporate BrdU, which is a measure of DNA synthesis and thus start of the S-phase. However, in these same mice, the primary degeneration of cerebellar Purkinje cells does not require this loss of control at the G_1-S transition *(228)*. Apart from the question of how these particular differences between degenerating neurons come to life, it still remains to be explored if expression of some G_1-related protein causes apoptosis by initiating the cell cycle or by some other way.

How would this correlate with the notion of Heintz mentioned above? We share the view of Ross *(229)* that some "selected cell cycle proteins are likely to be specific for particular initiating events leading the apoptotic program, depending on the proliferative status of the cell." This would in our opinion best describe the sometimes contradictory results from the various fields of research and in the different experimental paradigms used today, be it that it still reveals the enormous lack in our understanding of the phenomenon.

2.2.4. Is Therapeutic Manipulation Feasible?

If one thing has become clear in the text above it is that thus far no simple substrate has been identified on which pharmacological strategies can be based. A central death effector that would unify all types of cell death will perhaps be found in the near future. When this molecule's sole function is to induce death, it may turn out to be an ideal target for intervention. Alternatively, if some elements of the cell cycle machinery appear critical in specific cases of cell death, they may form a substrate for pharmacological modulation. Supporting these kind of strategies is a recent study by the Farinelli and coworkers in which they show that apoptosis in NGF-deprived PC12 cells can be prevented by several G_1-S inhibitors, but not by S–, G_2–, and M-phase inhibitors *(230)*. Whether this offers new strategies for in vivo use remains to be determined. For now, we recommend studying every individual type of apoptotic cell death in itself and try and determine all parameters involved in this particular death type. Basing future experimental and therapeutic manipulation of the process in this way would exclude extrapolations of data from other experimental situations. These may not or may differently apply to your own paradigm. Hopefully, in the years to come fundamental research will give us tools to treat the many cases in which the process of cellular suicide goes awry (reviewed in ref. *231*).

3. VISUALIZATION OF NEURONAL DAMAGE

Various in vivo imaging techniques, including positron (PET) and single photon emission tomography (SPECT), computed tomography scanning (CT), and magnetic resonance imaging (MRI) are available for visualization of CVA or hemorrhage-induced cerebral damage in patients. Possibilities, advantages, and disadvantages of these in vivo imaging techniques are reviewed in Chapter 2.

In vivo MRI imaging and various histological methods may be employed for damage assessments in the animal stroke models (Chapter 11). Changes of the regional cerebral blood flow (rCBF) and formation of edematous tissue have been visualized with MRI imaging after focal brain ischemia in rats *(232)*. For histological damage assessments, we recommend the silver-impregnation techniques *(233)* because they are easy, reliable, and performed on paraformaldehyde-fixed tissue that allows immunocytochemical and molecular biological studies of the adjacent nonimpregnated sections *(234)*. Tetrazolium and hematoxylin-eosin (HE) histochemistry, and glial fibrillary acidic (GFAP) and heat shock protein (HSP72) immunocytochemistry may be alternative protocols for revealing hypoxic/ischemic neuronal damage. Histochemical stains that employ tetrazolium salts show oxidation-reduction enzyme activity in fresh, unfixed slices of the brain. In the viable neurons and glial cells, the tetrazolium salt is reduced by dehydrogenases to a purple-colored formazan. Degenerating cells lack such dehydrogenase activity and will remain colorless *(235)*. The tetrazolium method is rapid and reliable, but a serious disadvantage is that it must be performed on unfixed slices, which prohibits additional immunocytochemical or molecular biological studies of adjacent sections. Pale areas, so-called "dark-neurons" and neuronal swelling characterize the injured areas and/or cells in the HE stains. The GFAP is a marker protein specific for astrocytes that may be used to reveal glial scar formation after hypoxic/ischemic injury *(234)*. The stress protein HSP72 is rapidly expressed after hypoxia/ischemia both in neuronal and glial cells, however, predominantly in the zone adjacent to the infarct; the so-called "penumbra." In the ischemic core, the HSP72 expression is restricted to the viable vascular elements. This HSP72 immunocytochemical method, therefore, may be a valuable tool for studying effects of treatments aimed at reducing the hypoxic/ischemic damage in the penumbra or ischemic tolerance development *(164)*.

For revealing mechanisms of programmed cell death, various immunocytochemical methods have become available that utilize antibodies directed against cell cycle regulating proteins (Cdks), immediate early genes, or growth factors that either inhibit or promote the cell death program. Examples of the latter group of immediate early genes and growth factors that may switch on the death program are c-myc, FAS, p53, TNF, and ICE. Typical examples of survival promoters are the members of the Bcl-2 family *(236)*.

4. TREATMENT FROM A MECHANISTIC POINT OF VIEW

The neuropathology of cerebral ischemia and stroke is characterized by three fundamental events: early, acute cell death resulting directly from failure of the blood supply; secondary bystander necrotic cell death in the penumbra; and delayed cell death of the apoptotic type. Because there is no warning signal for stroke, the acute damage probably cannot be prevented in first-ever stroke patients. In patients of the known risk groups (previous stroke, hypertension, and TIA) one could attempt prophylactic treatments, for example, with calcium entry-blockers like nimodipine (Chapter 4) and/or monoamine oxidase inhibitors *(135)*, or antioxidants. The preclinical studies of mechanisms of cell death after hypoxic/ischemic injury reviewed above, have shown that processes leading to bystander cell death in the penumbra are initiated within hours after the insult. Therapies started thereafter cannot reverse or stop the process. More or less the same lessons have been learned from the clinical studies. Thus far, no therapeutic approach tested for cerebral damage reduction after stroke has been reported to be very effective. The main reason may be that most of the therapies were initiated late, between 12–24 h after the hypoxic/ischemic event. Some preclinical studies have presented evidence now that delayed neuronal death of the apoptotic type can be initiated within minutes to hours after the occurrence of the hypoxic/ischemic event *(84,237)*. This implies that beneficial effects on the development of the cerebral damage may be expected only of therapies that have been initiated within the first hour(s) after the insult. Therefore, the future of acute stroke treatment will most likely include the administration of agents for metabolic intervention by paramedics "in the field" or the general practitioner, so that no valuable time is lost.

REFERENCES

1 Pulsinelli, W., The ischemic penumbra in stroke. *Sci. Am. Sci. Med.,* **2** (1995) 16–25.

2 Ricci, S., Celani, M. G., Guercini, G., Rucireta, P., Vitali, R., La Rosa, F., Duca, E., Ferraguzzi, R., Paolotti, M., and Seppolini, D., First-year results of a community-based study of stroke incidence in Umbria, Italy, *Stroke,* **20** (1989) 853–857.

3 Ricci, S., Celani, M. G., La Rosa, F., Vitali, R., Duca, E., Ferraguzi, R., Paolotti, M., Seppolini, D., Caputo, N., and Chiurulla, C., SEPIVAC: a community based study of stroke incidence in Umbria, Italy, *J. Neurol. Neurosurg. Psychiatry,* **54** (1991) 695–698.

4 Ricci, S., Celani, M. G., La Rosa, F., Vitali, R., Duca, E., Ferraguzi, R., Paolotti, M., Seppolini, D., Caputo, N., and Chiurulla, C., A community-based study of incidence and risk factors and outcome of transient ischaemic attacks in Umbria, Italy, *J. Neurol.,* **238** (1991) 87–90.

5 Bergman, L., Van der Meulen, J. H. P., Limburg, M., and Habbema, J. D. F., Costs of medical care after first-ever stroke in the Netherlands, *Stroke,* **26** (1995) 1830–1836.

6 Knollema, S., Knigge, M. F., Jansen, H. M. L., Ter Horst, G. J., Korf, J., Minderhoud, J. M., and Meyboom-De Jong, B., Het cerebrovasculair accident in de praktijk, *Patient Care,* **22** (1995) 21–32.

7 Anderson, C. S. Jamrozik, K. D., Burvill, P. W., Chakera, T. M., Jonhson, G. A., and Stewart-Wynne, E. G., Ascertaining the true incidence of stroke: experience from the Perth community stroke study, *Med. J. Aust.,* **158** (1993) 80–84.

8 Scott, W. G. and Scott, H., Ischaemic stroke in New Zealand: an economy study, *N. Z. Med. J.* **107** (1994) 443–446.

9 Itoh, T., Matsumoto, M., Handa, N. , Maeda, H., Hougaku, H., Tsukamoto, Y., Kondo, H., Tanouchi, J., and Kamada, T., Paradoxial embolism as a cause of ischemic stroke of uncertain etiology. A transcranial Doppler sonographic study, *Stroke,* **25** (1994) 771–775.

10 Corelei, A., Marini, C., Ferranti, E., Frontini, M., Prencipe, M., and Fieschi, C., A prospective study of cerebral ischemia in the young. Analysis of pathogenic determinants, *Stroke,* **24** (1993) 362–367.

11 Futrell, N. and Millikan, C., Frequency, etiology, and prevention of stroke in patients with systemic lupus erythematosus, *Stroke,* **20** (1989) 583–591.

12 Herrschaft, H., Heart diseases as a cause of cerebral symptoms and syndromes, *Fortschr. Neurol. Psychiatr.,* **58** (1990) 287–300.

13 Grau, A. J., Buggle, F., Heindl, S., Steichen-Wiehn, C., Banerjee, T., Maiwald, M., Rohlfs, M., Suhr, H., Fiehn, W., Becher, H., and Hacke, W., Recent infection as a risk factor for cerebrovascular ischemia, *Stroke,* **26** (1995) 373–379.

14 Grau, A. J., Buggle, F., Steichen-Wiehn, C., Heindl, S., Banerjee, T., Seitz, R., Winter, R., Forsting, M., Werle, E., Bode, C., Nawroth, P. P., Becher, H., and Hacke, W., Clinical and biochemical analysis in infection-associated stroke, *Stroke,* **26** (1995) 1520–1526.

15 Lindenstrom, E., Boysen, G., and Nyboe, J., Risk factors for stroke in Copenhagen, Denmark. I. Basic demographic and social factors, *Neuroepidemiology,* **12** (1993) 37–42.

16 Laloux, P., Ossemann, M., and Jamart, J., Stroke subtypes and risk factors associated with silent infarctions in patients with first-ever stroke or transient ischemic attack, *Acta Neurol. Belg.,* **94** (1994) 17–23.

17 Kane-Carlsen, P. A., Transient ischemic attacks: clinical features, pathophysiology and management, *Nurse Pract.,* **15** (1990) 9–14.

18 Lyrer, P., Prognosis of and basis for decision making in transient ichemic attacks, *Schweiz. Med. Wochenschr.,* **125** (1995) 1299–1306.

19 Rouhart, F., Zagnoli, F., Goas, J. Y., and Mocquard, Y., Cerebral ischemic arterial accidents in young adults, *Rev. Neurol.* (Paris), **149** (1993) 547–553.

20 Ebrahim, S., *Clinical Epidemiology of Stroke,* Oxford University Press, Oxford, 1990.

21 MacMahon, S., Peto, R., Cutler, J., Collins, R., Sorlie, P., Neaton, J., Abbott, R., Godwin, J., Dyer, A., and Stamler, J., Blood pressure, stroke, and coronary heart disease, *Lancet,* **335** (1990) 765–774.

22 Stamler, J., Dietary salt and blood pressure. Nutrition and cardiocerebrovascular diseases, *Ann. NY Acad. Sci.,* **676** (1993) 122–156.

23 Post, D. and Smit, P. Th., Sociaal geneeskundige aspecten van het CVA, *Patient Care,* **22** (1995) 51–55.

24 Lai, S. M., Alter, M., Friday, G., Sobel, E., Gil-Peralta, A., McCoy, R. L., Levitt, L. P., and Isack, T., Transient ischemic attacks: their frequency in the Leigh Valley, *Neuro-epidemiology,* **9** (1990) 124–130.

25 Whisnant, J. P., Melton, L. J., Dvis, P. H., O'Fallon, W. M., Nishimura, K. and Schoenberg, B. S., Comparisons of case ascertainment by medical record linkage in a cohort follow-up to determine incidence rates for transient ischemic attacks and stroke, *J. Clin. Epidemiol.,* **43** (1990) 791–797.

26 Ueda, K., Kyohara, Y., and Hasua,Y., Transient ischemic attacks in a Japanese community, *Stroke,* **18** (1987) 844–847.

27 Van der Werf, Y. D., De Jongste, M. J. L., and Ter Horst, G. J., The immune system mediates blood-brain barrier damage: possible implications for pathophysiology of neuropsychiatric illnesses, *Acta Neuropsychiat.,* **7** (1995) 114–121.

28 Barron, S. A., Rogovski, Z., and Hemli, J., Autonomic consequences of cerebral hemisphere infarction, *Stroke,* **25** (1994) 113–116.

29 Yamamoto, Y., Akiguchi, I., Oiwa, K., Satoi, H., and Kimura, J., Diminished nocturnal blood pressure decline and lesion site in cerebrovascular disease, *Stroke,* **26** (1995) 829–833.

30 Naver, H., Blomstrand, C., Ekholm, S., Jensen, C., Karlsson, T., and Wallin, B. G., Autonomic and thermal sensory symptoms and dysfunction after stroke, *Stroke,* **26** (1995) 1379–1385.

31 Demeurisse, G., Verhas, M., and Capon, A., Remote cortical dysfunction in aphasic stroke patients, *Stroke,* **22** (1991) 1015–1020.

32 Evans, J. G., Transient neurological dysfunction and risk of stroke in an elderly English population: the different significance of vertigo and non-rotatory dizziness., *Age. Ageing,* **19** (1990) 43–49.

33 Lane, R. D., Wallace, J. D., Petrosky, P. P., Schwartz, G. E., and Gradham, A. H., Supraventricular tachycardia in patients with right hemisphere strokes, *Stroke,* **23** (1992) 362–366.

34 Ter Horst, G. J., Van den Brink, A., Homminga, S. A., Hautvast, R. W. M., Rakhorst, G., Mettenleiter, T. C., De Jongste, M. J. L., Lie, K. I., and Korf, J., Transneuronal viral labelling of rat heart left ventricle controlling pathways, *NeuroReport,* **4** (1993) 1307–1311.

35 Standish, A., Enquist, L. W., and Schwaber, J., S., Innervation of the heart and its central medullary origin defined by viral tracing, *Science,* **263** (1994) 232–234.

36 Ter Horst, G. J., Hautvast, R. W. M., De Jongste, M. J. L., and Korf, J., Neuroanatomy of cardiac activity regulating circuitry; a transneuronal retrograde viral tracing study in the rat, *Eur. J. Neuroscience,* **8** (1996) 101–113.

37 Oppenheimer, S. M., Gelb, A., Girvin, J. P., and Hachinski, V. C., Cardiovascular effects of human insular cortex stimulation, *Neurology,* **42** (1992) 1727–1732.

38 Oppenheimer, S. M., The anatomy and physiology of cortical mechanisms of cardiac control, *Stroke,* **24** (1993) 13–15.

39 Buja, L. M., Eigenbrodt, M. L., and Eigenbrodt, E. H., Apoptosis and necrosis. Basic types and mechanisms of cell death, *Arch. Pathol. Lab. Med.,* **117** (1993) 1208–1214.

40 Kerr, J. F. R. and Harmon, B. V., Definition and incidence of apoptosis: an historical perspective. In L. D. Tomei and F. O. Cope (eds.) *Apoptosis: The Molecular Basis of Cell Death,* Cold Spring Harbor Laboratory, Cold Spring Harbor, NY, 1991, pp. 5–29.

41 Server, A. C. and Mobley, W. C., Neuronal cell death and the role of apoptosis. In L. D. Tomei and F. O. Cope (eds.) *Apoptosis: The Molecular Basis of Cell Death,* Cold Spring Harbor Laboratory, Cold Spring Harbor, NY, 1991, pp. 263–279.

42 Baethmann, A., Schurer, L., Unterberg, A., Wahl, W., Staub, F., and Kempski, O. S., Mediator substances of brain edema in cerebral ischemia, *Arzneimittelforsch.,* **41** (1991) 310–315.

43 Nowak, T. S., Tomida, S., Pluta, R., Xu, S., Kozuka, M., Vass, K., Wagner, H. G., and Klatzo, I., Cumulative effect of repeated ischemia on brain edema in the gerbil, *Adv. Neurol.,* **52** (1990) 1–9.

44 Stummer, W., Baethmann, A., Murr, R., Schurer, L., and Kempski, O. S., Cerebral protection against ischemia by locomotor activity in Gerbils. Underlying mechanisms, *Stroke,* **26** (1995) 1423–1430.

45 Ames, A., Wright, R. L., Kowada, M., Thurston, J. M., and Majno, G., Cerebral ischemia, II. The no-reflow phenomenon, *Am. J. Pathol.,* **52** (1968) 437–453.

46 Chiang, J., Kowada, M., Ames, A., Wright, R. L., and Majno, G., Cerebral ischemia, III. Vascular changes, *Am. J. Pathol.,* **52** (1968) 455–476.

47 Dougherty, J. H., Levy, D. E., and Weksler, B., B. Experimental cerebral ischemia produces platelet aggregation, *Neurology,* **29** (1979) 1460–1465.

48 Hossmann, V., Hossmann, K.-A., and Tagaki, S., Effect of intravascular platelet aggregation on blood recirculation following prolonged ischemia in the cat brain, *J. Neurol.,* **222** (1980) 159–170.

49 Hallenbeck, J. M., Dutka, A. J., Tanishima, T., Kochanek, P. M., Kumaroo, K. K., Thompson, C., Obrenovitch, T. P., and Contreas, T. J., Polymorphonuclear leukocyte accumulation in brain regions with low flow during the early post-ischemic period, *Stroke,* **17** (1986) 246–253.

50 Grogaard, B., Schurer, L., Gerdin, B., and Arfors, K. E., Delayed hypoperfusion after incomplete forebrain ischemia in the rat: the role of polymorphonuclear leukocytes, *J. Cereb. Blood Flow Metab.,* **9** (1989) 500–505.

51 Moncada, S. and Vane, J. R., Arachidonic acid metabolites and interactions between platelets and blood vessel walls, *N. Engl. J. Med.,* **300** (1979) 1142–1147.

52 Black, K. L., Hoff, J. T., and Deshmukh, G. D., Eicosapentaenoic acid: effect on brain prostaglandins, cerebral blood flow and edema in ischemic gerbils, *Stroke,* **15** (1984) 65–69.

53 Stevens, M. K., Yaksh, T. L., Hansen, R. B., and Anderson, R. E., Effect of preischemic cyclo-oxygenase inhibition by zomepirac sodium on reflow, cerebral autoregulation, and EEG recovery in the cat after global ischemia, *J. Cereb. Blood Flow Metab.,* **6** (1986) 691–702.

54 Pettigrew, L. C., Grotta, J. C., Rhoades, H. M., and Wu, K. K., Effect of thromboxane inhibition on eicosanoid levels and blood flow in ischemic rat brain, *Stroke,* **20** (1989) 627–632.

55 Nakagomi, T., Sasaki, T., Kirino, T., Tamura, A., Noguchi, M., Saito, I., and Takakura, K., Effects of cyclo-oxygenase and lipo-oxygenase inhibitors on delayed neuronal death in the gerbil hippocampus, *Stroke,* (1989) **20** 925–929.

56 Wang, X., Tian-Li, Y., Barone, F. C., and Feuerstein, G. Z., Demonstration of increased endothelial-leukocyte adhesion molecule-1 mRNA expression in rat ischemic cortex, *Stroke,* **26** (1995) 1665–1669.

57 Zhang, R. L., Chopp, M., Jiang, N. , Tang, W. X., Prostak, J., Manning, A. M., and Anderson, D. C., Anti-intercellular adhesion molecule-1 antibody reduces ischemic cell damage after transient but not permanent middle cerebral artery occlusion in the wistar rat, *Stroke,* **26** (1995) 1438–1443.

58 Beekhuizen, H. and Furth, R., Monocyte adherence to human vascular endothelium, *J. Leukocyte Biol.,* **54** (1993) 363–378.

59 Weller, A., Isenmann, S., and Vestwever, D., Cloning of the mouse endothelial selectins: expression of both E- and P-selectin is induced by tumor necrosis factor, *J. Biol. Chem.,* **267** (1992) 15,176–15,183.

60 Zhang, R. L., Chopp, M., Li, Y., Zaloga, C., Jiang, M., Jones, M., Miyasaka, M., and Ward, P., Anti-ICAM-1 antibody reduces ischemic cell damage after transient middle cerebral artery occlusion in the rat, *Neurology,* **44** (1994) 1747–1751.

61 Flower, R. J., Lipocortin and the mechanism of action of the glucocorticoids, *Br. J. Pharmacol.,* **94** (1988) 987–1015.

62 Flower, R. J. and Rothwell, N. J., Lipocortin-1: cellular mechanisms and clinical relevance, *TIPS,* **15** (1994) 71–76.

63 Go, K. G., Zuiderveen, F., De Ley, L., Ter Haar, J. G., Parente, L., Solito, E., and Molenaar, W. M., Effects of steroids on brain lipocortin immunoreactivity, *Acta Neurochir.,* **60** (1994) 101–103.

64 Voermans, P. H., Go, K. G., Ter Horst, G. J., Ruiters, M. H. J., Solito, E., and Parente, L., Induction of annexin-1 mRNA and annexin-1 in rat brain by methylprednisolone and the 21-aminosteroid U74389F, *Med. Inflamm.,* submitted.

65 Parente, L. and Solito, E., Association between glucocorticosteroids and lipocortin-1, *TIPS,* **15** (1994) 362–365.

66 Russo-Marie, F., Lipocortins: an update, *Prostagl. Leukotr. Ess. Fatty Acids,* **42** (1991) 83–89.

67 Relton, J. K., Strijbos, P. J., O'Shaughnessy, C. T., Carey, F., Forder, R. A., Tilders, F. J., and Rothwell, N. J., Lipocortin-1 is an endogenous inhibitor of ischemic damage in the rat brain, *J. Exp. Med.,* **174** (1991) 305–310.

68 McKenna, J. A., Lipocortin-1 in apoptosis. Mammary regression, *Anat. Rec.,* **242** (1995) 1–10.

69 Siesjo, B. K., Zhao, Q., Pahlmark, K., Sicsjo, P., Katsura, K., and Folbergrova, J., Glutamate, calcium and free radicals as mediators of ischemic brain damage, *Ann. Thorac. Surg.* **59** (1995) 1316–1320.

70 Miljanich, G. P. and Ramachandran, J., Antagonists of neuronal calcium channels: structure, function and therapeutic implications, *Annu. Rev. Pharmacol. Toxicol.,* **35** (1995) 707–734.

71 Alps, B. J., Drugs acting on calcium channels: potential treatment for ischaemic stroke, *Br. J. Clin. Pharmacol.,* **34** (1992) 199–206.

72 Ginsberg, M. D., Lin, B., Morikawa, E., Dietrich, W. D., Busto, R., and Globus, M. Y., Calcium antagonists in the treatment of experimental cerebral ischemia, *Arzneimittelforschung,* **41** (1991) 334–337.

73 Meyer, F. B., Calcium, neuronal hyperexcitability and ischemic injury, *Brain Res. Reg.,* **14** (1989) 227–243.

74 Morley, P., Hogan, M. J., and Hakim, A. M., Calcium-mediated mechanisms of ischemic injury and protection, *Brain Pathol.,* **4** (1994) 37–47.

75 Bloom, S., Magnesium deficiency cardiomyopathy, *Am. J. Cardiovasc. Pathol.,* **2** (1988) 7–17.

76 Maki, A., Berezesky, I. K., Fargnoli, J., Holbrook, N. J., and Trump, B. T., Role of $[Ca^{2+}]i$ in induction of c-*fos,* c-*jun* and c-*myc* mRNA in rat PTE after oxidative stress, *FASEB J.,* **6** (1992) 919-924.

77 Csermely, P., Schnaider, T., and Szanto, I., Signalling and transport through the nuclear membrane, *Biochem. Biophys. Acta,* **1241** (1995) 425–452.

78 Paoletti, P., Neyton, J., and Ascher, P., Glycine-independent and subunit-specific potentiation of NMDA responses by extracellular Mg^{2+}, *Neuron,* **15** (1995) 1109–1120.

79 Traynelis, S. F., Hartley, M., and Heinemann, S., Control of proton sensitivity of NMDA receptor by RNA splicing and polyamines, *Science,* **268** (1995) 873–876.

80 Choi, D. W. and Rothman, S. M., The role of glutamate neurotoxicity in hypoxic-ischemic neuronal death, *Ann. Rev. Neurosci.,* **13** (1990) 171–182.

81 Meldrum, B. and Garthwaite, J., Excitatory amino acid neurotoxicity and neuro-degenerative disease, *Trends Pharmacol. Sci.,* **11** (1990) 379–387.

82 Kure, S., Tominaga, T., Yoshimoto, T., Tada, K., and Narisawa, K., Glutamate triggers internucleosomal DNA cleavage in neuronal cells, *Biochem. Biophys. Res. Commun.,* **179** (1991) 39–45.

83 MacManus, J. P., Hill, I. E., Huang, Z. G., Rasquinha, I., Xue, D., and Buchan, A. M., DNA damage consistent with apoptosis in transient focal ischaemic cortex, *NeuroReport,* **5** (1994) 493–496.

84 Ankarcrona, M., Dypbukt, J. M., Bonfoco, E., Zhivotovsky, B., Orrenius, S., Lipton, S. A., and Nicotera, P., Glutamate induced neuronal death: a succession of necrosis or apoptosis depending on mitochondrial function, *Neuron,* **15** (1995) 961–973.

85 Wood, E. R., Bussey, T. J., and Phillips, A. G., A glycine antagonist 7-chlorokynurenic acid attenuates ischemia-induced learning deficits, *NeuroReport,* **4** (1993) 151–154.

86 Pellegrini-Giampietro, D. E., Cozzi, A., and Moroni, F., The glycine antagonist and free radical scavenger 7-Cl-thio-kynurnate reduces CA1 ischemic damage in the gerbil, *Neuroscience,* **63** (1994) 701–709.

87 Patel, M., Day, B. J., Crapo, J. D., Fridovich, I., and McNamara, J. O., Requirement for superoxide in excitotoxic cell death, *Neuron,* **16** (1996) 345–355.

88 Hatfield, R. H., Gill, R., and Brazell, C., The dose-response relationship and therapeutic window for dizocilpine (MK-801) in a rat focal ischemia model, *Eur. J. Pharmacol.,* **216** (1992) 1–7.

89 Hewitt, K. and Corbett, D., Combined treatment with MK-801 and nicardipine reduces global ischemic damage in the gerbil, *Stroke,* **23** (1992) 82–86.

90 Uematsu, D., Araki, N. , Greenberg, J. H., Sladky, J., and Reivich, M., Combined therapy with MK-801 and nimodipine for protection of ischemic brain damage, *Neurology,* **41** (1991) 88–94.

91 Lyden, P. D. and Lonzo, L., Combination therapy protects ischemic brain in rats. A glutamate antagonist plus a gamma-aminobutyric acid agonist, *Stroke,* **25** (1994) 189–196.

92 Javitt, D. C. and Zukin, S. R., Recent advances in the phencyclidine model of schizophrenia, *Am. J. Psychiat.,* **148** (1991) 1301–1308.

93 Kinouchi, H., Sharp, F. R., Chan, P. H., Mikawa, S., Kamii, H., Arai, S., and Yoshimoto, T., MK-801 inhibits the induction of immediate early genes in cerebral cortex, thalamus, and hippocampus, but not in the substantia nigra following middle cerebral artery occlusion, *Neurosci. Lett.,* **179** (1994) 111–114.

94 Vendrell, M., Curran, T., and Morgan, J. I., Glutamate, immediate early genes, and cell death in the central nervous system, *Ann. NY Acad. Sci.,* **679** (1993) 132–141.

95 Swanson, R. A., Yu, A. C. H., Chan, P. H., and Sharp, F. R., Glutamate increases glycogen content and reduces glucose utilization in primary astrocyte cultures, *J. Neurochem.,* **54** (1990) 490–496.

96 Elekes, O., Venema, K., Postema, F., Dringen, R., Hamprecht, B., and Korf, J., Glial contribution of rat hippocampus lactate as assessed with microdialysis and stress, *Neurosci. Lett.* (1996), in press.

97 Cornell-Bell, A. H., Finkbeiner, S. M., Cooper, M. S., and Smith, S. J., Glutamate induces calcium waves in cultured astrocytes: long range glial signaling, *Science,* **247** (1990) 470–473.

98 Glaum, S. R., Holzwarth, J. A., and Miller, R. J., Glutamate receptors activate Ca^{2+} mobilization and Ca^{2+} influx in astrocytes, *Proc. Natl. Acad. Sci. USA,* **87** (1990) 3454–3458.

99 Rosenberg, P. A. and Aizenman, E., Hundred-fold increase in neuronal vulnerability to glutamate toxicity in astrocyte-poor cultures of rat cerebral cortex, *Neurosci. Lett.,* **103** (1989) 162–168.

100 Rosenberg, P. A., Accumulation of extracellular glutamate and neuronal death in astro-cyte-poor cortical cultures exposed to glutamine, *Glia,* **4** (1991) 91–100.

101 Volterra, A., Trotti, D., Cassutti, P., Tromba, C., Salvaggio, A., Melcangi, R. C., and Racagni, G., High sensitivity of glutamate uptake to extracellular free arachidonic acid levels in rat cortical synaptosomes and astrocytes, *J. Neurochem.,* **59** (1992) 600–606.

102 Dugan, L. L., Bruno, V. M. G., Amagasu, S. M., and Giffard, R. G., Glia modulate the response of murine cortical neurons to excitotoxicity: glia exacerbate AMPA neurotoxic-ity. *J. Neurosci.,* **15** (1995) 4545–4555.

103 Eng, L. F., Glial fibrillary acidic protein (GFAP): the major protein of glial intermediate filaments in differentiated astrocytes, *J. Neuroimmunol.,* **8** (1985) 203–214.

104 Lindsay, R. M., Reactive gliosis. In S. Federoff and A. Vernadakis (eds.) *Astrocytes: Cell, Biology and Pathology of Astrocytes,* vol. 3, Academic Press, Orlando, FL, 1986, pp. 231–262.

105 Wood, P. L., Microglia as a unique cellular target in the treatment of stroke: potential neurotoxic mediators produced by activated microglia, *Neurol. Res.,* **17** (1995) 242–248.

106 Schwartz, J. P., Sheng, J. G., Mitsuo, K., Shirabe, S., and Nishiyama, N., Trophic factor production by reactive astrocytes in injured brain, *Ann. NY Acad. Sci. NY,* **679** (1993) 226–234.

107 Morrison, R. S., Sharma, A., De Vellis, J., and Bradshaw, R. A., Basic fibroblast growth factor supports the survival of cerebral cortical neurons in primary culture, *Proc. Natl. Acad. Sci. USA,* **83** (1986) 7537–7541.

108 Maiese, K., Boniece, I., DeMeo, D., and Wagner, J. A., Peptide growth factors protect against ischemia in culture by preventing nitric oxide toxicity, *J. Neurosci.* **13** (1993) 3034–3040.

109 Pechan, P. A., Chowdhury, K., Gerdes, W., and Seifert, W., Glutamate induces the factors NGF, bFGF, the receptor FGF-R1 and c-fos mRNA expression in rat astrocytes in culture, *Neurosci. Lett.,* **153** (1993) 111–114.

110 Mattson, M. P. and Scheff, S. W., Endogenous neuroprotection factors and traumatic brain injury: mechanisms of action and implications for therapy, *J. Neurotrauma,* **11** (1994) 3–33.

111 Hsu, C. Y., An, G., Liu, J. S., Xue, J. J., He, Y. Y., and Lin, T. N., Expression of immedi-ate early gene and growth factor mRNAs in a focal cerebral ischemia model in the rat, *Stroke,* **24** (1993) 178–181.

112 Takeda, A., Onodera, H., Sugimoto, A., Kogure, K., Obinata, M., and Shibahara, S., Coordinated expression of messenger RNAs for nerve growth factor, brain derived neurotrophic factor and neurotrophin-3 in the rat hippocampus following transient fore-brain ischemia, *Neuroscience,* **55** (1993) 23–31.

113 Dragunow, M., Beilharz, E., Sirimanne, E., Lawlor, P., Williams, C., Bravo, R., and Gluckman, P., Immediate early gene protein expression in neurons undergoing delayed death, but not necrosis, following hypoxic-ischemic injury to the young rat brain, *Mol. Brain Res.,* **25** (1994) 19–33.

114 Koh, J. Y., Gwag, B. J., Lobner, D., and Choi, D., W. Potentiated necrosis of cultured cortical neurons by neurotrophins, *Science,* **268** (1995) 573–575.

115 Halliwell, B. and Gutteridge, J. M. C., *Free Radicals in Biology and Medicine,* Oxford University Press, London, 1985.

116 Patt, A., Xanthine oxidase-derived hydrogen peroxide contributes to ischemia reperfusion-induced edema in gerbil brains, *J. Clin. Invest.,* **81** (1988) 1556–1562.

117 Siesjo, B. L., Agardh, C. D., and Bengtson, F., Free radicals and brain damage, *Cereb. Brain Metab. Rev.,* **1** (1989) 165–211.

118 Agardh, C. D., Zhang, H., Smith, M. L., and Siesjo, B. L., Free radical production and ischemic brain damage: influence of post-ischemic oxygen tension, *Int. J. Dev. Neurosci.,* **9** (1991) 127–138.

119 Damsma, G., Boisvert, D. P., Mudrick, L. A., Wenkstern, D., and Fibiger, H. C., Effects of transient forebrain ischemia and pargyline on extracellular concentrations of dopamine, serotonine, and their metabolites in the rat striatum as determined by in vivo microdialysis, *J. Neurochem.,* **54** (1990) 801–808.

120 Gutteridge, J. M. C., Hydroxyl radicals, iron, oxidative stress and neurodegeneration, *Ann. NY Acad. Sci.,* **738** (1994) 201–213.

121 Palmer, R. M. J., Ashton, D. S., and Moncada, S., Vascular endothelial cells synthesize nitric oxide from l-arginine, *Nature,* **333** (1988) 664–666.

122 Palmer, R. M. J., Ferrige, A. G., and Moncada, S., Nitric oxide release account for the biological activity of endothelium-derived relaxing factor, *Nature,* **327** (1987) 524–526.

123 Ischiropoulos, H., Zhu, L., and Beckman, J. S., Peroxynitrite formation from macrophage-derived nitric oxide, *Arch. Biochem. Biophys.,* **208** (1992) 446–451.

124 Ter Horst, G. J., Knollema, S., Stuiver, B., Hom, H. W., Yoshimura, S., Ruiters, M. H. J., and Korf, J., Differential glutathione peroxidase mRNA up-regulations in rat forebrain areas after transient hypoxia/ischemia, *Ann. NY Acad. Sci. USA,* **738** (1994) 329–333.

125 Moreno, S. and Muganini, E., Immunocytochemical localization of catalase in rat brain, *Soc. Neurosci. Abstr.,* **18** (1992) 1604.

126 Knollema, S., Hom, H. W., Schirmer, H., Korf, J., and Ter Horst, G. J., Immunolocalization of glutathione reductase in the murine brain, *J. Comp. Neurol.* (1996), in press.

127 Mahadik, S. P., Makar, T. K., Murthy, J. N., and Karplak, S. E., Temporal changes in superoxide dismutase, glutathione peroxidase and catalase levels in primary and peri-ischemic tissue: monosialoganglioside (GM1) treatment effects, *Mol. Chem. Neuropathol.,* **18** (1993) 1–14.

128 Matsuyama, T., Michishita, H., Nakamura, N. , Tsuchiyama, M., Shimizu, S., Watanabe, K., and Sugita, M., Induction of copper-zinc superoxide dismutase in gerbil hippocampus after ischemia, *J. Cereb. Blood Flow Metab.,* **13** (1993) 135–144.

129 Kirino, T., Delayed neuronal death in the gerbil hippocampus following ischemia, *Brain Res.,* **239** (1982) 57–69.

130 Truelove, D., Shuaib, A., Ijaz, S., Ishaqzay, R., and Kalra, J., Neuronal protection with superoxide dismutase in repetitive forebrain ischemia in gerbils, *Free Radic. Biol. Med.,* **17** (1994) 445–450.

131 He, Y. Y., Hsu, C. Y., Ezrin, A. M., and Miller, M. S., Polyethylene glycol-conjugated superoxide dismutase in focal cerebral ischemia-reperfusion, *Am. J. Physiol.,* **265** (1993) H252–256.

132 Knollema, S., Elting, J. W., Korf, J., and Ter Horst, G. J., Ebselen (PZ-51) protects the caudate putamen against hypoxia/ischemia induced neuronal damage, *Neurosci. Res. Comm.* (1996), in press.

133 Yue, T. L., Gu, J. L., Lysko, P. G., Cheng, H. Y., Barone, F. C., and Feuerstein, G., Neuroprotective effects of phenyl-t-butyl-nitrone in gerbil global brain ischemia and in cultured rat cerebellar neurons, *Brain Res.,* **574** (1992) 193–197.

134 Kinouchi, H., Epstein, C. J., Mizui, T., and Chan, P. H., Attenuation of focal cerebral ischemic injury in transgenic mice overexpressing Cu/Zn-superoxide dismutase, *Proc. Natl. Acad. Sci. USA,* **88** (1991) 11,158–11,162.

135 Knollema, S., Aukema, W., Hom, H. W., Korf, J., and Ter Horst, G. J., L-Deprenyl reduces brain damage in rats exposed to transient hypoxia/ischemia, *Stroke,* **26** (1995) 1883–1888.

136 Sivenius, J., Kuhmonen, J., Miettinen, R., Baapalinna, A., and Riekkinen, P. J., Effect of selegeline on the hippocampal CA1 layer following transient ischemia in gerbils, *Soc. Neurosci. Abstr.,* **20** (1994) 188.

137 Carillo, M. C., Kanai, S., Nokubo, M., and Kitani, K., (–)Deprenyl induces activities of both superoxide dismutase and catalase but not glutathione peroxidase in the striatum of young male wistar rats, *Life Sci.,* **48** (1991) 517–521.

138 Kochanek, P. M. and Hallenbeck, J. M., Polymorphonuclear leukocytes and monocytes/macrophages in the pathogenesis of cerebral ischemia and stroke, *Stroke,* **23** (1992) 1367–1379.

139 Knollema, S., Van de Witte, S. V., Korf, J., and Ter Horst, G. J., The reaction of microglia and astrocytes in relation to the development of regional neuronal damage after ischemia, *J. Comp. Neurol.,* submitted.

140 Morioka, T., Kalehua, T., and Streit, W. J., Characterization of microglial reaction after middle cerebral artery occlusion in rat brain, *J. Comp. Neurol.,* **327** (1993) 123–132.

141 Jander, S., Kraemer, M., Schroeter, M., Witte, O. W., and Stoll, G., Lymphocyte infiltration and expression of intercellular adhesion molecule 1 in photochemically induced ischemia of the rat cortex, *J. Cereb. Blood Flow Metabol.,* **15** (1995) 42–51.

142 Korematsu, K., Goto, S., Nagahiro, S., and Ushio, Y., Microglial response to transient focal cerebral ischemia: an immunocytochemical study on rat cerebral cortex using anti-phosphotyrosine antibody, *J. Cereb. Blood Flow Metabol.,* **14** (1994) 825–830.

143 Wang, X., Yue, T. L., Barone, F. C., and Feuerstein, G. Z., Demonstration of increased endothelial-leukocyte adhesion molecule-1 mRNA expression in rat ischemic cortex, *Stroke,* **26** (1995) 1665–1669.

144 Guilian, D., Baker, T. J., Shih, L., and Lachman, L. B., Interleukin-1 of the central nervous system is produced by ameboid microglia, *J. Exp. Med.,* **164** (1987) 594–604.

145 Sawada, M., Kondo, N. , Suzumura, A., and Marunouchi, T., Production of tumor necrosis factor-alpha by microglia and astrocytes in culture, *Brain Res.,* **491** (1989) 394–397.

146 Woodroofe, M. N. , Sarna, G. S., Wadhwa, M., Hayes, G. M., Loughlin, A. J., Tinker, A., and Cuzner, M. L., Detection of interleukin-1 and interleukin-6 in adult rat brain, following mechanical injury, by in vivo microdialysis: evidence of a role for microglia in cytokine production, *J. Neuroimmunol.,* **33** (1991) 227–236.

147 Wong, G. H. W., Protective roles of cytokines against radiation: Induction of mitochondrial MnSOD, *Biochem. Biophys. Acta,* **1271** (1995) 205–209.

148 Cheng, B., Christakos, S., and Mattson, M. P., Tumor necrosis factor protect neurons against metabolic-excitotoxic insults and promote maintenance of calcium homeostasis, *Neuron,* **12** (1994) 139–153.

149 Tchelingerian, J. L., Quinonero, J., Booss, J., and Jacque, C., Localization of TNFa and IL-1a immunoreactivities in striatal neurons after surgical injury to the hippocampus, *Neuron,* **10** (1993) 213–224.

150 Sharkey, J. and Butcher, S. P., Immunophilins mediate the neuroprotective effects of FK506 in focal cerebral ischemia, *Nature,* **371** (1994) 336–339.

151 Minami, M., Kuraishi, Y., Yabuuchi, K., Yamazaki, K., and Satoh, M., Induction of interleukin-1b mRNA in rat brain after transient forebrain ischemia, *J. Neurochem.,* **58** (1992) 390–394.

152 Rothwell, N. J. and Strijbos, P. J. L. M., Cytokines in neurodegeneration and repair, *Int. J. Devl. Neurosci.,* **13** (1995) 179–185.

153 Lindholm, D., Heumann, R., Meyer, M., and Thoenen, H., Interleukin-1 regulates synthesis of nerve growth factor in non-neuronal cells of rat sciatic nerve, *Nature,* **230** (1987) 658,659.

154 Wong, D. and Dorovini-Zis, K., Upregulation of intercellular adhesion molecule-1 (ICAM-1) expression in primary cultures of human brain microvessel endothelial cells by cytokines and lipopolysaccharide, *J. Neuroimmunol.,* **39** (1992) 11–22.

155 Rothwell, N. J. and Relton, J. K., Involvement of cytokines in acute neurodegeneration in th˒ CNS, *Neurosci. Behav. Res.,* **17** (1993) 217–227.

156 Qauglirello, V. J., Wispelwey, B., Long, W. J., and Scheld, W. M., Recombinant human interleukin-1 induces meningitis and blood-brain barrier injury in the rat, *J. Clin. Invest.,* **87** (1991) 1360–1366.

157 Fan, L., Young, P. R., Barone, F. C., Feuerstein, G. Z., Smith, D. H., and McIntosh, T. K., Experimental brain injury induces expression of interleukin-1b mRNA in the rat brain, *Mol. Brain Res.,* **30** (1995) 125–130.

158 Tracey, K. J. and Cerami, A., Tumor necrosis factor, other cytokines and disease, *Ann. Rev. Cell Biol.,* **9** (1993) 317–343.

159 Oehm, A., Behrmann, I., Falk, W., Pawlita, M., Maier, G., Klas, C., Li-Weber, M., Richards, S., Dhein, J., and Trauth, B. C., Purification and molecular cloning of the APO-1 cell surface antigen, a member of the tumor necrosis factor/nerve growth factor family, *J. Biol. Chem.,* **267** (1992) 10,709–10,715.

160 Dickson, D. W., Apoptosis in the brain, *Am. J. Pathol.,* **146** (1995) 1040–1044.

161 Matsuyama, T., Hata, R., Tagaya, M., Yamamoto, Y., Nakajima, T., Furuyama, J., Wanake, A., and Sugita, M., Fas antigen mRNA induction in postischemic murine brain, *Brain Res.,* **657** (1994) 342–346.

162 Weller, M., Frei, K., Groscurth, P., Krammer, P. H., Yonekawa, Y., and Fontana, A., Anti-Fas/APO-1 antibody mediated apoptosis of cultured human glioma cells. Induction and modulation of sensitivity by cytokines, *J. Clin. Invest.,* **94** (1994) 954–964.

163 D'Souza, S. D., Antel, J. P., and Freedman, M. S., Cytokine induction of heat shock protein expression in human oligodendrocytes: an interleukin-1 mediated mechanism, *J. Neuroimmunol.,* **50** (1994) 17–24.

164 Nowak, T. S., Synthesis of heat shock/stress protein during cellular injury, *Ann. NY Acad. Sci.,* **679** (1993) 142–156.

165 Kerr, J. F. R., Wyllie, A. H., and Currie, A. R., Apoptosis: a basic biological phenomenon with wide-ranging implications in tissue kinetics, *Br. J. Cancer,* **26** (1972) 239–257.

166 Martin, D. P., Schmidt, R. E., DiStefano, P. S., Lowry, O. H., Carter, J. G., Johnson, E. M., Jr., Inhibitors of protein synthesis and RNA synthesis prevent neuronal death caused by nerve growth factor deprivation, *J. Cell Biol.,* **106** (1988) 829–844.

167 Jacobsen, M. D., Burne, J. F., and Raff, M. C., Programmed cell death and Bcl-2 protection in the absence of a nucleus, *EMBO J.,* **13** (1994) 1899–1910.

168 Steller, H., Mechanisms and genes of cellular suicide, *Science,* **267** (1995) 1445–1449.

169 Wyllie, A. H., Kerr, J. F. R., and Currie, A. R., Cell death: the significance of apoptosis, *Int. Rev. Cytol.,* **68** (1980) 251–306.

170 Smeyne, R. J., Vendrell, M., Hayward, M., Baker, S. J., Miao, G. G., Schilling, K., Robertson, L. M., Curran, T., and Morgan, J. I., Continuous c-fos expression precedes programmed cell death in vivo, *Nature,* **363** (1993) 166–169.

171 Estus, S., Zaks, W. J., Freeman, R. S., Gruda, M., Bravo, R., and Johnson, E. M., Jr., Altered gene expression in neurons during programmed cell death: identification of c-jun as necessary for neuronal apoptosis, *J. Cell Biol.,* **127** (1994) 1717–1727.

172 Ham, J., Babij C., Whitfield J., Pfarr, C. M., Lallemand, D., Yaniv, M., and Rubin, L. L., A c-*jun* dominant negative mutant protects sympathetic neurons against programmed cell death, *Neuron,* **14** (1995) 927–939.

173 Debbas, M. and White, E., Wild-type p53 mediates apoptosis by E1A, which is inhibited by E1B, *Genes & Dev.,* **7** (1993) 546–554.

174 Yonish-Rouach, E., Resnitzki, D., Lotem, J., Sachs, L., Kimchi, A., and Oren, M., Wild-type p53 induces apoptosis of myeloid leukemic cells that is inhibited by interleukin 6, *Nature,* **352** (1991) 345–347.

175 Freeman, R. S., Estus, S., Johnson, E. M., Jr., Analysis of cell cycle related gene expression in postmitotic neurons, *Neuron,* **12** (1994) 343–355.

176 Shi, L., Nishioka, W. K., Th'ng, J., Bradbury, E. M., Litchfield, D. W., and Greenberg, A. H., Premature p34cdc2 activation required for apoptosis, *Science,* **263** (1994) 1143–1145.
177 Evan, G. I., Wyllie, A. H., Gilbert, C. S., Littlewood, T. D., Land, H., Brooks, M., Waters, C. M., Penn, L. Z., Hancock, D. C., Induction of apoptosis in fibroblasts by c-myc protein, *Cell,* **69** (1992) 119–128.
178 Bissonnette, R. P., Echeverri, F., Mahboubi, A., and Green, D. R., Apoptotic cell death induced by c-myc is inhibited by bcl-2, *Nature,* **359** (1992) 552–553.
179 Rubin, L. L., Gatchalian, C. L., Rimon, G., and Brooks, S. F., The molecular mechanisms of neuronal apoptosis, *Curr. Opin. Neurobiol.,* **4** (1994) 696–702.
180 Yuan, J. and Horvitz, H. R., The *Caenorhabitis elegans* cell death gene ced-4 encodes a novel protein and is expressed during the period of extensive programmed cell death, *Development,* **116** (1992) 309.
181 Thornberry, N. A., Bull, H. G., Calaycay, J. R., Chapman, K. T., Howard, A. D., Kostura, M. J., Miller, D. K., Milineaux, S. M., Weidner, J. R., and Aunins, J., A novel heterodimeric cysteine protease is required for interleukin 1-β processing in monocytes, *Nature,* **356** (1992) 768–774.
182 Yuan, J., Shaham, S., Ledoux, S., Ellis, H. M., and Horvitz, H. R., The *C. elegans* cell death gene *ced*-3 encodes a protein similar to mammalian Interleukin 1-β Converting Enzyme, *Cell,* **75** (1993) 641–652.
183 Wang, L., Miura, M., Bergeron, L., Zhu, H., and Yuan, J., Ich-1 an Ice/ced-3-related gene, encodes both positive and negative regulators of programmed cell death, *Cell,* **78** (1994) 739–750.
184 Fernandez-Alnemri, T., Litwack, G., and Alnemri, E. S., CPP32 a novel human apoptotic protein with homology to *Caenorhabitis elegans* cell death protein ced-3 and mammalian Interleukin 1-β Converting Enzyme, *J. Biol. Chem.,* **269** (1994) 30,761–30,764.
185 Miura, M., Zhu, H., Rotello, R., Hartwieg, E. A., and Yuan, J., Induction of apoptosis in fibroblasts by Il-1-β-Converting Enzyme, a mammalian homolog of the *C. elegans* cell death gene *ced*-3, *Cell,* **75** (1993) 653–660.
186 Gagliardini, V., Fernandez, P. A., Lee, R. K. K., Drexler, H. C. A., Rotello, R. J., Fishman, M. C., and Yuan, J., Prevention of vertebrate neuronal death by the *crm*A gene, *Science,* **263** (1994) 826–828.
187 Hengartner, M. O., Ellis, R. E., and Horvitz, H. R., *Caenorhabitis elegans* gene *ced*-9 protects cells from programmed cell death, *Nature,* **356** (1992) 494–499.
188 Hengartner, M. O. and Horvitz, H. R., *C. elegans* cell survival gene encodes a functional homolog of the mammalian proto-oncogene *bcl*-2, *Cell,* **76** (1994) 665–676.
189 Batistatou, A., Merry, D. E., Korsmeyer, S. J., and Greene, L. A., Bcl-2 affects survival but not neuronal differentiation of PC12 cells, *J. Neurosci.,* **13** (1993) 4422–4428.
190 Mah, S. P., Zhong, L. T., Liu, Y., Roghani, A., Edwards, R. H., and Bredesen, D. E., The proto-oncogen *bcl*-2 inhibits apoptosis in PC12 cells, *J. Neurochem.,* **60** (1993) 1183–1186.
191 Garcia, I., Martinou, I., Tsujimoto, Y., and Martinou, J. C., Prevention of programmed cell death of sympathetic neurons by the bcl-2 proto-oncogene, *Science,* **258** (1992) 302–304.
192 Allsopp, T. E., Wyatt, S., Paterson, H. F., and Davies, A. M., The proto-oncogene bcl-2 can selectively rescue neurotrophic factor-dependent neurons from apoptosis, *Cell,* **73** (1993) 295–307.
193 Martinou, J. C., Dubois-Dauphin, M., Staple, J. K., Rodriguez, I., Frankowski, H. Misotten, M., Albertini, P., Talabot, D., Catsicas, S., and Pietra, C., Overexpression of Bcl-2 in transgenic mice protects neurons from naturally occurring cell death and experimental ischaemia, *Neuron,* **13** (1994) 1017–1030.
194 Clem, R. J., Fechheimer, M., and Miller, L. K., Prevention of apoptosis by a Baculovirus gene during infection of insect cells, *Science,* **254** (1991) 1388–1390.

195 Sugimoto, A., Friesen, P. D., and Rothman, J. H., Baculovirus p35 prevents developmentally programmed cell death and rescues a ced9 mutant in the nematode *Caenorhabitis elegans*, *EMBO J.*, **13** (1994) 2023–2028.

196 Rabizadeh, S., LaCount, D. J., Friesen, P. D., and Bredesen, D. E., Expression of the Baculovirus p35 gene inhibits mammalian neural cell death, *J. Neurochem.*, **61** (1993) 2318–2321.

197 Davies, A. M., The Bcl-2 family of proteins and the regulation of neuronal survival, *Trends Neurosci.*, **18** (1995) 355–358.

198 Oltvai, Z. N. , Milliman, C. L., and Korsmeyer, S. J., Bcl-2 heterodimerizes in vivo with a conserved homolog, Bax, that accelerates programmed cell death, *Cell,* **74** (1993) 609–619.

199 Yin, X. M., Oltvai, Z. N., and Korsmeyer, S. J., BH1 and BH2 domains of Bcl-2 are required for ihibition of apoptosis and heterodimerization with Bax, *Nature,* **369** (1994) 321–323.

200 Sato, T., Hanada, M., Bodrug, S., Jrie, S., Jwama, N., Boise, L. H., Thompson, C. B., Golemis, E., Fong, L., and Wang, H. G., Interactions among members of the Bcl-2 protein family analyzed with a yeast two-hybrid system. *Proc. Natl. Acad. Sci. USA,* **91** (1994) 9238–9242.

201 Barinaga, M., Cell suicide: by ICE, not fire, *Science,* **263** (1994) 754–756.

202 Boise, L. H., Gonzalez, G. M., Postema, C. E., Ding, L., Lindsten, T., Turka, L. A., Mao, X., Nunez, G., and Thompson, C., B. bcl-x, a bcl-2-related gene that functions as a dominant regulator of apoptotic cell death, *Cell,* **74** (1993) 597–608.

203 Motoyama, N. , Wang, F., Roth, K. A., Sawa, H., Nakayama, K., Nakayama, K.,Negishi, I., Senju, S., Zhang, Q., Fujii, S., and Loh., D. Y., Massive cell death of immature hematopoietic cells and neurons in Bcl-x deficient mice, *Science,* **267** (1995) 1506–1510.

204 Krajewski, S., Mai., J. K., Krajewska, M., Sikorska, M., Mossakowski, M. J., and Reed, J. C., Upregulation of Bax protein levels in neurons following cerebral ischaemia, *J. Neurosci.* **15** (1995) 6364–6376.

205 Honkaniemi, J., Massa, S. M., and Sharp, F. R., Sequential induction of specific apoptosis-associated genes in gerbil CA1 pyramidal hippocampal neurons following global ischaemia, *J. Cereb. Blood Flow Metab.,* **15** (1995) S148.

206 Krajewski, S., Tanaka, S., Takayama, S., Schibler, M. J., Fenton, W., and Reed, J. C., Investigations of the subcellular distribution of the *bcl-2*oncoprotein: residence in the nuclear envelope, endoplasmatic reticulum, and outer mitochondrial membranes, *Cancer Res.,* **53** (1993) 4701–4714.

207 Lam, M., Dubyak, G., Chen, L., Nunez, G., Miesfeld, R. L., and Distelhorst, D. W., Evidence that Bcl-2 represses apoptosis by regulating endoplasmatic reticulum-associated Ca^{2+} fluxes, *Proc. Natl. Acad. Sci. USA,* **91** (1994) 6569–6573.

208 Hockenberry, D., Oltvai, Z., Yin, X. M., Milliman, C., and Korsmeyer, S. J., Bcl-2 functions in an antioxidant pathway to prevent apoptosis, *Cell,* **75** (1993) 241–251.

209 Kane, D. J., Sarafin, T. A., Auton, S., Hahn, H., Gralla, F. B., Valentine, J. C., Ord, T., and Bredesen, D. E., Bcl-2 inhibition of neural cell death: decreased generation of reactive oxygen species, *Science,* **262** (1993) 1274–1276.

210 Fernandez-Sarabia, M. and Bischoff, J. R., Bcl-2 associates with the *ras*-related protein R-*ras* p23, *Nature,* **366** (1993) 274,275.

211 Buttyan, R., Zakeri, Z., Lockshin, R., and Wohlgemuth, D., Cascade induction of c-fos, c-myc and heat shock 70 K transcripts during regression of the rat ventral prostate gland, *Mol. Endocrinol.,* **2** (1988) 650–657.

212 Dragunow, M. and Preston, K., The role of inducible transcription factors in apoptotic nerve cell death, *Brain Res. Rev.,* **21** (1995) 1–28.

213 DeFelipe, C. and Hunt, S. P., The differential control of c-jun expression in regenerating sensory neurons and their associated glial cells, *J. Neurosci.,* **14** (1994) 2911–2923.

214 Anderson, A. J., Pike, C. J., and Cotman, C. W., Differential induction of immediate early gene proteins in cultured neurons by β-Ameloid (Aβ): association of c-jun with Aβ-induced apoptosis, *J. Neurochem.,* **65** (1995) 1487–1498.

215 Wiessner, C., Cyclin D1—a gene associated with programmed cell death in sympathetic ganglion neurons—is expressed in the postischemic rat brain, *J. Cereb. Blood Flow Metab.,* **15** (1995) S146.

216 Schreiber, S. S., Sakhi, S., Dugich-Djordjevic, M. M., and Nichols, N. R., Tumor suppressor p53 induction and DNA damage in hippocampal granule cells after adrenalectomy, *Exp. Neurol.,* **130** (1994) 368–376.

217 Yonish-Rouach E., Wilder, S., Kimchi, A., May, E., Lawrence, J. J., May, P., and Oren, M., p53-mediated cell death: relationship to cell cycle control, *Mol. Cell Biol.,* **13** (1993) 1415–1423.

218 El-Deiry, W. S., Tokino, T., Velculescu, V. E., Levy, D. S., Parsons, R., Trent, J. M., Lin, D. T., Mercer, W. E., Kinzler, K. W., and Vogelstein, B., WAF1 a potential mediator of p53 tumor suppression, *Cell,* **75** (1993) 817–825.

219 Harper, J. W., Adami, G. R., Wei, N. , Keyomarsi, K., and Elledge, S. J., The p21 cdk-interacting protein Cip1 is a potent inhibitor of G1 cyclin-dependent kinases, *Cell,* **75** (1993) 805–816.

220 Sherr, C. J. and Roberts, J. M., Inhibitors of mammalian G1 cyclin-dependent kinases, *Genes & Dev.,* **9** (1995) 1149–1163.

221 Davis, R. J., MAPKs: new JNK expands the group, *Trends Biochem. Sci.,* **19** (1994) 470–473.

222 Nobes, C. D., Reppas, J. B., Markus, A., and Tolkovsky, A. M., Active p21Ras is sufficient for rescue of NGF-dependent rat sympathetic neurons, *Neuroscience,* **70** (1996) 1067–1079.

223 Xia, Z., Dickens, M., Raingeaud, J., Davis, R. J., and Greenberg, M. E., Opposing effects of ERK and JNK-p38 MAR kinases on apoptosis, *Science,* **270** (1995) 1326–1331.

224 White, K., Grether, M. E., Abrams, J. M., Young, L., Farrell, K., and Steller, H., Genetic control of programmed cell death in *Drosophila, Science,* **264** (1994) 677–683.

225 Heintz, N., Cell death and the cell cycle: a relationship between transformation and neurodegeneration?, *Trends Biochem. Sci.,* **18** (1993) 157–159.

226 Feddersen, R. M., Ehlenfeldt, R., Yunis, W. S., Clark, H. B., and Orr, H. T., Disrupted cerebellar cortical development and progressive degeneration of purkinje cells in SV40 T antigen transgenic mice, *Neuron,* **9** (1992) 955–966.

227 Lee, E. Y., Chang, C. Y., Hu, N. , Wong, Y. C., Lai, C. C., Herrup, K., Lee, W. H., and Bradley, A., Mice deficient for Rb are nonviable and show defects in neurogenesis and hemotopoiesis, *Nature,* **359** (1992) 288–294.

228 Herrup, K. and Busser, J. C., The induction of multiple cell cycle events precedes target-related neuronal death, *Development,* **121** (1995) 2385–2395.

229 Ross, M. E., Cell division and the nervous system: regulating the cycle from neural differentiation to death, *Trends Neurosci.,* **19** (1996) 62–68.

230 Farinelli, S. E. and Greene, L. A., Cell cycle blockers Mimosine, Ciclopirox, and Deferoxamine prevent the death of PC12 cells and postmitotic sympathetic neurons after removal of trophic support, *J. Neurosci.,* **16** (1996) 1150–1162.

231 Thompson, C. B., Apoptosis in the pathogenesis and treatment of disease, *Science,* **267** (1995) 1456–1462.

232 Verheul, H. B., Balazs, R., Berkerbach van Sprenkel, J. W., Tulleken, C. A. F., Nicolay, K., Tamminga, K. S., and Van Lookeren Campagne, M., Comparison of diffusion-weighted MRI with changes in cell volume in a rat model of brain injury, *NMR in BioMed.,* **7** (1994) 96–100.

233 Ter Horst, G. J., Knollema, S., Knigge, M. F., Krugers, H. J., Van de Witte, S. V., Postema, F., and Hom, H. W., Silver staining of traumatized neurons: application of a Gallyas procedure in experimental cerebral hypoxia/ischemia research, *Neurosci. Prot.,* **050-02** (1995) 1–13.

234 Krugers, H. J., Medema, R. M., Postema, F., and Korf, J., Induction of glial fibrillary acidic protein (GFAP)-immunoreactivity in the rat dentate gyrus after adrenalectomy: comparison with neurodegenerative changes using silver impregnation, *Hippocampus,* **4** (1994) 307–314.

235 Izumi, Y., Pinard, E., Roussel, S., and Seylaz, J., Insulin protects brain tissue against focal brain ischemia in rats, *Neurosci. Lett.,* **144** (1992) 121–123.

236 Hockenbery, D., Nunez, G., Milliman, C., Schreiber, R. D., and Korsmeyer, S. J., Bcl-2 is an inner mitochondrial membrane protein that blocks programmed cell death, *Nature,* **348** (1990) 334–336.

237 Li, Y., Chopp, M., Jiang, N. , Zhang, M. G., and Zalago, C., Induction of DNA fragmentation after 10 to 120 minutes of focal cerebral ischemia in rats, *Stroke,* **26** (1995) 1252–1258.

2

Imaging of Stroke

Henk Stevens, Hugo M. L. Jansen, Jacques DeReuck, and Jakob Korf

1. INTRODUCTION

In vivo demonstration of stroke-related pathological processes has evolved from scintigraphical and angiographical techniques to computer-assisted tomographic technologies. Today, computer tomography (CT) and magnetic resonance imaging (MRI) are widely available *(1–5)*. CT is mainly used to distinguish between hemorrhagic and nonhemorrhagic pathology in the acute phase of stroke *(6)*. Both MRI and CT visualize morphological changes in the brain; MRI, however, has a far better spatial resolution and allows visualization of other aspects of the pathology as well. To detect functional changes in both the normal and the pathological brain, additional procedures can be used such as visualization of regional cerebral blood flow with positron emission tomography (PET) or single photon emission tomography (SPECT). Both hypoperfusion of the core of the infarcted brain area and the penumbra and hyperperfusion in more peripherally located brain regions (luxury perfusion) can readily be shown *(7–18)*. PET has the advantages of a higher resolution and absolute quantitative information on flow and metabolism of brain pathology compared to SPECT, but SPECT is cheaper and more accessible.

The merging of PET and MRI not only supports the previous anatomical (CT and MRI) technologies, but offers additional functional possibilities as well. Both MR- and PET-based technologies are still rapidly developing with applications in the clinical and experimental stroke research. In this chapter, we will present some recent developments and applications of imaging techniques in stroke. The usefulness of these technologies to visualize pathological changes during the development of ischemic stroke will be considered.

2. TIME-COURSE OF CEREBRAL ISCHEMIA

The possibility of distinguishing between reversible and irreversible damaged tissue (penumbra vs core of the infarct) is of main clinical interest, as reversibility could allow (potential) therapeutic interventions. It is not our purpose to describe the course of stroke pathology in detail, since other reviews deal with that subject more intensively. However, it is important to emphasize the major phases of the development of ischemia to infarction as they could eventually be distinguished by the various neuroimaging techniques.

From: Clinical Pharmacology of Cerebral Ischemia *Edited by: G. J. Ter Horst and J. Korf Humana Press Inc., Totowa, NJ*

Stroke is most often caused by arterial stenosis or obstruction owing to a (thrombo) embolic process with temporary or permanent obstruction of blood flow in the downstream territory. In the initial phase, there is a decrease in blood flow in a well-defined brain area, in which the lack of energy supply causes immediate (in experimental animals within minutes) depolarization of nerve cells and presumably of other cells as well. Such depolarizations can be compensated for a few minutes by anaerobic glycolysis, but because the energy reserves of the brain are limited, most cells will soon become depolarized, and irreversibly damaged, leading to cell death if blood flow remains impaired. During this process of cell death, the ion gradient over the cellular membrane disappears and water enters the cell, so the extracellular space, comprising approx 20–25% of the normal human brain, is diminished by 50%. In the tissue surrounding the death core, blood flow may remain insufficient to maintain normal function, but cells are still polarized, because energy supply is sufficient to maintain ion gradient over the membrane; subsequently the extracellular space remains normal. Part of the energy required to maintain membranous gradients is of anaerobic origin, and lactate may become an important energy substrate for neurons. These events are thought to occur within the first 12 h poststroke, with the predominant features being hypoperfusion and cellular edema.

As soon as the damage has become irreversible, inflammatory processes start, such as the activation of the microglia cells and leukocyte infiltration *(19–25)*. The activation of microglia is an early event after (any) brain damage. Microglia comprise 5–12% of the cells in the brain and are considered as resident macrophages. These cells can be activated, and start producing nitric oxide, excitatory amino acids, proteases, and cytokines. Microglia can also change from an activated state to a phagocytic state and start to remove cellular debris. Soon after the infarction, an infiltration of T-lymphocytes is also seen in both the infarcted and the surrounding regions. T-cell infiltration may precede microglia activation as has recently been shown in a photochemical stroke model *(20)*. Such infiltration mostly occurs when cellular adhesion molecules (such as ICAM-1) are expressed upon stimulation with cytokines (interferon-γ, interleukin-1, and tumor necrosis factor-α) and by opening of the blood-brain barrier (BBB).

Inflammatory events take place in the core of the infarct but extend further into the penumbra regions. The role of these processes is unclear in the pathogenesis of stroke. The inflammatory response is maximal after 1–3 wk, and approx 6 wk after stroke onset, the response starts to decline to normality *(24)*. In this case, the core of the infarct becomes partially cystic, and the extracellular space is increased. Permanent degenerative changes such as loss of specific neurons and other cells may have occurred, whereas concomitant compensory changes, including regenerative processes, may be still underway. Neuroimaging can be a tool to visualize these time-dependent changes. The potency to visualize these aspects of pathology is highly dependent upon the imaging modality. The time course of major events following stroke is shown in Fig. 1.

3. CT

CT scanning is an accurate diagnostic tool in patients with cerebrovascular diseases. CT is especially used in the acute phase of stroke to differentiate between hemorrhagic and nonhemorraghic events *(6,26)*. Within the first 8–12 h, the majority of nonhemor-

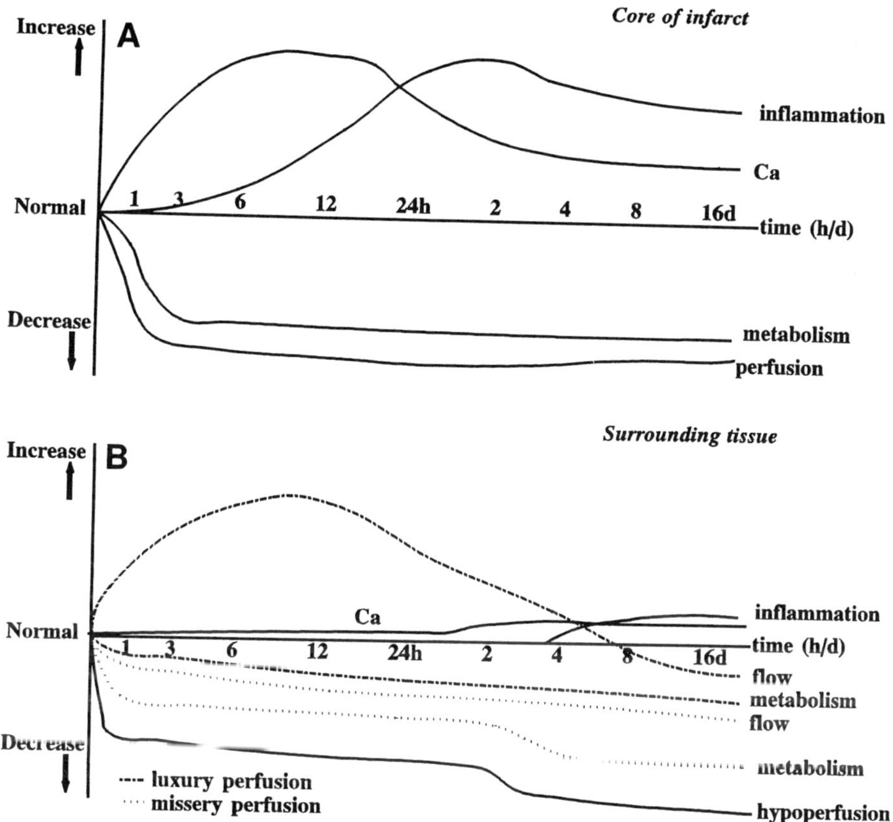

Fig. 1. Time-course of metabolic and inflammatory processes after stroke in the core (**A**) and the surrounding brain tissue (**B**).

rhagic infarcts escape CT demonstration because the density values of these infarcts correspond, at this time, with those of normal brain tissue. Larger infarcts can sometimes be suspected in the earlier phase by the presence of brain swelling. Demarcation of the infarct begins between 8–24 h, and becomes definite in relation to the particular vascular territories after 2–3 d. After 3–5 d it becomes possible to visualize contrast enhancement in the infarction, starting at the margin of the infarcted area and extending during the next 1–2 wk. This enhancement of contrast in the infarcted zone is owing to hyperemia and is particularly important at 10–20 d after stroke, as CT examination made at that time may not reveal the infarct zone without contrast. At this time the density values that are measured in the infarcted region (with reparative processes, edema, and increased perfusion) correspond to those of healthy brain tissue (fogging effect).

4. MRI

MRI has already proven to be of great practical value in diagnosing stroke, using T_1, T_2-images and T_1-images after iv injection of contrast (gadolinium). MRI visualizes body tissue using a static, oscillating magnetic field that excites protons. When the oscillating magnetic field is stopped, excited protons relax and induce a radio frequency in a receiver coil. The signal intensity is reconstructed on gray scale images on which the different body tissues can be recognized *(27,28)*. With MRI, cerebral ischemia can be visualized earlier than CT. Because of its better resolution, lacunar infarcts, territorial infarcts of the posterior fossa, especially of the brainstem are more frequently demonstrated by MRI than by CT. Old hemorrhages, and infarcts with a hemorrhagic component, can be diagnosed because of the characteristic signal behavior of the iron ions in the hemosiderin deposits in the infarct. The integrity of the BBB can be showed by contrast enhancement in brain areas where the BBB is disrupted. In stroke, it appears that both intact and disrupted BBB can be seen. In the initial phase, the BBB is often intact, whereas after several days or weeks, when the damage and debris is maximal, the BBB becomes disrupted *(1,3–5,29)*. At this time, lymphocyte infiltration is also maximal.

More recent and advanced technologies based on MRI, like magnetic resonance spectroscopy (MRS) and diffusion and perfusion MRI can partially support the previous anatomical (CT and MRI) technologies and offer additional functional possibilities as well. MRS can noninvasively measure metabolites such as N-acetylaspartate (NAA) and lactate in the human brain during normal and pathologic conditions such as after stroke. The NAA is almost exclusively located in neurons and is widely used as a neuronal marker in MRS studies of brain infarction *(15,30–39)*. Several (partial serial) studies, with MRS in patients with stroke show a reduced NAA level and an increased level of lactate within the infarcted area during the acute stage of stroke. During the chronic stage, a remarkably reduced level of NAA in large middle cerebral artery infarctions and, in some studies, elevated levels of lactate (up to 23 mo) are found *(13,40)*. Presence of lactate in the acute stage of brain infarction probably reflects the degree of ischemia. In the chronic stage, lactate may be produced by inflammatory and phagocytic cells and by glia cells that metabolize glucose mainly to lactate even under normoxic conditions *(13,14)*. MRS could be used in the future for the demonstration of biochemical markers in the early stage of stroke to provide guidance in prognosis and treatment planning.

Diffusion and perfusion MR have been used in clinical and experimental stroke studies *(12)*. Whereas T_1 and T_2 weighed MRI is usually not very sensitive to visualize early events, both diffusion and perfusion MRI have more potential in this regard. Diffusion MR visualizes changes in the diffusion characteristics of interstitial water, apparent diffusion coefficients (ADC_w) *(41)*. Changes in ADC_w are seen within minutes after the onset of stroke, concomitant with shrinkage of the extracellular space, as determined by whole tissue impedance measurements. Since change in ADC_w is one of the earliest changes seen after cession of blood flow, it is possible to visualize ischemic brain tissue within 1 h after stroke onset *(42)*. Welch et al. (1995) histopathologically rated tissue changes according to the results obtained in the rat with a MCA-stroke-model with T_2 and ADC_w *(41)*. Whereas T_2 values began to increase after about 16 h poststroke, the ADC_w had already decreased during that previous period. After 24 h, the changes in both parameters were increased and followed each other closely. The changes are illustrated in Fig. 2.

Regional cerebral blood flow (CBF) can be assessed with MR angiography (based on the difference between excitation and recording of MRI, as excited blood moves out of the imaging plane); on clearance studies MRI-detectable tracers are used (using MRI-detectable tracers and A-V differences); on bolus track imaging (measuring tran-

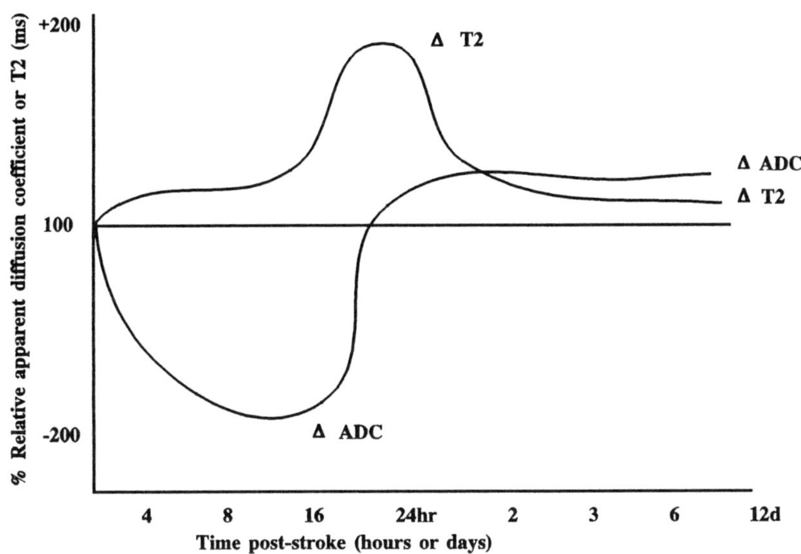

Fig. 2. Changes of T_2 and apparent diffusion coefficient of water in the core of an ischemic infarct, as derived from an experimental animal stroke model. The figure is reconstructed from data of ref. *41*.

sit time of a paramagnetic substance, such as Gd-DPTA); or on MRI based on blood oxygenation level-dependent imaging (functional MRI). Functional MRI is used to estimate regional CBF and metabolism making use of different signal behavior of oxidized hemoglobin and deoxyhemoglobin. Functional MRI is especially used for nonpathological brain studies. In pathological conditions when tissue oxygen is decreased or absent, the possible changes in deoxyhemoglobin can be masked and not well predicted or related to rCBF. Under ischemic conditions, the bolus track imaging is best suited. Bolus track perfusion has been used experimentally in a few studies to identify brain regional hypoperfusion *(43)*.

5. PET

Radiolabeled markers (positron emitters) enable us to get a better view of the metabolism of the brain in both normal circumstances and pathological states. The most common radiolabels for stroke in PET are $^{15}O_2$, $H_2^{15}O$, $C^{15}O_2$, and $C^{15}O$ *(7–10,17,44)*. With both $H_2^{15}O$ and $C^{15}O_2$ the regional cerebral blood flow (rCBF) can be measured. $C^{15}O$ can be used to visualize blood volume (CBV); regional cerebral metabolic rate for oxygen (rCRMO$_2$), and oxygen extraction factor (OEF) can be measured with $^{15}O_2$ and ^{11}F-deoxyglucose is used to study regional cerebral metabolic rate for glucose (rCRM$_{glu}$). Besides these labels, today ^{55}Co is used as a calcium analog to visualize ischemic lesions in experimental studies *(16,45,46)*.

6. BLOOD FLOW AND OXYGEN METABOLISM

Both $H_2^{15}O$ and $C^{15}O_2$ can be used to quantify the degree of ischemia after stroke in the core of infarction and in the border zone. During the acute phase of stroke (up to 4 d), irreversible damage has taken place in the infarcted core that is characterized by a rCBF below 12 mL/100 g/min and with a $rCMRO_2$ of 1.5 mL/100 g/min. Compared with the contralateral mirror region, this is respectively below 45% and 65%. The CMR_{glu} and OEF are reduced compared to the contralateral mirror region, whereas CBV remains unchanged. During the first 2 wk after onset of symptoms $CMRO_2$, CMR_{glu}, CBV, and OEF do not change significantly while flow increases slightly *(7,9)* in the core of infarct. This slight increase in flow does not affect the metabolic state of the necrotic tissue as indicated by the reduced OEF. In the border zone of ischemia there may still be viable tissue that can be subdivided into tissue with a reduced rCBF (12–18 mL/100 g/min), tissue with increased OEF (misery perfusion), and tissue with high rCBF and low $rCMRO_2$ (luxury perfusion) *(7,8)*. Studies in stroke patients showed that tissue with a rCBF between 12–18 mL/100 g/min in the acute phase, the rCBF, $CMRO_2$, and $rCMR_{glu}$ decrease during the first week, turning this tissue into infarction. Misery perfusion is seen in 45–57% of cases studied in the first 4 d after stroke and is accompanied by a slightly decreased $rCMRO_2$ and CMR_{glu}, whereas CBF is slightly reduced. During the following 2 wk, with a few exceptions, this tissue will deteriorate into necrotic tissue when the $rCMRO_2$ declines and the blood flow stays the same *(7,8)*. Luxury perfusion in the ischemic area is characterized by initially high CBF and low $CMRO_2$. In a study of 22 patients, the role of differences in perfusion in the acute phase and the clinical outcome was studied *(47)*. This study proved that patients with hyperperfusion in the peri-infarct area fared slightly better after 1 yr than patients without hyperperfusion in the peri-infarct area.

This viable peri-infarct tissue may form a substrate for potential therapeutics of ischemic stroke, but the therapeutic routines usually applied today cannot yet prevent all the metabolic derangement and progression to necrosis of this viable peri-infarct tissue.

7. ⁵⁵CO-PET IN VISUALIZING CEREBRAL ISCHEMIA

Since massive calcium influx takes place during ischemia, cobalt as a calcium analog has been postulated to be a suitable isotope for visualizing this mechanism. Animal experiments have already shown that cobalt acts like a calcium analog in kainic acid lesions in the brain *(45)*. Therefore, ⁵⁵Co has been used to visualize cerebral ischemia. Jansen et al. studied six patients with a middle cerebral artery stroke with ⁵⁵Co-PET in the first 2 wk after onset of symptoms *(16)*. This study showed Co uptake in the infarcted area of the brain independently of the integrity of the BBB and a positive correlation with clinical prognosis as defined on the Orgogozo scale. Figure 3 shows a picture of a patient with a middle cerebral artery stroke in which in the first week after onset of symptoms, Co-PET was made in combination with $^{15}O_2$, $C^{15}O_2$, and $C^{15}O$ PET-scan and MRI. In this study, it showed that Co uptake in the acute phase of stroke mainly takes place in the penumbra region and in the later phase, in patients with major stroke, in the infarcted area. The cobalt showed to have a partial overlap with CBF and Gd–DTPA of the MRI *(46)*.

8. SPECT

Most clinical applications of SPECT concern regional CBF measurement with tracers such as iodine-123-isopropyl iodocamphetamine (IA), Technetium-99m hexamethylpropyleneamine oxime (⁹⁹ᵐTcHMPAO), or ¹³³Xe *(47,49,50)*. The IA is not

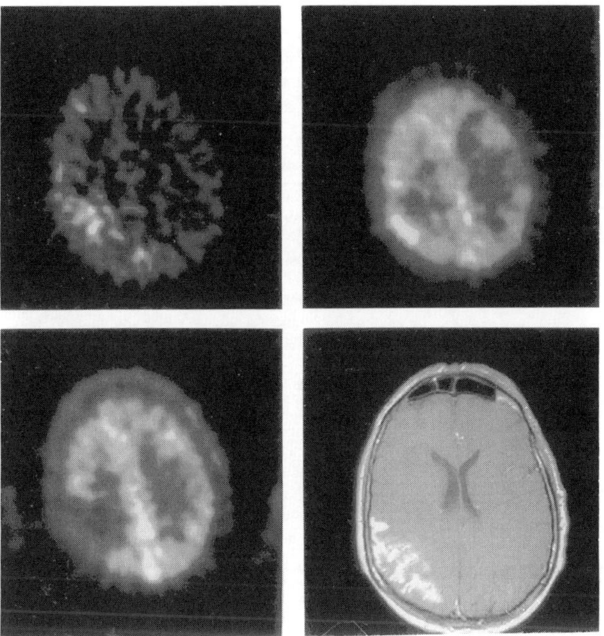

Fig. 3. PET scans of a patient with an MCA infarction in the parietal occipital right cerebral cortex made 4 d after onset of the symptoms. Left upper scan: ^{55}Co, right upper scan: $C^{15}O_2$, left bottom: $^{15}O_2$, right bottom: GdMRI.

useful, since a considerable redistribution of the tracer invalidates the tracer in stroke *(51)*; so not only the infarct but also the surrounding tissue are similarly radioactive. For clinical routine, ^{99m}Tc-HMPAO is the most suitable tracer in SPECT for visualizing CBF *(11)*. In the acute stroke, SPECT is very accurate for visualization of the extent of the hypoperfusion in stroke (with high sensitivity). Generally >90% of stroke patients exhibit detectable perfusion deficits. A few days after stroke onset (e.g., after 5 d) luxury perfusion may occur and mask hypoperfusion in the infarct area. There is uncertainty if SPECT scans made within the first 12 h after stroke can predict clinical outcome. Leukocyte infiltration has been shown both with In-oxime and Tc-HMPAO-labeled leukocytes *(21)*. In general, SPECT can visualize after major strokes, most significantly after 2 wk. For the detection of inflammation, recent developments may lead to significant improvements *(52)*. Thallium 201 SPECT has mainly been used to detect malignancies of the brain, but more recently it has been shown that with this technique stroke can be detected, presumably in a later phase only *(29)*. The uptake may be related to the breakdown of the blood-brain barrier and may reflect K^+ uptake or exchange in degenerative neuronal tissue. Similar to PET, cobalt has been explored as a tracer for SPECT in stroke *(53)*, using ^{57}Co as a calcium analog (Fig. 4). In SPECT, the role of cobalt accumulation during the inflammatory reaction after stroke is now studied using Co-SPECT in comparison with ^{99}Tc-HMPAO-labeled leukocyte SPECT. Results indicate that Co uptake and leukocyte accumulation tend to concur in the infarction *(46)*. The disadvantages of SPECT compared with PET diminish since it is possible in

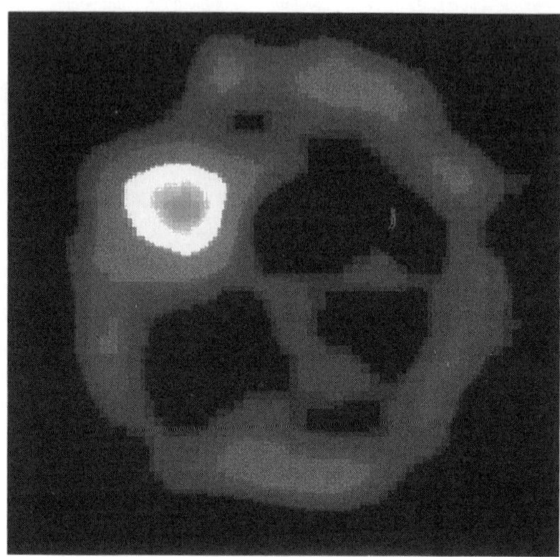

Fig. 4. Co-SPECT of patient with an MCA infarction in the right hemisphere, made 12 d after onset of symptoms.

SPECT to quantify flow in terms of absolute perfusion values and the new generation of SPECT cameras show almost the same spatial resolution as PET.

ACKNOWLEDGMENTS

Part of the work described is supported by the Netherlands Heart Foundation (92.305) and the Dutch Technology Foundation (GGN 222741).

REFERENCES

1 Bogousslavsky, J., Fox, A. J., Barnett, H. J. M., Hachinski, V. C., Vinitski, S., and Carey, L. S., Clinicotopographic correlation of small vertebro-basilar infarct using magnetic resonance imaging, *Stroke,* **17** (1986) 929–938.

2 Brant-Zawadzki, M., Solomon, M., Newton, T. H., Weinstein, P., Schmidley, J., and Norman, D., Basic principles of magnetic resonance imaging in cerebral ischemia and initial clinical experience, *Neuroradiology,* **27** (1985) 517–520.

3 Ramadan, N. M., Deveshwar, R., and Levine, S. R., Magnetic resonance and clinical cerebrovascular disease. An update, *Stroke,* **20** (1989) 1279–1283.

4 Ross, M. A., Biller, J., Adams, H. P. Jr., and Dunn, V., Magnetic resonance imaging in Wallenberg's lateral medullary syndrome, *Stroke,* **17** (1986) 542–545.

5 Rothrock, J. F., Lyden, P. D., Hesselink, J. R., Brown, J. J., and Healy, M. E., Brain magnetic resonance imaging in the evaluation of lacunar stroke, *Stroke,* **18** (1987) 781–788.

6 Koudstaal, P. J., Van Gijn, J., Lodder, J., Frenken, C. W. G. M., Vermeulen, M., Franke, C. L., Hijdra, A., and Bulens, C., Transient ischemic attacks with and without a relevant infarct on computed tomographic scans cannot be distinguished clinically, *Arch. Neurol.,* **48** (1991) 916–920.

7 Heiss, W.-D. and Herholz, K., Assessment of pathophysiology of stroke by positron emission tomography, *Eur. J. Nucl. Med.,* **21** (1994) 455–465.

8   Heiss, W.-D., Huber, M., Fink, G. R., Herholz K., Pietrzyk, U., Wagner, R., and Wienhard, K., Progressive derangement of periinfarct viable tissue in ischemic stroke, *J. Cereb. Blood Flow Metab.*, **12** (1992) 193–201.

9   Heiss, W.-D., Fink, G. R., Huber, M., and Herholz, K., Positron emission tomography imaging and the therapeutic window, *Stroke,* **24(Suppl 1)** (1993) 150–153.

10 Heiss, W.-D., Emunds, H.-G., and Herholz, K., Cerebral glucose metabolism as a predictor of rehabilitation after ischemic stroke, *Stroke,* **21** (1993) 1784–1788.

11 Holman, B. L. and Devous, M. D., Functional brain SPECT: the emergence of a powerful clinical method, *N. Nucl. Med.,* **33** (1992) 1888–1904.

12 Hossmann, K.-A. and Hoehn-Berlage, M., Diffusion and perfusion MR imaging of cerebral ischemia, *Cerebrov. Brain Metab. Rev.,* **7** (1995) 187–217.

13 Houkin, K., Kamada, K., Kamiyama, H., Iwasaki, Y., Abe, H., and Kashiwaba, T., Longitudinal changes in proton magnetic resonance spectroscopy in cerebral infarction, *Stroke,* **24** (1993) 1316–1321.

14 Hugg, J. W., Duijn, J. H., Matson, G. B., Maudsley, A. A., Tsuruda, J. S., Gelinas, D. F., and Weiner, M. W., Elevated lactate and alkalosis in chronic human brain infarction observed by 1H and 31P MR spectroscopic imaging, *J. Cereb. Blood Flow Metab.,* **12** (1992) 724–744.

15 Husted, C. A., Duijn, J. H., Matson, G. B., Maudsley, A. A., and Weiner, M. W., Molar quantitation of in vivo proton metabolites in human brain with 3D magnetic resonance spectroscopic imaging, *Magn. Reson. Imaging,* **12** (1994) 661–667.

16 Jansen, H. M. L., Pruim, J., Van der Vliet, A. M., Paans, A. M. J., Hew, J. M., Franssen, E. J. F., De Jong, B. M., Kosterink, J. G. W., Haaxma, R. and Korf, J., Visualization of damage brain tissue after ischemic stroke with 55Co positron emission tomography, *J. Nuclear Med.,* **35** (1994) 456–460.

17 Powers, W. J., Grubb, R. L., Darriet, D., and Raichle, M. E., Cerebral blood flow and cerebral metabolic rate of oxygen requirements for cerebral function and viability in humans, *J. Cerebr. Blood Flow Metab.,* **5** (1985) 600–608.

18 Wise, R. J. S., Bernardi, S., Frackowiak, S. J., Legg, N. J., and Jones, T., Serial observations on the pathophysiology of acute stroke, the transition from ischaemia to infarction as reflected in regional oxygen extraction, *Brain,* **106** (1983) 197–222.

19 Fassbender, K., Mössner, R., Motsch, L., Kischka, U., Grau, A., and Hennerici, M., Circulating selectin- and immunoglobulin-type adhesion molecules in acute ischemic stroke, *Stroke,* **26** (1995) 1361–1364.

20 Jander, S., Kraemer, M., Schroeter, M., Witte, O. W., and Stoll, G., Lymphocytic infiltration and expression of intercellular adhesion molecule-1 in photochemically induced ischemia of the rat cortex, *J Cerebr. Blood Flow Metab.* **15** (1995) 42–51.

21 Kao, C. H., Wang, P. Y., Wang, Y. L., Chang, L., Wang, S. J., and Yeh, S. H., A new prognostic index—leucocyte infiltration—in human cerebral infarcts by ^{99}Tc^{m}-HMPAO-labelled white blood cell brain SPECT, *Nucl. Med. Comm.,* **12** (1991) 1007–1012.

22 Kochanek, P. M. and Hallenbeck, J. M., Polymorphonuclear leukocytes and monocytes/macrophages in the pathogenesis of cerebral ischemia and stroke, *Stroke,* **23** (1992) 1367–1379.

23 Silvestrini, M., Pietroiusti, A., Magrini, A., Matteis, M., Carta, S., Bernardi, G., and Galante, A., Leukocyte aggregation in patients with a previous cerebral ischemic event, *Stroke,* **25** (1994) 1390–1392.

24 Wang, P.-Y., Kao, C.-H., Mui, M.-Y., and Wang, S.-J., Leukocyte infiltration in acute hemispheric ischemic stroke, *Stroke,* **24** (1993) 236–240.

25 Wood, P. L., Microglia as a unique cellular target in the treatment of stroke: potential neurotoxic mediators produced by activated microglia, *Neurol. Res.,* **17** (1995) 242–248.

26 Hacke, W., Hennerici, M., Gelmers, H. J., and Krämer, G. (eds.), *Cerebral Ischemia,* Springer Verlag, Berlin, 1991.

27 Bushong, S. C., *Magnetic Resonance Imaging Physical and Biological Principles,* Mosby, St. Louis, 1988, pp. 84–105.

28 Kent, D. L. and Larson, E. B., Magnetic resonance imaging of the brain and spine, is clinical efficacy established after the first decade? *Ann. Intern. Med.,* **108** (1988) 402–424.

29 Bernat, I., Toth, G., and Kovács, L., Tumour-like thallium-201 accumulation in brain infarcts, an unexpected finding on single–photon emission tomography, *Eur. J. Nucl. Med.,* **21** (1994) 191–195.

30 Duijn, J. H., Matson, G. B., Maudsley, A. A., Hugg, J. W., and Weiner, M. W., Human brain infarction: proton MR spectroscopy, *Radiology,* **183** (1992) 711–718.

31 Felber, S., Luef, G., and Aichner, F., 1-H-Localized magnetic resonance spectroscopy: preliminary observations in transients ischemic attacks, *J. Neurol. Sci.,* **108** (1992) 35–38.

32 Felber, S. R., Aichner, F. T., Sauter, R., and Gerstenbrand, F., Combined magnetic resonance imaging and proton magnetic resonance spectroscopy of patients with acute stroke, *Stroke,* **23** (1992) 1106–1110.

33 Fenstermacher, M. J. and Narayana, P. A., Serial proton magnetic resonance spectroscopy of ischemic brain injury in humans, *Invest. Radiol.,* **9** (1990) 1034–1039.

34 Ford, C. C., Griffey, R. H., Matwiyoff, N. A., and Rosenberg, G. A., Multivoxel 1H-MRS of stroke, *Neurology,* **42** (1992) 1408–1412.

35 Gideon, P., Henriksen, O., Sperling, B., Christiansen, P, Olsen, T. S., Jorgensen, H. S., and Arlien-Soborg, P., Early time course of N-acetyl-aspartate, creatine and phosphocreatine, and compounds containing choline in the brain after acute stroke: a proton magnetic resonance spectroscopy study, *Stroke,* **23** (1992) 1566–1572.

36 Graham, G. D., Blamire, A. M., Howseman, A. M., Rothman, D. L., Fayad, P. B., Brass, L. M., Petroff, O. A. C., Shulman, R. G., and Prichard, J. W., Proton magnetic resonance spectroscopy of cerebral lactate and other metabolites in stroke patients, *Stroke,* **23** (1992) 333–340.

37 Henriksen, O., Gideon, P., Sperling, B., Olsen, T. S., Jorgensen, H. S., and Arlien-Soborg, P., Cerebral lactate production and blood flow in patients with acute stroke, *Magn. Reson. Imaging,* **2** (1992) 511–517.

38 Petroff, O. A. C., Graham, G. D., Blamire, A. M., Al-Rayess, M., Rothman, D. L., Fayad, P. B., Brass, L. M., Shulman, R. G., and Prichard, J. W., Spectroscopic imaging of stroke in humans: histopathology correlates of spectral changes, *Neurology,* **42** (1992) 1349–1354.

39 Sappey-Marinier, D., Calabrese, G., Hetherington, H. P., Fisher, S. H., Deicken, R., Van Dyke, C., Fein, G., and Weiner, M. W., Proton magnetic resonance spectroscopy of human brain: applications to normal white matter, chronic infarction, and MRI white matter hyperintensities, *Magn. Reson. Med.,* **26** (1992) 313–317.

40 Gideon, P., Sperling, B., Arlien-Soborg, P., Olsen T. S., Henriksen, O., Long-term follow-up of cerebral infarction patients with proton magnetic resonance spectroscopy, *Stroke,* **25** (1994) 967–973.

41 Welch, K. M. A., Windham, J., Knight, R. A., Vijaya, N., Hugg, J. W., Jacobs, M., Peck, D., Booker, P., Dereski, M. O., and Levine, S. R., A model to predict the histopathology of human stroke using diffusion and T2-weighted magnetic resonance imaging, *Stroke,* **26** (1995) 1983–1989.

42 Warach, S., Chien, D., Li, W., Ronthal, M., and Edelman, R. R., Fast magnetic resonance diffusion-weighted imaging of acute human stroke, *Neurology,* **42** (1992) 1717–1723.

43 Wittlich, F., Kohno, K., Mies, G., Norris, D. G., and Hoehn-Berlage, M., Quantitative measurement of regional blood flow with gadolinium diethylenetriaminepentaacetate bolus track NMR imaging in cerebral infarcts in rats: validation with the iodo(14C)antipyrine technique, *Proc. Natl. Acad. Sci. USA,* **92** (1995) 1846–1850.

44 Weiller, C., Recovery from motor stroke: human positron emission tomography studies, *Cerebrovasc. Dis.,* **5** (1995) 282–291.

45 Gramsbergen, J. B. P., Veenma-van der Duin, L., Loopuijt, L. D., Paans, A. M. J., Vaalburg, W., and Korf, J., Imaging of the degeneration of neurons and their processes in rat or cat brain by $^{45}CaCl_2$ autoradiography or $^{55}CoCl_2$ positron emission tomography, *J. Neurochem.,* **50** (1988) 1798–1807.

46 Stevens, H., Vd Wiele, Ch., De Reuck, J., Jansen, H. M. L., Dierckx, R. A., and Korf, J., Co^{57} and Tc^{99}-m HMPAO labeled leucocytes SPECT in visualizing cerebral ischemia. *J. Neurol.,* **243 (Suppl.)** (1996) S107.

47 Fujinuma, K., Kitai, N., Sakai, S., Iizuka, T., and Kanda, T., IN-111-labeled leukocyte brain SPECT imaging in acute embolic stroke in man, *J. Cerebr. Blood Flow Metab.* **15** (1995) S679.

48 Linde, R., Laursen, H., and Hansen, A. J., Is calcium accumulation post–injury an indicator of cell damage? *J. Cerebr. Blood Flow Metab.,* **15** (1995) S725.

49 Buell, U., Stirner, H., Braun, H., Kreiten, K., and Ferbert, A., SPECT with $^{99}Tc^m$-HMPAO and $^{99}Tc^m$-pertechnetate to assess regional cerebral blood flow (rCBF) and blood volume (rCBV). Preliminary results in cerebrovascular disease and interictal epilepsy, *Nucl. Med. Comm.,* **8** (1987) 519–524.

50 Lassen, N. A., Imaging brain infarcts by single-photon emission tomography with new tracers, *Eur. J. Nucl. Med.,* **21** (1994) 189,190.

51 Gupta, S., Bushnell, D. L., Mlcoch, A., Eastman, G., Barnes, W. E., and Fisher, S. G., Utility of late N-Isopropyl-p-(Iodine-123)-Iodoamphetamine brain distribution in predicting outcome following cerebral infarction, *Stroke,* **22** (1991) 1512–1518.

52 Datz, F. L. and Morton, K. A., New radiopharmaceuticals for detecting infection, *Invest. Radiol.,* **28** (1993) 356–365.

53 Stevens, H., Knollema, S., Piers, D. A., Akkermans, G., Venema, L., Kool, W., De Jager, A. E. J., and Korf, J., $^{57}Co^{2+}$ as a SPECT tracer in visualizing ischaemic infarcts, *J. Neurol.,* **243 (Suppl.)** (1996) S105.

3

Clinical Aspects of Stroke
and Therapeutic Strategies

Keith W. Muir, Elkan Gamzu, and Kennedy R. Lees

1. INTRODUCTION

"Stroke" is defined by the World Health Organization as "rapidly developing clinical signs of focal (or global) disturbance of cerebral function, with symptoms lasting 24 h or longer or leading to death, with no apparent cause other than of vascular origin." Ultimately, stroke results from the death of neurons in the central nervous system (CNS), and improved understanding of the pathological processes that mediate neuronal death at the cellular and neurochemical level has led to the development of therapeutic strategies that promise to improve the outcome after acute stroke. Many pathological processes may cause stroke, and many disease processes may mimic stroke. Advances in the therapeutic possibilities in stroke require physicians and clinical trialists alike to appreciate the clinical spectrum of stroke disease, and the problems that are peculiar to it.

2. EPIDEMIOLOGY AND RISK FACTORS

Acute stroke is the third largest cause of mortality in the Western world, ranking behind malignant disease and coronary heart disease *(1)*, and it is the single largest medical cause of disability in adults. In the Western world, approx 85% of strokes are caused by atherothrombotic occlusion of a blood vessel, and 15% by intracranial hemorrhage *(2)*.

Of ischemic strokes, the principal mechanisms include:

1. Thrombotic embolism from the heart.
2. Atherothrombotic embolism from large artery atherosclerosis.
3. *In situ* thrombotic occlusion of small perforating arteries (e.g., in diabetes).

Rarer causes include vasculitic processes of the medium or small arteries, embolism from vasculitic processes in large arteries (e.g., giant cell arteritis or Takayasu's arteritis), vasogenic conditions of uncertain pathogenesis such as migraine or Moya-Moya disease, or coagulopathic conditions such as stroke associated with systemic lupus erythematosus.

In individuals, risk factors for stroke may be thought of in terms of modifiable and unmodifiable factors. Unmodifiable risk factors include age, sex (risk of stroke is

From: Clinical Pharmacology of Cerebral Ischemia *Edited by: G. J. Ter Horst and J. Korf Humana Press Inc., Totowa, NJ*

43

greater in men, although lifetime risk of death from stroke is higher in women *(1)*, and race (higher incidence in blacks). The incidence at 30 yr of age is 0.3/100,000 per annum, rising logarithmically to 300/100,000 by age 85 *(3)*. Genetic factors may play a role, but are difficult to separate from a family history of other established risk factors with a hereditary component such as diabetes mellitus and hypertension *(4)*.

Major modifiable risk factors are coexisting cardiac disease *(5)*, especially atrial fibrillation *(6–8)*, which predisposes to thromboembolism, and myocardial infarction; chronic diastolic *(9)* or isolated systolic hypertension *(4,10)*; diabetes mellitus *(11)*; and cigaret smoking *(12,13)*. A further major risk is a history of previous cerebral ischemic events, that may either be completed strokes or transient ischemic attacks (TIAs). A TIA is defined as an event conforming to the WHO definition of stroke (*see* Section 1.) but less than 24 h in duration. In practice, full recovery from a TIA takes place in 1 h or less in at least 60% of cases, and the probability of a cerebral ischemic event resolving entirely within the 24-h period decreases rapidly with its duration *(14)*. The longer the duration of a focal deficit, the greater the probability of an associated infarct being identified on computed tomography (CT) scan *(15)*. The implications of this for clinical trials are discussed at greater length later. A TIA is a valuable marker of cardiovascular risk: following a TIA, there is a 7–12% risk of stroke in the first year *(16,17)*, reducing to 3.4% per annum thereafter, and a 3.1% annual risk of coronary arterial events. Risk factors for stroke combine multiplicatively rather than additively *(18,19)*.

Factors that have a less consistent profile of risk for stroke include excess alcohol consumption in binges, hypercholesterolemia *(20,21)*, or increased lipoprotein(a) *(22)* concentration. Minor risk may be conferred by elevated plasma homocysteine levels *(23)*, increased circulating fibrinogen levels *(24,25)*, obesity, and recent infection.

Primary prevention of stroke is a major public health goal. Treatment of modifiable risk factors, notably hypertension *(26)*, valvular heart disease, and nonvalvular atrial fibrillation produces significant reductions in the incidence of stroke. Lowering of both diastolic *(27,28)* and isolated systolic *(29)* hypertension with either thiazide diuretics or beta adrenoceptor antagonist drugs reduces the incidence of stroke in line with epidemiological predictions *(26)*, with the greatest absolute treatment effect evident among elderly patients *(27,30)* in whom the risk of stroke is greatest. Calculations of relative risk for stroke from 17 studies involving over 47,000 patients *(31)* show reduction after 3 yr of treatment of 38%. Newer antihypertensive drugs such as calcium antagonists and angiotensin converting enzyme inhibitors have yet to be tested adequately but are expected to confer similar benefit. Since there is no identified threshold of blood pressure that confers increased risk of stroke, but rather a continuous linear increase in stroke risk with blood pressure, population measures such as reduction of dietary salt intake (which will reduce blood pressure) may also reduce stroke incidence *(32)*.

Anticoagulation with warfarin for patients with nonvalvular atrial fibrillation has been shown in five large randomized clinical trials *(33–37)* to reduce the relative risk of stroke by around 60%. Again, the greatest benefit for anticoagulant therapy is seen among the elderly. Younger patients (<60 yr) may derive insufficient absolute benefit to warrant anticoagulation, but may derive some useful risk reduction from aspirin *(38,39)*. Carotid endarterectomy is of uncertain value in primary prevention of stroke *(40–42)*.

Several modalities of treatment are effective in secondary prevention of stroke. Antithrombotic therapy (aspirin or ticlopidine, and possibly dipyridamole *[43]*) is effective in prevention of strokes in patients who have suffered a TIA *(44)*. Anticoagulation for nonrheumatic atrial fibrillation *(45)* is also effective in secondary prevention of stroke. Carotid endarterectomy has been shown to be superior to antithrombotic drugs alone for patients with symptomatic (i.e., in patients who have suffered either TIA or stroke) carotid stenosis of greater than 70% *(46,47)*, reducing the relative risk of subsequent ipsilateral disabling or fatal stroke by 40–80%. Reduction of blood pressure in patients with previous cerebrovascular disease has been specifically studied in only a small number of patients comparatively, but appears to reduce the relative risk of subsequent stroke *(48–51)*. Confidence limits for risk reduction are wide, however, and the true magnitude of any effect is unclear. Blood pressure reduction in normotensive or mildly hypertensive stroke survivors is currently being tested in a large randomized trial.

In younger patients with stroke, other risk factors have been identified: Recognized etiological factors include the presence of congenital or acquired cardiac abnormalities *(52)* (e.g., patent foramen ovale, atrial septal aneurysm, mitral valve prolapse, atrial myxoma), genetic disorders (e.g., the syndromes of mitochondrial encephalomyopathy lactic acidosis and stroke [MELAS] and cerebral autosomal dominant acute subcortical infarct and leukoaraiosis [CADASIL]), vasculitides, inherited or acquired prothrombotic states (e.g., hereditary antithrombin III or protein C deficiency, factor V Leiden, systemic lupus erythematosus), and use of estrogen-containing oral contraceptives *(53–55)*. These mechanisms account for only a small proportion of all stroke.

3. THE HUMAN CEREBRAL CIRCULATION

The human brain is supplied by four main vessels. The anterior circulation consists of paired internal carotid arteries (ICAs) derived from the common carotid arteries (CCAs). The ICAs enter the skull base and form the anterior part of the circle of Willis, giving off the arteries that supply the main part of the cerebral hemispheres, the middle cerebral arteries (MCAs) and the anterior cerebral arteries (ACAs). The posterior circulation is derived from paired vertebral arteries that unite on the ventral surface of the brainstem to form the basilar artery. Terminal branches of the vertebral and basilar arteries are given off in pairs to supply the brainstem and cerebellum. The basilar artery anastamoses with the anterior circulation via the posterior communicating arteries forming the posterior part of the circle of Willis. The posterior cerebral arteries (PCAs) arise from the terminal portion of the basilar artery and supply the posterior part of the cerebral hemispheres.

4. CLINICAL PRESENTATIONS

Approximately 80% of strokes occur in the territory of the anterior circulation, and the majority of these affect the territory of the middle cerebral arteries (MCAs), that supply the lateral part of the cerebral hemispheres, including the brain regions responsible for the motor, sensory, language (in the dominant hemisphere), and visuospatial function (in the nondominant hemisphere).

A large number of clinical syndromes has been described, but in approx 70% of strokes there is limb weakness (although this is likely to reflect recognition and referral bias). Limb weakness may result from stroke affecting any part of the motor pathways from the cerebral cortex (prefrontal gyrus) to the internal capsule, to the pyramidal (corticospinal) tracts as they traverse the brainstem. The associated features of a stroke assist in the localization of the lesion and may also provide a guide to the extent of the stroke—e.g., isolated hemiparesis usually arises from subcortical infarction affecting the internal capsule because of small vessel occlusive disease (lacunar stroke), but the combination of hemianopia, hemisensory loss, and dysphasia with hemiparesis signifies a proximal MCA occlusion with extensive infarction of the dominant cerebral hemisphere.

The Oxfordshire Community Stroke Project (OCSP) *(56)* devised a system of clinical differentiation of stroke subtypes with prognostic and etiological implications that did not seek to quantify degree of impairment. This system divides strokes into total anterior circulation strokes (TACS), partial anterior circulation strokes (PACS), lacunar strokes (LACS), and posterior circulation strokes (POCS), entirely on clinical criteria (Table 1).

5. INVESTIGATION OF ACUTE STROKE

Investigation of the acute stroke patient should have three goals:

1. Exclusion of nonstroke pathology.
2. Distinction of pathological basis of stroke—infarct vs hemorrhage.
3. Identification of risk factors and etiology of stroke.

Stroke may be misdiagnosed in up to 20% of patients presenting to a hospital medical service *(57)*. The recognition of treatable disorders that may present as hemiparesis is essential, hypoglycemia being the most obvious example of a condition that is entirely reversible if recognized.

Computed tomography (CT) scanning of the brain is essential in all patients, preferably on admission to hospital, since this imaging modality is able to distinguish infarction from primary intracerebral hemorrhage (PICH). Despite several attempts to codify clinical impressions into scoring systems to separate infarct from hemorrhage, clinical diagnosis remains unreliable *(58)*. CT will also identify nonstroke pathology in up to 5% of cases: Subarachnoid hemorrhage, brain tumors, cerebral abscess, and HIV-related infections such as toxoplasmosis may all present as stroke. CT may also assist in defining the etiology of ischemic stroke since certain patterns of infarction are associated with different pathological mechanisms. In particular, venous rather than arterial infarct may be identified on the basis of infarct pattern on CT (hemorrhagic infarcts that do not conform to an arterial territory) and confirmed by dynamic contrast CT; boundary zone infarcts may signify global hypoperfusion, or unilateral carotid artery occlusion; and large subcortical MCA territory infarcts are usually seen with embolic occlusion of the proximal MCA.

Age and clinical status will dictate the requirement for many of the etiological investigations, whereas local availability of neuroimaging will limit investigation for many. Even intensive investigation yields a definitive cause in only 50–60% of patients.

MRI scanning improves identification and localization of infarcts, particularly lacunar strokes or those in the posterior circulation, and has advantages over CT in the imaging of abnormalities characteristic of more diffuse vascular disease processes such as stroke associated with vasculitides, HIV infection, mitochondrial cytopathy

Table 1
Oxfordshire Community Stroke Project Clinical Classification of Stroke Syndromes

Total Anterior Circulation Syndrome (TACS)	All three of the following: Hemiparesis Hemianopia Higher cortical dysfunction
Partial Anterior Circulation Syndrome (PACS)	Any two of three under TACS *or* isolated higher cortical dysfunction
Lacunar Syndrome (LACS)	Pure motor stroke Pure sensory stroke Sensorimotor stroke Dysarthria-clumsy hand syndrome Ataxic hemiparesis
Posterior Circulation Syndrome (POCS)	Isolated hemianopia *or* combinations of the following: Limb sensory or motor loss Cranial nerve palsies Ataxia

(MELAS syndrome) or hereditary abnormalities (CADASIL). Newer MRI techniques such as perfusion scanning, which visualizes cerebral blood flow noninvasively, and diffusion weighted imaging (DWI), which may allow early definition of the volume of tissue that is ischemic but not yet infarcted, may have a clinical role in the future, but at present remain research tools. Although advances in MRI technology have great promise, for most stroke patients, improved access to CT scanning is of greater relevance.

Advances in MRI and ultrasound have improved the early identification of etiological factors. Magnetic resonance angiography (MRA) promises noninvasive imaging of carotid and vertebrobasilar systems (particularly with echo-planar imaging), and may identify intracranial vascular malformations. Currently, conventional angiography remains the investigation of choice to grade carotid artery stenosis or define the intracranial circulation in detail since MRA is subject to artifact. Ultrasound imaging of the carotid circulation is well established, and many surgeons consider duplex scanning to be sufficiently accurate to permit carotid surgery without confirmatory angiography. Transcranial Doppler ultrasound has yet to find its niche in routine stroke management, but embolus detection by TCD is of potential value in defining the responsible source of stroke in patients with dual pathology.

6. ORGANIZATION OF STROKE CARE

Until recently, therapeutic nihilism has led to low priority being given to stroke patients at every stage of care, from primary care assessment, to transport to hospital services, to in-hospital care, and ultimately to rehabilitation. Increased access of stroke patients to acute hospital services has been driven by the demonstration that properly organized stroke services significantly improve mortality and disability, and reduce the duration of hospital stay *(59)*. It is also increasingly recognized that safe and optimal secondary prevention requires accurate diagnosis, and therefore at least CT scan within 2 wk of onset, and preferably the suggested mandatory investigations outlined in Table 2.

Studies that define the factors responsible for the efficacy of stroke units are awaited. At present, evidence suggests that improved outcome is achieved by attention to basic care, including routine ascertainment of swallowing reflexes, alternative hydration and feeding for patients unable to swallow, early mobilization, antibiotic therapy for pulmonary infections, and prophylaxis for deep vein thrombosis (DVT). If these results are borne out by further study, then it will be an indication of the poverty of routine hospital care for stroke patients.

A valuable by-product of stroke units should be an infrastructure that enables large randomized, controlled clinical trials similar to those conducted in acute myocardial infarction after the establishment of coronary care units.

7. COMPLICATIONS OF ACUTE STROKE

Outcome after acute stroke is heterogeneous. The 30-d case fatality rate for first strokes is between 17 and 34%, depending on the population structure *(3,60)*. Only 50% of deaths within 30 d result directly from the stroke *(60)*. Stroke causes death either by cerebral edema with attendant raised intracranial pressure and transtentorial herniation, or less commonly by infarction of medullary cardiorespiratory centers in basilar artery occlusion. The pattern of mortality from stroke is that deaths directly consequent to the stroke occur in the first week, with a further peak of mortality from 2–4 wk owing to secondary pathologies related to immobility or dysphagia (pneumonia, sepsis, DVT, and pulmonary thromboembolism) *(61)*. The 1-yr case fatality rate is approx 40%, with the secondary causes of death representing a mixture of the consequences of immobility and recurrent cardiovascular events, particularly second strokes *(61)*. There is therefore a great variation in stroke outcome, that is largely independent of the immediate pathology.

8. TARGETS FOR THERAPEUTIC INTERVENTION

8.1. The Ischemic Penumbra

The development of infarction is dependent upon both the severity of ischemia and its duration *(62)*. Normal cerebral blood flow (CBF) is maintained at approx 60 mL/100 g tissue/min. Reversible loss of neuronal function is seen at around 16–20 mL/100 g/min *(63)*, and below this level of CBF, ischemia may result in death of tissue depending on duration of ischemia and type of tissue (glia and white matter being inherently more resistant to ischemia than gray matter, and some neuronal subpopulations being more vulnerable to ischemia than others).

Focal cerebral ischemia produces a gradient of CBF across the MCA territory, with severe ischemia of the striatum, and less severe ischemia of the cortex *(64,65)*. This appears to depend on collateral blood supply, and not on any functional changes induced by ischemia. The gradient of CBF reduction gives rise to the concept of the ischemic penumbra ("almost shadow"), the region with CBF between the threshold for reversible neuronal impairment and the threshold at which rapidly irreversible damage occurs. Since the penumbra progresses to infarction over a period of time, interruption of ischemia-induced processes may prevent death of penumbral tissue if applied sufficiently early, and if the processes are reversible.

Table 2
Investigation of Acute Stroke Patients

	Investigation	Suggested Priority
Mandatory	Blood glucose	Immediate
	Chest X-ray	Day of admission
	ECG	Day of admission
	Full blood count	Day of admission
	Electrolytes	Day of admission
	Erythrocyte sedimentation rate	Day of admission
	CT scan of brain	Day of admission
Selected Patients	Carotid duplex ultrasound	
	Cardiac ultrasound	
	Magnetic resonance imaging scan	
	Cerebral angiography	
	Transesophageal echocardiography	
	Thrombophilia screen	
	Autoantibodies	
	Genetic analysis (e.g., CADASIL, MELAS)	
Clinical use undefined	SPECT scan	
	Magnetic resonance angiography	
	Transcranial Doppler ultrasound	

Preclinical animal models of focal cerebral ischemia demonstrate that the duration of the time window in which intervention is useful depends upon the methods of defining the penumbra, the drug being studied, and the animal model. The window appears to be short, perhaps of the order of 2 h in a rat permanent MCA occlusion model. Translation of these observations to human stroke is difficult. By using radioisotope-based imaging modalities, positron emission tomography (PET; which images cerebral metabolism of glucose and oxygen), and single photon emission computed tomography (SPECT; which images relative regional cerebral blood flow), changes in blood flow and metabolism after stroke can be related to clinical outcome. Studies in human stroke have shown that clinical recovery is associated with early reperfusion *(66,67)*. The PET appearances of a penumbra persist in some patients for up to 48 h *(68)*, but there is wide interindividual variation. These observations support the feasibility of therapeutic intervention in human stroke.

8.2. Cellular Targets for Intervention

Ischemia produces rapid depletion of cellular energy stores with attendant depolarization of cell membranes caused directly by influx of sodium (Na^+) ions and outflow of potassium (K^+) ions. This depolarization leads to release of neurotransmitters, particularly the excitatory amino acid (EAA) glutamate, and also opening of postsynaptic neuronal ion channels. There is postsynaptic entry of Na^+ ions and more importantly, calcium (Ca^{2+}) ions, via both voltage-gated and ligand-gated channels (especially the glutamate operated N-methyl D-aspartate, or NMDA receptor).

Intracellular free Ca^{2+} overload causes pathological activation of many enzyme systems, notably phospholipase A_2, protein kinase C, nitric oxide, calpains, and endonucleases, with secondary generation of oxygen free radicals, and potentiation of EAA-mediated Ca^{2+} entry. Inflammatory mediators including interleukins 1β and 6 are generated soon after ischemic onset, and an inflammatory response develops over the hours following stroke. There are also changes in neuronal protein synthesis that may signify apoptosis.

Drug treatments have been developed that target every aspect of this ischemic cascade. Calcium channel antagonists, antagonists of the NMDA receptor, free radical scavenging agents, sodium channel antagonists, presynaptic glutamate release inhibitors, and antiinflammatory agents are all at various stages of clinical development at present.

9. THERAPEUTIC STRATEGIES

Therapeutic approaches to stroke have been centered on two distinct approaches, one primarily vascular and one primarily neuronal. Reperfusion is theoretically attractive since it is perfusion failure that underlies all ischemic stroke, and relief of the initiating event should prevent all consequences of neuronal ischemia. Strategies for reperfusion have included thrombolytic drugs to promote thrombus dissolution actively, anticoagulants to prevent propagation of thrombus, and a variety of therapies designed to increase regional cerebral blood flow or alter the rheological characteristics of blood. The alternative approach has sought to prolong the viability of neurons subjected to ischemia and is therefore known as neuroprotection.

Clinical trials have been conducted in stroke for almost 40 yr. The majority of trials were conducted before widespread availability of CT scanning, animal models that permitted preclinical testing of drugs, and before recognition of the ischemic penumbra. The methodology developed over the years with these trials forms a valuable background to current concepts of how to conduct stroke trials, but many of the studies fail to answer definitively whether treatments did or did not work, since numbers involved were insufficient to do so with adequate statistical power.

9.1. Reperfusion

9.1.1. Thrombolysis

Animal data have supported the concept of thrombolytic treatment of acute focal ischemia, but the earliest studies were conducted in humans. Case series have been reported intermittently since 1958, but patient numbers have been statistically inadequate, and, by modern standards, drugs were administered very late after stroke onset (up to 30 d in some cases). The principal risk of thrombolysis is provocation of *de novo* intracerebral hemorrhage, or exacerbation of unidentified preexisting hemorrhagic conversion of the infarct, which must be excluded prior to treatment by brain CT. Until 1992, only four randomized controlled trials had been conducted with pretreatment CT scanning. More recently, with developing understanding of free radicals, there was concern over reperfusion injury *(69)* and consequent cerebral edema. A meta analysis of the limited data available in 1992 *(70)* suggested possible benefit from thrombolysis in reduction of disability or death, although with very wide confidence intervals since only six randomized trials (two without CT) were included. An updated meta analysis of all trials is planned.

Thrombolytic therapy has now been tested in five large randomized controlled trials, three of which used streptokinase (SK) and two of which used recombinant tissue plasminogen activator (rtPA). All trials required CT prior to entry, and initiated treatment early after onset with windows between 90 min and 6 h.

All three trials of streptokinase (SK)—the Multicentre Acute Stroke Trial-Europe (MAST-E) *(71)*, MAST-Italy *(72)*, and the Australian Streptokinase Trial (ASK) *(73)*—were stopped prematurely by the relevant safety monitoring committees due to increased early mortality in the treatment arms. Adverse outcome was due entirely to an early excess of symptomatic intracerebral hemorrhage. Both trials of recombinant tissue plasminogen activator (rtPA)—the European Cooperative Acute Stroke Study (ECASS) *(74)* and the National Institute of Neurological Disorders and Stroke (NINDS) trial *(75)*—were completed as planned.

The dichotomy between the trials' results is striking: of the five trials, only NINDS found improved outcome with treatment, all others found increased mortality with thrombolysis. The NINDS trial used a 3-h time window, whereas the other trials used 4 (ASK) or 6 h (MAST-E, MAST-I, ECASS). Only ASK planned a separate analysis of 0–3 h data, and suggested possible benefit in this group although numbers were small (24% of recruited patients) and confidence intervals wide. There was a suggestion from both ECASS and MAST-I that survivors had improved outcome with respect to disability, but these findings were based on unplanned post hoc analyses, and any benefit was overshadowed by the substantial excess mortality.

The combined mortality data for thrombolytic trials in stroke are shown in Fig. 1 and Table 3. Treatment up to 3 h appears to have no excess mortality, whereas trials with longer time windows have significantly increased mortality. Formal meta analysis of the combined endpoint of death and disability is ongoing. The NINDS trial demonstrated that this combined outcome was improved significantly by early administration of rtPA using conservative measures of disability.

Restoration of blood flow by thrombolytic drugs to brain tissue that has been ischemic for 3 h or longer carries an unfavorable risk-benefit ratio. This appears to reflect an intrinsically poor therapeutic index for these drugs since administration of low mol-wt heparin up to 48 h after stroke may be beneficial *(76)*, and is not associated with increased mortality. Since safe thrombolysis will only be available to a small number of patients (probably fewer than 3% in the United Kingdom) owing to the short time window, there remains a requirement for safe and more widely applicable therapy.

9.1.2. Rheological Modification

Other strategies predominantly intended to restore or enhance perfusion have been assessed clinically with varying adequacy.

Isovolemic hemodilution by venesection and infusion of dextran 40 or hydroxyethyl starch alters the rheological characteristics of blood by reducing hematocrit to a target of approx 0.3, in order to increase cardiac output and thereby penumbral CBF. Randomized controlled trials involving 2605 patients have been reported. Several small trials using different regimens were inconclusive. Two large trials (the Scandinavian Stroke Study Group *(77)* and the Italian Acute Stroke Study Group *(78)* included 1640

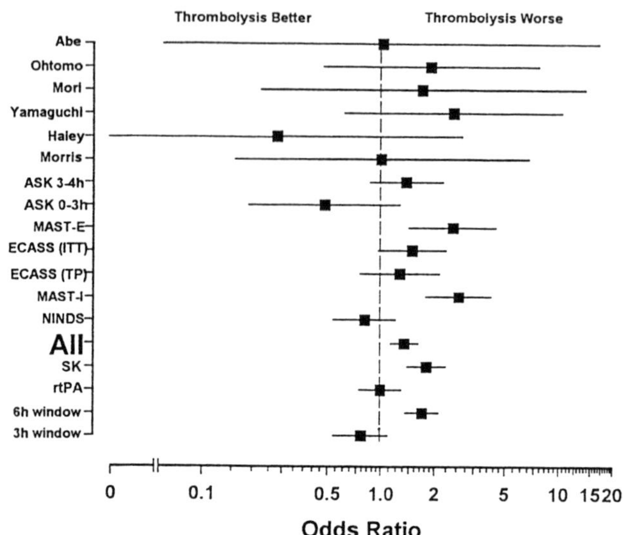

Fig. 1. Meta analysis of mortality for thrombolytic trials in stroke.

patients and found no benefit: Meta analysis of hemodilution trials similarly failed to show benefit *(79)*.

Naftidrofuryl is a modest in vitro vasodilator that may increase CBF after stroke. Naftidrofuryl was ineffective in four published trials that suffered from late administration of drug and small patient numbers. A trial including 620 patients within 48 h of stroke found no difference in duration of hospital stay on intention to treat analysis *(80)*.

Three randomized controlled trials of the prostacyclin analog vasodilator epoprostenol in 101 patients found no consistent clinical or hemodynamic benefits.

The drugs pentoxifylline and propentifylline are inhibitors of cAMP breakdown and inhibit adenosine reuptake. No significant benefits were found from two randomized trials of pentoxifylline (total of 407 patients) *(81,82)*; small trials of propentifylline (30 patients) and aminophylline (46 patients) have been reported.

Heparin, both unfractionated and low mol-wt, has been the subject of several randomized controlled trials in acute stroke. Early studies used heparin in progressing stroke without statistical power to detect benefit. A recent large trial of fraxiparine *(76)* found improved outcome at 3 mo after subcutaneous administration within 48 h of stroke: whether this effect is due to improvements in the initial infarct or prevention of secondary events such as DVT or recurrent stroke is uncertain. Subacute use (within 48 h) of heparin and aspirin is being studied in a very large randomized, controlled trial, the International Stroke Trial (IST).

9.2. Neuroprotection

9.2.1. Antiedema Agents

Since raised intracranial pressure associated with cerebral edema is a significant detrimental component of stroke, several early trials involved drugs intended to reduce edema formation.

Table 3
Thrombolytic Trials in Acute Stroke Conducted with CT Scan

Trial	Yr	Time window, d	Agent	Total n
Abe	1981	30	UK	107
Ohtomo	1985	5	UK	350
Mori	1991	6	rtPA	31
Yamaguchi	1992	6	rtPA	102
Haley	1992	3	rtPA	27
Glasgow	1992	6	SK	20
ASK	1995	4	SK	270
ASK 0-3h	1995	3	SK	70
MAST-E	1995	6	SK	270
ECASS ITT	1995	6	rtPA	620
ECASS TP	1995	6	rtPA	511
MAST-I	1995	6	SK	622
NINDS	1995	3	rtPA	624
All				3113
All SK				1252
All rtPA				1404
6-h window				1935
3-h window				721

SK, streptokinase; UK, urokinase; rtPA, recombinant tissue plasminogen activator; ITT, intention to treat; TP, target population.

Corticosteroids reduce the volume of edema surrounding cerebral tumors, and were accordingly considered potential neuroprotectants. They have a large number of adverse effects, such as immunosuppression, gastrointestinal bleeding, fluid retention, and hyperglycemia. Prospective trials using dexamethasone included only small numbers of patients (<200): Retrospective series have also been reported *(83)*. No useful conclusions can be drawn from the results.

Glycerol is a hyperosmolar agent also used in management of edema around tumors. Several trials in stroke included almost 500 patients, but long time windows and variable methodology have resulted in wide confidence intervals for the pooled results, which encompass both the possibilities that glycerol may be of significant benefit or of significant harm *(84)*. Hemolysis was noted in several trials to be a clinically significant adverse effect of glycerol infusion.

9.2.2. Calcium Channel Antagonists

Dihydropyridine Ca^{2+} antagonists block L-type voltage-gated Ca^{2+} channels, found in vascular smooth muscle and on postsynaptic neurons. In vitro neuroprotection and success in treatment of subarachnoid hemorrhage prompted clinical trials, despite an inability to demonstrate neuroprotection in any model with postischemic drug administration.

Nimodipine has been studied extensively, although smaller studies of nicardipine, flunarizine, and PY 108-068 in stroke have been reported. After initially positive results with oral nimodipine *(85,86)*, no benefit was found in 14 further trials. Meta analysis *(87)* of 9 of the available trials (which included 3360 patients) suggested that there may have been some benefit in those patients treated within 12 h, but its methodology was seriously flawed, and it excluded several relevant trials. Most of the oral nimodipine trials suffered from long time windows (typically 48 h) and late treatment administration. Two trials of iv nimodipine have also been conducted, INWEST *(88)*, and a still unpublished American trial. Both trials found poorer outcome with nimodipine. Outcome in INWEST was clearly related to systemic blood pressure, with disability increasing with the degree of drug-induced hypotension.

9.2.3. Excitatory Amino Acid Antagonists

The experimental profile of drugs that influence EAA release or binding is remarkably consistent, with significant reductions of infarct size in animal models with all drugs able to block transmission via the NMDA receptor *(89)*. There are concerns regarding the tolerability of these agents in clinical practice *(90)*, but several are currently undergoing testing in major clinical trials.

Magnesium ions are responsible for the voltage-dependent block of the NMDA ion channel, relief of which seems to be a critical event in NMDA-mediated toxicity. Increasing extracellular magnesium concentration behaves pharmacologically like a noncompetitive ion channel blocking drug *(91)*, and may also have beneficial vascular effects. Magnesium sulfate reduces the risk of eclamptic seizures in preeclampsia *(92)*, terminates eclamptic seizures more effectively than other anticonvulsants *(93)*, and has been given in several small clinical trials in stroke.

Hyperpolarization of the neuronal membrane by various strategies is neuroprotective. Benzodiazepines hyperpolarize the membrane by increasing chloride conductance via the $GABA_A$ receptor. The short acting benzodiazepine chlormethiazole is neuroprotective in preclinical models and is currently undergoing clinical trials.

9.2.4. Free Radical Scavengers

Several agents that prevent free radical-mediated damage are neuroprotective, the most widely tested being the 21-aminosteroids (lazaroids), a series of steroid-based molecules which lack gluco- or mineralocorticoid activity. The 21-aminosteroid tirilazad (U74006F) failed to show any effect on outcome in two trials *(94,95)* of 6 mg/kg/d of tirilazad given within 6 h of onset to 1072 stroke patients.

9.2.5. Gangliosides

Ganglioside GM_1 (monosialoganglioside) probably inhibits protein kinase C translocation. Several large clinical trials *(96–98)* including over 1000 patients, failed to show clinical benefit. Use of gangliosides has been banned in Europe after reports of an excess incidence of Guillain-Barré syndrome possibly attributable to these drugs.

10. STROKE TRIAL METHODOLOGY

The clinical development of drug therapies requires a series of clinical trials that must explore drug safety, tolerability, pharmacokinetics, and ultimately efficacy (Table 4). All modern pharmaceutical development follows a scheme outlined in Table 5.

Trial methods will vary according to the requirements of individual drug development. Whereas the methodology of phase I and early phase II trials is similar in all important respects to other clinical situations, trials that seek to include outcome mea-

Table 4
Current Clinical Trials of Neuroprotective Drugs in Stroke

Site of action	Agent	Clinical development
Sodium channel blockade	Lubeluzole	Phase III
	619C89	Phase II (terminated)
	Lifarizine	Phase II (terminated)
NMDA antagonists	Selfotel (CGS 19755)	Phase III (suspended)
	Eliprodil	Phase III (suspended)
	Magnesium sulfate	Phase III
	Aptiganel HCl (CNS 1102)	Phase II/III
	Remacemide HCl	Phase II
Glycine site antagonists	GV 150526A	Phase II
Free radical scavengers	Tirilazad	Phase III (recommended at higher dose)
Other mechanisms	Chlormethiazole	Phase III
	GM1 ganglioside	Phase III (terminated)

Table 5
Clinical Trial Definitions in Stroke

Trial type	Population	Aims
Phase I	Healthy volunteers	Safety and tolerability Pharmacokinetics
Phase II	Patients	Safety and tolerability Pharmacokinetics ? Surrogate markers of therapeutic effect
Phase III	Patients	Efficacy

sures are beset with methodological uncertainties and have yet to adhere to a common framework.

10.1. Design Issues

The substantial difficulties with outcome measurement in stroke trials, addressed below, mean that the sample size for phase III stroke trials is typically of the order of 500–800 patients. The size and duration of a clinical trial is therefore influenced significantly by the choice of entry criteria and the length of follow-up.

Most trials base entry criteria upon several factors, typically:

1. Time since stroke onset ("time window").
2. Stroke pathology (infarct vs hemorrhage).
3. Stroke severity.

The most important factor in stroke trial design is the time window chosen. Animal focal ischemia models suggest penumbral viability for a few hours after onset of ischemia, and, whereas human PET studies show that in some individuals this may extend to 48 h *(68)*, it is likely that the magnitude of infarct reduction by stroke therapies will be greatest when treatments are commenced soon after the onset of ischemia. The statistical power of phase III trials should therefore be enhanced by early initiation of treatment. Most phase III clinical trials opt for a 6-h time window, although some extend this to 12 h. The reasons for choosing a 6-h window are largely historic, there being no evidence of such a cut-off being of relevance to clinical practice: the 6-h window represents a compromise between the desirable (very short) window that is assumed to maximize therapeutic effect and the limits that very short time windows place on recruitment to trials *(99)*. Concern that short time windows may lead to inclusion of significant numbers of patients with TIA rather than stroke are probably unfounded: the definition of a TIA based upon a 24-h cut-off is arbitrary and, in the era of stroke trials, perhaps in need of revision. Retrospective analysis of patients with TIAs has found that 60% of events have resolved entirely within 1 h. In the NINDS rtPA trial, only 2% of 620 patients presenting within 3 h of symptom onset had complete resolution of their deficit by 24 h. Whereas a TIA has proved a useful epidemiological risk marker, a limit of 1 h may be a more practical boundary for current clinical practice, suggesting a time window of 1–6 h after symptom onset.

In phase II trials designed to assess tolerability or pharmacokinetic parameters, it is unnecessary to recruit patients within a very short time window, and studies with windows of up to 24 h may be conducted.

Reliance on a clinical diagnosis of stroke will inevitably mean inclusion of patients with both intracerebral hemorrhage and cerebral infarction *(58)*, as well as occasional patients with nonstroke pathology. Preinclusion CT scanning to exclude patients with diagnoses other than ischemic infarction will result in a more homogeneous study population, but may delay trial treatment administration significantly, and may also not reflect the reality of clinical practice. The ECASS trial of rtPA required strict CT criteria for entry, and retrospectively excluded over 100 subjects out of 600 randomized due to disagreements over CT interpretation (principally relating to the extent of visible infarction). Some trials include an assessment of the probable aetiology of stroke, the most widely used criteria originating in the Trial of ORG 10172 in Acute Stroke Treatment (TOAST) *(100)*. The TOAST criteria, although robust with respect to interrater reliability, depend entirely upon the results of investigations, and cannot therefore be useful at the time of trial entry.

Demonstration of a treatment effect will be more difficult if the sample includes significant numbers of patients with a very good prognosis, and probably also those with a very poor prognosis. Patients with minor degrees of weakness, those with lacunar syndromes, or rapidly resolving deficits are all commonly excluded, as are patients who are comatose on admission. Since the standard outcome measure is the Barthel disability scale, which is weighted heavily in favor of physical strength and mobility, there is a bias towards inclusion of patients with significant limb weakness in clinical trials.

Phase II trials may be slowed considerably by unnecessarily rigorous entry criteria, which are seldom of relevance in the assessment of pharmacokinetics or tolerability, but will restrict recruitment rates. From a clinician's standpoint, phase III trials in stroke probably also suffer from excessively narrow entry criteria, the main driving force for which is the expense of pharmaceutical development. Large trials with simpler but more clinically relevant end-points are likely to have greater weight with clinicians.

10.2. Outcome Measures for Stroke Trials

Outcome after stroke encompasses everything from complete resolution of all symptoms and signs within a few hours to rapid death. Survivors of stroke may have neurological deficits that give rise to physical disabilities of varying degree, but often have abnormalities of cognition, language, or visuospatial function that are much less readily detected by routine clinical assessment.

Since a significant proportion of survivors of stroke will be left disabled and dependent, stroke trialists regard reduction in mortality alone as an inadequate end-point. Stroke trials must therefore attempt to assess the degree of dependence of patients as well as determining mortality. Assessment of dependence and disability is difficult, and has led to the development of means of measuring functional impairment by stroke scales. Outcome assessments have been extended to surrogates of the patient's clinical state such as duration of hospitalization, place of care, or requirement for rehabilitative therapy. However, these indicators vary among hospitals and countries as a result of differences in the process of care, and differences that result entirely from drug treatments may be difficult to separate.

The timing of assessment has varied among clinical trials, with most opting for 3 mo after stroke, and some (e.g., MAST-E) extending this to 6 or 12 mo. The time taken to reach a plateau in recovery from stroke depends both on time and on stroke severity *(101)*. Recovery from mild strokes plateaus after 1 mo or less, whereas it plateaus between 2 and 3 mo after more severe strokes. By choosing three months, it will be certain that any treatment effect is sustained, but the effect may be diluted by the accumulation of secondary complications that contribute significantly to morbidity and mortality but are not necessarily related to the stroke. It may be relevant to consider shorter follow-up periods, since few patients move from dependence to independence between 1 and 3 mo.

10.3. Stroke Scales

The World Health Organization defines a hierarchical scheme of impairment, disability, and handicap *(102)*. This defines impairment as the physical problem (e.g., limb weakness), which gives rise to disability (e.g., inability to walk or to climb stairs independently), which in turn gives rise to handicap (e.g., inability to return to work). Whereas impairments may be described relatively objectively, disability depends on other aspects of health and on home circumstances. Assessment of handicap includes social and environmental interaction by the patient: Such abstract concepts ensure that assessment of handicap is highly subjective.

Stroke scales exist only for use in clinical trials, the first widely used scale being that of Mathew *(103)* from a trial of glycerol. The Mathew scale employed arbitrary test items of uncertain interrater reliability or prognostic value, and with weighting of the test item scores to favor motor assessments. This scale was subsequently employed in other trials but interrater reliability was only assessed some 16 yr later *(104)* when it was found to be so poor that the scale was deemed unusable.

Further scales have been developed for specific clinical trials (e.g., the Scandinavian Stroke Scale for hemodilution *[105]*, the National Institutes of Health stroke scale for the NINDS rtPA trial *[106]*, or the European Stroke Scale *[107]* for lubeluzole). No single scale has yet become the standard, and trials may be compromised by collecting assessments on multiple scales.

Stroke scales are ordinal scales—i.e., they assign arbitrary scores to predefined functional levels. The scores are discontinuous and cannot be analyzed by parametric statistics, a limitation that is still poorly appreciated. Analysis of the mathematical and conceptual basis *(108)* of stroke scales has led to a more scientific approach to design *(109,110)*, particularly with respect to selection of test items, evaluation of interrater reliability, and development of appropriate statistical tools for analysis *(111)*. Some investigators have modified analysis by mathematical models that include assessments from multiple scales *(75)*.

Most scales are heavily weighted toward the assessment of motor dysfunction by limb strength testing, a comparatively robust test item. Interrater reliability is poorer for cognitive or higher cerebral functions. Many scales have also included items of doubtful relevance such as muscle tone, plantar reflexes, and conjugate eye deviation.

Some acute stroke scales (e.g., the Canadian Neurological Scale and the Scandinavian Stroke Scale) are designed for serial assessment of patients, and are of proven prognostic relevance. Other scales of similar construction have been less rigorously assessed (e.g., the Middle Cerebral Artery Neurological Score or the European Stroke Scale); some, such as the widely used "Unified Scale" lack even published methodology or adequate validation. The NIH scale differs significantly from Canadian or European scales in design and weighting, and cannot be converted into other scales. The NIH scale is, however, of prognostic value.

The nonparametric statistical analysis of ordinal impairment scales means that analytical methods are insensitive. Power calculations based on the proportion of patients improving by 4 points on the NIH stroke scale found that 1600 patients per treatment group would be required to detect a 10% treatment effect *(112)*. Since a shift of 4 points on an ordinal scale may carry very different implications for individuals (a 4-point shift on the NIH scale could signify complete recovery of facial weakness only, with arm strength marginally better, or it could signify complete recovery of a severe motor deficit affecting arm and leg), clinicians have expressed concern about basing treatment decisions on changes in numbers that do not have clear functional relevance *(113)*.

10.4. Disability and Handicap Measurement

Outcome assessment has been much more uniform than acute impairment measurement. Most trials assess disability with the Barthel score *(114)*, a scale that rates activities of daily living (ADL). The Barthel scale has the advantages of extensive validation for interrater reliability, ability to score patients without relying on their own answers (which may be affected by cognitive or language dysfunction), and utility in predicting dependence or independence. Dichotomizing the Barthel around a cut-off of 60/100 has been shown to predict dependence or independence following rehabilitation (115), although the use of more rigid criteria to define favorable clinical outcomes may be preferable: The NINDS rtPA trial used a Barthel of 95 or 100 to define good outcome. The Barthel score has been criticized for a concentration on motor activities to the exclusion of other important items and for "top-end" insensitivity—i.e., many patients with significant handicap score 100 on the Barthel scale. It has significant advantages, however, in ease of use, familiarity to medical and nursing staff, and proven reliability in telephone interview *(116)*. More complex evaluations of a patient's functional status require longer to complete, are more dependent on the rater's familiarity with the scale and also correlate closely with the Barthel score.

The Functional Independence Measure (FIM) scores both physical and cognitive function *(117)*, and has been validated in rehabilitation of stroke patients. It may predict dependence and prognosis even from an early stage after stroke *(118)*. The complexity of the FIM (full scoring involves grading of 18 separate items on a 7-point ordinal scale) necessitates specific training, however. The scale weights some measures (e.g., transfers from wheelchair) more heavily than the Barthel, and includes assessments of cognitive function that are absent from the Barthel. It is unclear at present whether the additional time and effort required for scoring the FIM has any significant advantage over the Barthel score.

Other outcome scales combine elements of disability assessment with handicap assessment. The Rankin Index *(119)* was a five-point scale originally intended to rate disability, but modified to rate handicap and extended to include death and complete normality in a 0–6 score: In most respects the Glasgow Outcome scale *(120)*, originally developed for head injury, is similar. Inclusion of handicap assessment entails a significant element of subjective judgment, which is more dependent on the rater than disability scales *(121)*, although in practice the Rankin score may behave as a simplified Barthel *(122)*. Addition of the Rankin score to all those scoring 100 on the Barthel has been advocated as a means of reducing "top-end" insensitivity.

There are well-argued objections to the use of scales as trial end-points *(113)*. These indicate that simple outcome measures (e.g., dead or alive, disabled or able to conduct normal activities) have superior interrater reliability, and are more relevant to patient and physician. Further, many trials convert results back into broad patient groups, which defeats the purpose of using complex (and less reliable) scoring systems.

10.5. Surrogate End-Points

At present, the absence of clinically proven treatments for most stroke patients renders the use of surrogate end-points entirely speculative. Once an effective drug treatment has been developed, useful surrogate markers may become available.

Drug doses that are neuroprotective in animals cannot be translated into humans with any certainty. At present, investigators have opted variously for pharmacokinetic (e.g., equivalent areas under the plasma concentration-time profile in rat and human) or pharmacodynamic markers (e.g., CNS side-effects, such as electroencephalographic changes suggestive of CNS penetration). The use of brain and cerebrospinal fluid drug concentrations may be possible, although not in the stroke population *(123)*.

For drugs acting on excitatory amino acid systems, imaging by SPECT with radioligands of the NMDA receptor may give some indication of drug effect upon a pharmacologically relevant system. Direct CSF measurement of glutamate concentrations may also prove helpful, although normal ranges for glutamate have not yet been established.

Newer imaging techniques, notably DWI and perfusion MRI, allow repeated imaging of stroke patients, and may visualize ischemic tissue that is potentially salvageable. Image quality is inadequate at present to calculate tissue volumes with precision, and the wide interindividual variation in stroke size means that the role of new MRI technologies will remain uncertain in the immediate future.

Changes in stroke scale scores in the early period after stroke are crude, prone to interrater error, and of dubious clinical significance.

When a clinical trial is able to show drug effects with established clinically relevant end-points, then all of the above methods may prove to be applicable.

11. CONCLUSION

Once a truly effective and safe treatment for stroke becomes available, the utility of various outcome measures will be determined with certainty. The ECASS and NINDS trial results may not have a substantial impact on clinical practice, but they show that the existing outcome measures, however flawed, are sensitive. They also indicate that a time window of 3–6 h is realistic for demonstration of drug effects. Less dangerous therapies may also have considerably longer therapeutic windows, and it is likely that drugs that modify metabolic processes in the ischemic penumbra will prove to be effective for the treatment of stroke in the near future.

REFERENCES

1 Bonita, R., Epidemiology of stroke, *Lancet,* **339** (1992) 342–344.

2 Bamford, J., Sandercock, P., Dennis, M., Burn, J., and Warlow, C. A., prospective study of acute cerebrovascular disease in the community: the Oxfordshire Community Stroke Project—1981–1986. 2. Incidence, case fatality rates and overall outcome at one year of cerebral infarction, primary intracerebral and subarachnoid haemorrhage, *J. Neurol. Neurosurg. Psych.,* **53** (1990) 16–22.

3 Bonita, R., Beaglehole, R., and North, J. D., Event, incidence and case fatality rates of cerebrovascular disease in Auckland, New Zealand, *Am. J. Epidemiol.,* **120** (1984) 236–243.

4 Boysen, G., Nyboe, J., Appleyard, M., Sorensen, P. S., Boas, J., Somnier, F., Jensen, G., and Schnohr, P., Stroke incidence and risk factors for stroke in Copenhagen, Denmark, *Stroke,* **19** (1988) 1345–1353.

5 Bikkina, M., Levy, D., Evans, J. C., Larson, M. G., Benjamin, E. J., Wolf, P. A., and Castelli, W. P., Left ventricular mass and risk of stroke in an elderly cohort. The Framingham Heart Study (see comments), *JAMA,* **272** (1994) 33–36.

6 Flegel, K. M., Shipley, M. J., and Rose, G., Risk of stroke in non-rheumatic atrial fibrillation, *Lancet,* **1** (1987) 526–529.

7 Wolf, P. A., Abbott, R. D., and Kannel, W. B., Atrial fibrillation: a major contributor to stroke in the elderly. The Framingham Study, *Arch. Intern. Med.,* **147** (1987) 1561–1564.

8 Wolf, P. A., Abbott, R. D., and Kannel, W. B., Atrial fibrillation as an independent risk factor for stroke: the Framingham Study, *Stroke,* **22** (1991) 983–988.

9 MacMahon, S., Peto, R., Cutler, J., Collins, R., Sorlie, P., Neaton, J., Abbott, R., Godwin, J., Dyer, A., and Stamler, J., Blood pressure, stroke, and coronary heart disease. Part 1, Prolonged differences in blood pressure: prospective observational studies corrected for the regression dilution bias, *Lancet,* **335** (1990) 765–774.

10 Rutan, G. H., Kuller, L. H., Neaton, J. D., Wentworth, D. N., McDonald, R. H., and Smith, W. M., Mortality associated with diastolic hypertension and isolated systolic hypertension among men screened for the Multiple Risk Factor Intervention Trial, *Circulation,* **77** (1988) 504–514.

11 Fuller, J. H., Shipley, M. J., Rose, G., Jarrett, R. J., and Keen, H., Mortality from coronary heart disease and stroke in relation to degree of glycaemia: the Whitehall study, *Br. Med. J. Clin. Res. Ed.,* **287** (1983) 867–870.

12 Colditz, G. A., Bonita, R., Stampfer, M. J., Willett, W. C., Rosner, B., Speizer, F. E., and Hennekens, C. H., Cigarette smoking and risk of stroke in middle-aged women, *N. Engl. J. Med.,* **318** (1988) 937–941.

13 Wolf, P. A., D'Agostino, R. B., Kannel, W. B., Bonita, R., and Belanger, A. J., Cigarette smoking as a risk factor for stroke. The Framingham Study, *JAMA,* **259** (1988) 1025–1029.

14 Levy, D. E., How transient are transient ischemic attacks? *Neurology,* **38** (1988) 674–677.

15 Koudstaal, P. J., van Gijn, J., Frenken, C. W., Hijdra, A., Lodder, J., Vermeulen, M., Bulens, C., and Franke, C. L., TIA, RIND, minor stroke: a continuum, or different subgroups? Dutch TIA Study Group, *J. Neurol. Neurosurg. Psych.,* **55** (1992) 95–97.

16 Hankey, G. J., Slattery, J. M., and Warlow, C. P., The prognosis of hospital-referred transient ischaemic attacks, *J. Neurol. Neurosurg. Psych.,* **54** (1991) 793–802.

17 Dennis, M., Bamford, J., Sandercock, P., and Warlow, C., Prognosis of transient ischemic attacks in the Oxfordshire Community Stroke Project, *Stroke,* **21** (1990) 848–853.

18 Truelsen, T., Lindenstrom, E., and Boysen, G., Comparison of probability of stroke between the Copenhagen City Heart Study and the Framingham Study, *Stroke,* **25** (1994) 802–807.

19 van Latum, J. C., Koudstaal, P. J., Venables, G. S., van Gijn, J., Kappelle, L. J., and Algra, A., Predictors of major vascular events in patients with a transient ischemic attack or minor ischemic stroke and with nonrheumatic atrial fibrillation. European Atrial Fibrillation Trial (EAFT) Study Group, *Stroke,* **26** (1995) 801–806.

20 Lindenstrom, E., Boysen, G., and Nyboe, J., Influence of total cholesterol, high density lipoprotein cholesterol, and triglycerides on risk of cerebrovascular disease: the Copenhagen City Heart Study, *Br. Med. J.,* **309** (1994) 11–15.

21 Harmsen, P., Rosengren, A., Tsipogianni, A., and Wilhelmsen, L., Risk factors for stroke in middle-aged men in Goteborg, Sweden, *Stroke,* **21** (1990) 223–229.

22 Shintani, S., Kikuchi, S., Hamaguchi, H., and Shiigai, T., High serum lipoprotein(a) levels are an independent risk factor for cerebral infarction, *Stroke,* **24** (1993) 965–969.

23 Verhoef, P., Hennekens, C. H., Malinow, M. R., Kok, F. J., Willett, W. C., and Stampfer, M. J., A prospective study of plasma homocyst(e)ine and risk of ischemic stroke, *Stroke,* **25** (1994) 1924–1930.

24 Qizilbash, N., Jones, L., Warlow, C., and Mann, J., Fibrinogen and lipid concentrations as risk factors for transient ischaemic attacks and minor ischaemic strokes, *Br. Med. J.,* **303** (1991) 605–609.

25 Stout, R. W. and Crawford, V., Seasonal variations in fibrinogen concentrations among elderly people, *Lancet,* **338** (1991) 9–13.

26 MacMahon, S., Cutler, J. A., and Stamler, J., Antihypertensive drug treatment. Potential, expected, and observed effects on stroke and on coronary heart disease, *Hypertension,* **13** (1989) I45–50.

27 MacMahon, S. and Rodgers, A., The effects of blood pressure reduction in older patients: an overview of five randomized controlled trials in elderly hypertensives, *Clin. Exp. Hypertens.,* **15** (1993) 967–978.

28 MRC Working Party, Medical Research Council trial of treatment of hypertension in older adults: principal results, *Br. Med. J.,* **304** (1992) 405–412.

29 SHEP Cooperative Research Group, Prevention of stroke by antihypertensive drug treatment in older persons with isolated systolic hypertension. Final results of the Systolic Hypertension, in the Elderly Program (SHEP), *JAMA,* **265** (1991) 3255–3264.

30 Dahlof, B., Hansson, L., Lindholm, L. H., Schersten, B., Ekbom, T., and Wester, P. O., Swedish Trial in Old Patients with Hypertension (STOP-Hypertension) analyses performed up to 1992, *Clin. Exp. Hypertens.,* **15** (1993) 925–939.

31 MacMahon, S. and Rodgers, A., Blood pressure, antihypertensive treatment and stroke risk, *J. Hypertens. Suppl.,* **12** (1994) S5–14.

32 Stamler, J., Rose, G., Stamler, R., Elliott, P., Dyer, A., and Marmot, M., INTERSALT study findings. Public health and medical care implications, *Hypertension,* **14** (1989) 570–577.

33 Petersen, P., Boysen, G., Godtfredsen, J., Andersen, E. D., and Andersen, B., Placebo-controlled, randomised trial of warfarin and aspirin for prevention of thromboembolic complications in chronic atrial fibrillation. The Copenhagen AFASAK study, *Lancet,* **1** (1989) 175–179.

34 The Boston Area Anticoagulation Trial for Atrial Fibrillation Investigators, The effect of low-dose warfarin on the risk of stroke in patients with nonrheumatic atrial fibrillation, *N. Engl. J. Med.,* **323** (1990) 1505–1511.

35 Stroke Prevention in Atrial Fibrillation Study, Final results, *Circulation,* **84** (1991) 527–539.

36 Connolly, S. J., Laupacis, A., Gent, M., Roberts, R. S., Cairns, J. A., and Joyner, C., Canadian Atrial Fibrillation Anticoagulation (CAFA) Study, *J. Am. Coll. Cardiol.,* **18** (1991) 349–355.

37 Ezekowitz, M. D., Bridgers, S. L., James, K. E., Carliner, N. H., Colling, C. L., Gornick, C. C., Krause, Steinrauf, H., Kurtzke, J. F., Nazarian, S. M., Radford, M. J., Rickles, F. R., Shabetai, R., and Dcykin, S., Warfarin in the prevention of stroke associated with nonrheumatic atrial fibrillation. Veterans Affairs Stroke Prevention in Nonrheumatic Atrial Fibrillation Investigators, *N. Engl. J. Med.,* **327** (1992) 1406–1412.

38 Risk factors for stroke and efficacy of antithrombotic therapy in atrial fibrillation. Analysis of pooled data from five randomized controlled trials. Atrial Fibrillation Investigators: Atrial Fibrillation, Aspirin, Anticoagulation Study; Boston Area Anticoagulation Trial for Atrial Fibrillation Study; Canadian Atrial Fibrillation Anticoagulation Study; Stroke Prevention in Atrial Fibrillation Study; Veterans Affairs Stroke Prevention in Nonrheumatic Atrial Fibrillation Study, *Arch. Intern. Med.,* **154** (1994) 1449–1457.

39 Warfarin versus aspirin for prevention of thromboembolism in atrial fibrillation: Stroke Prevention in Atrial Fibrillation II Study, *Lancet,* **343** (1994) 687-691.

40 The CASANOVA Study Group, Carotid surgery versus medical therapy in asymptomatic carotid stenosis, *Stroke,* **22** (1991) 1229–1235.

41 Mayo Asymptomatic Carotid Endarterectomy Study Group, Results of a randomized controlled trial of carotid endarterectomy for asymptomatic carotid stenosis, *Mayo Clin. Proc.,* **67** (1992) 513–518.

42 Hobson, R. W., 2d, Weiss, D. G., Fields, W. S., Goldstone, J., Moore, W. S., Towne, J. B., and Wright, C. B., Efficacy of carotid endarterectomy for asymptomatic carotid stenosis. The Veterans Affairs Cooperative Study Group, *N. Engl. J. Med.,* **328** (1993) 221–227.

43 Matias Guiu, J., Davalos, A., Pico, M., Monasterio, J., Vilaseca, J., and Codina, A., Low-dose acetylsalicylic acid (ASA) plus dipyridamole versus dipyridamole alone in the prevention of stroke in patients with reversible ischemic attacks, *Acta Neurol. Scand.,* **76** (1987) 413–421.

44 Warlow, C., Secondary prevention of stroke, *Lancet,* **339** (1992) 724–727.

45 EAFT (European Atrial Fibrillation Trial) Study Group, Secondary prevention in non-rheumatic atrial fibrillation after transient ischaemic attack or minor stroke, *Lancet,* **342** (1993) 1255–1262.

46 European Carotid Surgery Trialists' Collaborative Group, MRC European Carotid Surgery Trial: interim results for symptomatic patients with severe (70–99%) or with mild (0–29%) carotid stenosis, *Lancet,* **337** (1991) 1235–1243.

47 North American Symptomatic Carotid Endarterectomy Trial Collaborators, Beneficial effect of carotid endarterectomy in symptomatic patients with high-grade carotid stenosis, *N. Engl. J. Med.,* **325** (1991) 445–453.

48 Carter, A. B., Hypotensive therapy in stroke survivors, *Lancet,* **1** (1970) 485–489.

49 Hypertension-Stroke Cooperative Study Group. Effect of antihypertensive treatment on stroke recurrence. *JAMA,* **229** (1974) 409–418.

50 The Dutch TIA Trial Study Group, Trial of secondary prevention with atenolol after transient ischemic attack or nondisabling ischemic stroke, *Stroke,* **24** (1993) 543–548.

51 Eriksson, S., Olofsson, B. O., Wester, P. O., and TEST Study Group, Atenolol in secondary prophylaxis after stroke and TIA. A report from the TEST study. Proceedings of the Second International Conference on Stroke, **2** (1992) (Abstract).

52 Mas, J. L. and Zuber, M., Recurrent cerebrovascular events in patients with patent foramen ovale, atrial septal aneurysm, or both and cryptogenic stroke or transient ischemic attack, *Am. Heart J.,* **130** (1995) 1083–1088.

53 Petitti, D. B., Wingerd, J., Pellegrin, F., and Ramcharan, S., Risk of vascular disease in women. Smoking, oral contraceptives, noncontraceptive estrogens, and other factors, *JAMA,* **242** (1979) 1150–1154.

54 Stampfer, M. J., Willett, W. C., Colditz, G. A., Speizer, F. E., and Hennekens, C. H., A prospective study of past use of oral contraceptive agents and risk of cardiovascular diseases, *N. Engl. J. Med.,* **319** (1988) 1313–1317.

55 Lidegaard, O., Oral contraception and risk of a cerebral thromboembolic attack: results of a case-control study, *Br. Med. J.,* **306** (1993) 956–963.

56 Bamford, J., Sandercock, P., Dennis, M., Burn, J., and Warlow, C., Classification and natural history of clinically identifiable subtypes of cerebral infarction, *Lancet,* **337** (1991) 1521–1526.

57 Libman, R. B., Wirkowski, E., Alvir, J., and Rao, T. H., Conditions that mimic stroke in the emergency department. Implications for acute stroke trials, *Arch. Neurol.,* **52** (1995) 1119–1122.

58 Weir, C. J., Murray, G. D., Adams, F. G., Muir, K. W., Grosset, D. G., and Lees, K. R., Poor accuracy of stroke scoring systems for differential clinical diagnosis of intracranial haemorrhage and infarction, *Lancet,* **344** (1994) 999–1002.

59 Langhorne, P., Williams, B. O., Gilchrist, W., and Howie, K., Do stroke units save lives? *Lancet,* **342** (1993) 395–398.

60 Bamford, J., Dennis, M., Sandercock, P., Burn, J., and Warlow, C., The frequency, causes and timing of death within 30 days of a first stroke: the Oxfordshire Community Stroke Project, *J. Neurol. Neurosurg. Psych.,* **53** (1990) 824–829.

61 Oppenheimer, S. and Hachinski, V., Complications of acute stroke, *Lancet,* **339** (1992) 721–724.

62 Jones, T. H., Morawetz, R. B., Crowell, R. M., Marcoux, F. W., Fitzgibbon, S. J., DeGirolami, U., and Ojemann, R. G., Thresholds of focal cerebral ischemia in awake monkeys, *J. Neurosurg.* **54** (1981) 773–782.

63 Sundt, T. M., Sharbrough, P. W., and Anderson, R. E., Cerebral blood flow and electroencephalograms during carotid endarterectomy, *J. Neurosurg.* **41** (1974) 310–320.

64 Tamura, A., Graham, D. I., McCulloch, J., and Teasdale, G. M., Focal cerebral ischaemia in the rat:1. Description of technique and early neuropathological consequences following middle cerebral artery occlusion, *J. Cereb. Blood Flow Metab.,* **1** (1981) 53–60.

65 Tamura, A., Graham, D. I., McCulloch, J., and Teasdale, G. M., Focal cerebral ischaemia in the rat:2. Regional cerebral blood flow determined by [14C]iodoantipyrine autoradiography following middle cerebral artery occlusion, *J. Cereb. Blood Flow Metab.,* **1** (1981) 61–69.

66 Marchal, G., Serrati, C., Rioux, P., Petit-Taboue, M. C., Viader, F., de la Sayette, V., le Doze, F., Lochon, P., Derlon, J. M., Orgogozo, J. M., and Baron, J. C., PET imaging of cerebral perfusion and oxygen consumption in acute ischaemic stroke: relation to outcome, *Lancet,* **341** (1993) 925–927.

67 Baird, A. E. and Donnan, G. A., Prognostic value of reperfusion during first 48 hours of ischaemic stroke, *Lancet,* **342** (1993) 236.

68 Heiss, W. D., Huber, M., Fink, G. R., Herholz, K., Pietrzyk, U., Wagner, R., and Wienhard, K., Progressive derangement of periinfarct viable tissue in ischemic stroke, *J. Cereb. Blood Flow Metab.,* **12** (1992) 193–203.

69 Koudstaal, P. J., Stibbe, J., and Vermeulen, M., Fatal ischaemic brain oedema after early thrombolysis with tissue plasminogen activator in acute stroke, *Br. Med. J.,* **297** (1988) 1571–1574.

70 Wardlaw, J. M. and Warlow, C. P., Thrombolysis in acute ischemic stroke: does it work? *Stroke,* **23** (1992) 1826–1839.

71 Hommel, M., Boissel, J. P., Cornu, C., Boutitie, F., Lees, K. R., Besson, G., Leys, D., Amarenco, P., and Bogaert, M., Termination of trial of streptokinase in severe acute ischaemic stroke, *Lancet,* **345** (1995) 57.

72 Multicentre Acute Stroke Trial-Italy (MAST-I) Group, Randomised controlled trial of streptokinase, aspirin, and combination of both in treatment of acute ischaemic stroke, *Lancet,* **346** (1995) 1509–1514.

73 Donnan, G. A., Davis, S. M., Chambers, B. R., Gates, P. C., Hankey, G. J., McNeil, J. J., Rosen, D., Stewart Wynne, E. G., and Tuck, R. R., Trials of streptokinase in severe acute ischacmic stroke (letter), *Lancet,* **345** (1995) 578–579.

74 Hacke, W., Kaste, M., Fieschi, C., Toni, D., Lesaffre, E., von Kummer, R., Boysen, G., Bluhmki, E., Hoxter, G., Mahagne, M. H., and Hennerici, M., Intravenous thrombolysis with recombinant tissue plasminogen activator for acute hemispheric stroke: The European Cooperative Acute Stroke Study (ECASS), *JAMA,* **274** (1995) 1017–1025.

75 The National Institute of Neurological Disorders and Stroke rt-PA Stroke Study Group, Tissue plasminogen activator for acute ischemic stroke, *N. Engl. J. Med.,* **333** (1995) 1581–1587.

76 Kay, R., Wong, K. S., Yu, Y. L., Chan, Y. W., Tsoh, T. H., Ahuja, A. T., Chan, F. L., Fong, K. Y., Law, C. B., Wong, A., and Woo, J., Low-molecular-weight heparin for the treatment of acute ischemic stroke, *N. Engl. J. Med.,* **333** (1995) 1588–1593.

77 Scandinavian Stroke Study Group, Multicenter trial of hemodilution in acute ischemic stroke. I. Results in the total patient population, *Stroke,* **18** (1987) 691–699.

78 Italian Acute Stroke Study Group, The Italian hemodilution trial in acute stroke, *Stroke,* **18** (1987) 670–676.

79 Asplund, K., Hemodilution in acute stroke. *Cerebrovasc. Dis.,* **1(Suppl 1)** (1991) 129–138.

80 Steiner, T. J., Naftidrofuryl in the treatment of acute cerebral hemisphere infarction, *Stroke,* **27** (1996) 195 (abstract)

81 Hsu, C. Y., Norris, J. W., Hogan, E. L., Bladin, P., Dinsdale, H. B., Yatsu, F. M., Earnest, M. P., Scheinberg, P., Caplan, L. R., and Karp, H. R., Pentoxifylline in acute non-hemorrhagic stroke. A randomized, placebo-controlled double-blind trial, *Stroke,* **19** (1988) 716–722.

82 Chan, Y. W. and Kay, C. S., Pentoxifylline in the treatment of acute ischaemic stroke—a reappraisal in Chinese stroke patients, *Clin. Exp. Neurol.,* **30** (1993) 110–116.

83 Candelise, L., Colombo, A., and Spinnler, H., Therapy against brain swelling in stroke patients. A retrospective clinical study on 227 patients, *Stroke,* **6** (1975) 353–356.

84 Sandercock, P., Important new treatments for acute ischaemic stroke? (editorial), *Br. Med. J. Clin. Res. Ed.,* **295** (1987) 1224,1225.

85 Gelmers, H. J., The effects of nimodipine on the clinical course of patients with acute ischemic stroke, *Acta Neurol. Scand.,* **69** (1984) 232–239.

86 Gelmers, H. J., Gorter, K., de Weerdt, C. J., and Wiezer, H. J., A controlled trial of nimodipine in acute ischemic stroke, *N. Engl. J. Med.,* **318** (1988) 203–207.

87 Mohr, J. P., Orgogozo, J. M., Harrison, M. J. G., Hennerici, M., Wahlgren, N. G., Gelmers, J. H., Martinez-Vila, E., Dycka, J., and Tettenborn, D., Meta-analysis of oral nimodipine trials in acute ischemic stroke, *Cerebrovasc. Dis.,* **4** (1994) 197–203.

88 Wahlgren, N. G., MacMahon, D. G., DeKeyser, J., Indredavik, B., and Ryman, T., Intravenous Nimodipine West European Stroke Trial (INWEST) of nimodipine in the treatment of acute ischaemic stroke, *Cerebrovasc. Dis.,* **4** (1994) 204–210.

89 McCulloch, J., Excitatory amino acid antagonists and their potential for the treatment of ischaemic brain damage in man, *Br. J. Clin. Pharmac.,* **34** (1992) 106–114.

90 Muir, K. W. and Lees, K. R., Clinical experience with excitatory amino acid antagonist drugs, *Stroke,* **26** (1995) 503–513.

91 Harrison, N. L. and Simmonds, M. A., Quantitative studies on some antagonists of N-methyl D-aspartate in slices of rat cerebral cortex, *Br. J. Pharmacol.,* **84** (1985) 381–391.

92 Lucas, M. J., Leveno, K. J., and Cunningham, F. G., A comparison of magnesium sulfate with phenytoin for the prevention of eclampsia, *N. Engl. J. Med.,* **333** (1995) 201–205.

93 The Eclampsia Trial Collaborative Group, Which anticonvulsant for women with eclampsia? Evidence from the Collaborative Eclampsia Trial, *Lancet,* **345** (1995) 1455–1463.

94 Peters, G. R., Hwang, L. J., Musch, B., Brosse, D. M., Orgogozo, J. M., Safety and efficacy of 6mg/kg/day tirilazad mesylate in patients with acute ischemic stroke (TESS Study), *Stroke,* **27** (1996) 195 (abstract).

95 RANTTAS Investigators, Randomized trial of tirilazad in acute stroke (RANTTAS), *Stroke,* **27** (1996) 195 (abstract).

96 Hoffbrand, B. I., Bingley, P. J., Oppenheimer, S. M., and Sheldon, C. D., Trial of ganglioside GM1 in acute stroke, *J. Neurol. Neurosurg. Psych.,* **51** (1988) 1213,1214.

97 Lenzi, G. L., Grigoletto, F., Gent, M., Roberts, R. S., Walker, M. D., Easton, J. D., Carolei, A., Dorsey, F. C., Rocca, W. A., Bruno, R., Patarnello, F., Fieschi, C., and the Early Stroke Trial Group, Early treatment of stroke with monosialoganglioside GM-1. Efficacy and safety results of the Early Stroke Trial, *Stroke,* **25** (1994) 1552–1558.

98 Alter, M., Bell, R., Brass, L., Gaines, K., Goldstein, L., Hollander, J., Jozefczyk, P., Kelley, R., Mayman, C., Miller, A., Pascuzzi, R , Ramirez Lassepas, M., Rosenbaum, D., Zachariah, S., Brennan, R., Chawluk, J., Furlan, A., and Mellits, D., Ganglioside GM1 in acute ischemic stroke: The SASS trial, *Stroke,* **25** (1994) 1141–1148.

99 Morris, A. D., Grosset, D. G., Squire, I. B., Lees, K. R., Bone, I., and Reid, J. L., The experiences of an acute stroke unit—implications for multicentre acute stroke trials, *J. Neurol. Neurosurg. Psych.,* **56** (1993) 352–355.

100 Albanese, M. A., Clarke, W. R., Adams, H. P., Jr., Woolson, R. F., and the TOAST Investigators, Ensuring reliability of outcome measures in multicenter clinical trials of treatments for acute ischemic stroke: the program developed for the Trial of ORG 10172 in Acute Stroke Treatment (TOAST), *Stroke,* **25** (1994) 1746–1751.

101 Jorgensen, H. S., Nakayama, H., Raaschou, H. O., ViveLarsen, J., Stoier, M., and Olsen, T. S., Outcome and time course of recovery in stroke. Part II: Time course of recovery. The Copenhagen Stroke Study, *Arch. Phys. Med. Rehab.,* **76** (1995) 406–412.

102 Anonymous, *International Classification of Impairments, Disabilities and Handicaps,* World Health Organisation, Geneva (1980).

103 Mathew, N. T., Meyer, J. S., Rivera, V. M., Charney, J. Z., and Hartmann, A., A double-blind evaluation of glycerol therapy in acute cerebral infarction, *Lancet,* **2** (1972) 1327–1329.

104 Gelmers, H. J., Gorter, K., de Weerdt, C. J., and Wiezer, H. J., Assessment of interobserver variability in a Dutch multicenter study on acute ischemic stroke, *Stroke,* **19** (1988) 709–711.

105 Scandinavian Stroke Study Group, Multicenter trial of hemodilution in ischemic stroke—background and study protocol, *Stroke,* **16** (1985) 885–890.

106 Brott, T., Adams, H. P., Jr., Olinger, C. P., Marler, J. R., Barsan, W. G., Biller, J., Spilker, J., Holleran, R., Eberle, R., Hertzberg, V., Rorick, M., Moomaw, C. J., and Walker, M., Measurements of acute cerebral infarction: a clinical examination scale, *Stroke,* **20** (1989) 864–870.

107 Hantson, L., De Weerdt, W., De Keyser, J., Diener, H. C., Franke, C., Palm, R., Van Orshoven, M., Schoonderwalt, H., De Klippel, N., Herroelen, L., and Feys, H., The European stroke scale, *Stroke,* **25** (1994) 2215–2219.

108 Lyden, P. D. and Lau, G. T., A critical appraisal of stroke evaluation and rating scales, *Stroke,* **22** (1991) 1345–1352.

109 Cote, R., Battista, R. N., Wolfson, C. M., and Hachinski, V., Stroke assessment scales: guidelines for development, validation, and reliability assessment, *Can. J. Neurol. Sci.,* **15** (1988) 261–265.

110 Orgogozo, J. M., Capildeo, R., Anagnostou, C. N., Juge, O., Pere, J. J., Dartigues, J. F., Steiner, T. J., Yotis, A., Rose, F. C., Development of a neurological score for the clinical evaluation of sylvian infarctions, *Presse Med.,* **12** (1983) 3039–3044.

111 Lesaffre, E., Scheys, I., Frohlich, J., and Bluhmki, E., Calculation of power and sample size with bounded outcome scores, *Stat. Med.,* **12** (1993) 1063–1078.

112 Goldstein, L. B., Sample size calculations for clinical trials using the NIH stroke scale (2), *Stroke,* **25** (1994) 1700,1701.

113 van Gijn, J., and Warlow, C. P., Down with stroke scales! *Cerebrovasc. Dis.,* **2** (1992) 244–246.

114 Mahoney, F. I., and Barthel, D. W., Functional evaluation: the Barthel index, *Maryland State Med. J.,* **14** (1965) 61–65.

115 Granger, C. V., Dewis, L. S., Peters, N. C., Sherwood, C. C., and Barrett, J. E., Stroke rehabilitation: analysis of repeated Barthel index measures, *Arch. Phys. Med. Rehabil.,* **60** (1979) 14–17.

116 Korner Bitensky, N., Wood Dauphinee, S., Siemiatycki, J., Shapiro, S., and Becker, R., Health-related information postdischarge: telephone versus face-to-face interviewing, *Arch. Phys. Med. Rehabil.,* **75** (1994) 1287–1296.

117 Keith, R. A., Granger, C. V., Hamilton, B. B., and Sherwin, F. S., The functional independence measure: a new tool for rehabilitation, *Adv. Clin. Rehabil.,* **1** (1987) 6–18X.

118 Oczkowski, W. J. and Barreca, S., The functional independence measure: its use to identify rehabilitation needs in stroke survivors, *Arch. Phys. Med. Rehabil.,* **74** (1993) 1291–1294.

119 Rankin, J., Cerebral vascular accidents in patients over the age of 60. 2: Prognosis, *Scott. Med, J.,* **2** (1957) 200–215.

120 Hall, K., Cope, D. N., and Rappaport, M., Glasgow Outcome Scale and Disability Rating Scale: comparative usefulness in following recovery in traumatic head injury, *Arch. Phys. Med. Rehabil.,* **66** (1985) 35–37.

121 Wolfe, C. D., Taub, N. A., Woodrow, E. J., and Burney, P. G., Assessment of scales of disability and handicap for stroke patients, *Stroke,* **22** (1991) 1242–1244.

122 De Haan, R., Limburg, M., Bossuyt, P., Van der Meulen, J., and Aaronson, N., The clinical meaning of Rankin 'handicap' grades after stroke, *Stroke,* **26** (1995) 2027–2030.

123 Steinberg, G. K., Perez-Pinzon, M. A., Maier, C. M., Sun, G. H., Yoon, E., Kunis, D. M., Bell, T. E., Powell, M., Kotake, A., and Giffard, R., CGS 19755: Correlation of in vitro neuroprotection, protection against experimental ischemia and CSF levels in cerebrovascular surgery patients. Proceedings of the Fifth International Symposium on Pharmacology of Cerebral Ischemia, GLU 4 (1994) (abstract).

4

Calcium Homeostasis, Nimodipine, and Stroke

**Paul G. M. Luiten, Bauke Stuiver, Gineke I. de Jong,
Csaba Nyakas, and Jacques H. A. De Keyser**

1. INTRODUCTION

Intracellular calcium as a divalent ion is the most common element involved in the transfer of signals in living cells. In cells at rest, Ca^{2+} levels will be very low, probably for a number of reasons. At low intracellular concentrations, as is the case in unstimulated cells, most Ca^{2+}-dependent intracellular processes will be at baseline levels necessary for the recovery stage upon activation after stimulation. Other reasons proposed for low resting levels of Ca^{2+} are the precipitation by Ca^{2+} of phosphates, the essential mediators of energy supply to the cell *(1)*.

In the cells of the nervous system, particularly in neurons which are the cells of primary interest of the current review, there is general consensus that the baseline levels of intracellular, free ionized Ca^{2+} is somewhere between 0.05-0.2 μM as measured with microfluorimetric methods. This low intracellular Ca^{2+} level *(2)* is about 10^4 lower than the Ca^{2+} concentration of 1.3 mM in the extracellular environment of the brain. It is obvious that such a large electrochemical gradient over the nerve cell membrane can only be maintained by a permanent active process. Spatiotemporal changes in intracellular calcium resulting from various stimuli are involved in the control of many cellular processes including growth, aging, adaptation, structural plasticity, and contraction, among others *(1)*.

The importance of understanding the behavior of intracellular calcium has rapidly gained impetus since it became clear that sustained elevation of intracellular calcium concentration $[Ca^{2+}]_i$ can activate degradative processes and lead to cell necrosis and programmed cell death. It is this role of $[Ca^{2+}]_i$ that is thought to underlie the devastating consequence of ischemic stroke that is the theme of the present survey. We will shortly review the mechanisms considered to regulate intracellular calcium homeostasis, the putative mechanism of distortion of $[Ca^{2+}]_i$ during stroke, and the pharmacological intervention by nimodipine as a representative of a group of compounds that specifically block part of the pathological influx of calcium. The influence of calcium blockade by nimodipine will be shortly reviewed in experimental models that mimic various aspects of ischemic stroke, followed by the outcome of nimodipine application in clinical stroke trials.

From: Clinical Pharmacology of Cerebral Ischemia *Edited by:* G. J. Ter Horst and J. Korf *Humana Press Inc., Totowa, NJ*

2. HOMEOSTATIC REGULATION OF INTRACELLULAR CALCIUM

2.1. Ca²⁺ Influx Through Ligand-Operated Ca²⁺ Channels

Within the framework of the current topic we will mainly confine our interest to cells of the nervous system, in which we may distinguish nonexcitable cells like the vascular endothelial cells from the neurons as an outspoken form of excitable cells. Since signal transduction between nerve cells is highly dependent on temporary and transient changes in local intracellular Ca^{2+}, a well controlled Ca^{2+} balance is required for adequate neuronal function *(1,3)*. Signals from external origin affect receptors on the neuronal surface in several ways. Short term effects are achieved when a released transmitter signal triggers changes in transmembranous receptor complexes that yield hydrophylic pores permitting ionic exchange. Owing to the electrochemical gradient over the neuronal membrane, short influxes of calcium are established as is notably the case through N-methyl-D-asparate (NMDA) receptor channels, amino-3-hydroxy-5-methyl-4-isoxazole propionic acid (AMPA), kainate, quisqualate glutamate receptors, and nicotinic receptors *(4)* (Fig. 1). MacDermott et al. *(5)* demonstrated in spinal cord neurons that the inward currents evoked by stimulation with NMDA are accompanied by an immediate and specific elevation of intracellular Ca^{2+} concentrations, which both characteristically could be blocked by Mg^{2+} ions in a voltage-dependent manner. In contrast to voltage-sensitive calcium channels, the NMDA channel itself is not dependent on changes in membrane potentials. However, in the presence of Mg^{2+} ions, NMDA-induced currents show a voltage sensitivity, that is the result of a voltage-dependent blockade by Mg^{2+} binding to the NMDA receptor complex *(6)*. It is notably the NMDA receptor that is highly permeable to calcium with a channel conductance of 50 pS, when compared to the other ionotropic receptor types. Studies comparing the contribution of calcium influx in depolarizing currents mediated by the various ligand-operated calcium channels all point to a predominant role of the NMDA channel *(7,8)*.

2.1.1. Ca²⁺ Increase via IP₃ Receptors and G-Proteins

More long-term impact of external messengers is mediated by Ca^{2+} release from intracellular stores. In excitable cells, this intracellular Ca^{2+} flux can be the result of activation of plasma membrane receptor complexes, which are coupled to G-proteins triggering the enzymatic activity of the β form of phospholipase C (PLC), hydrolysis of membrane phospholipids, and intracellular release of inositol trisphosphates (IP_3) *(9)*. The IP_3 binds to its receptors on the membranes of the endoplasmic reticulum, which initiates the release of Ca^{2+} into the cytosol (Fig. 1). Several receptor types sensitive to 'classical' neurotransmitters like acetylcholine, histamine, norepinephrine, serotonin, glutamate, and many neuropeptides influence the IP_3 signalling pathway leading to the increase of Ca^{2+} from intracellular origin.

2.1.2. Ca²⁺ Increase via Trk and IP₃ Pathway

The formation of IP_3 is not only mediated by G-protein-linked receptors. In several cell types, extracellular messages from a polypeptide family known as neurotrophins exert their cellular activation through tyrosine kinase (Trk)-linked receptors. Neurotrophins such as nerve growth factor and brain-derived neurotrophic factor and growth factors like platelet-derived growth factor and epidermal growth factor, stimulate Trk activity that leads to phosphorylation of the γ unit of PLC and subsequent production of

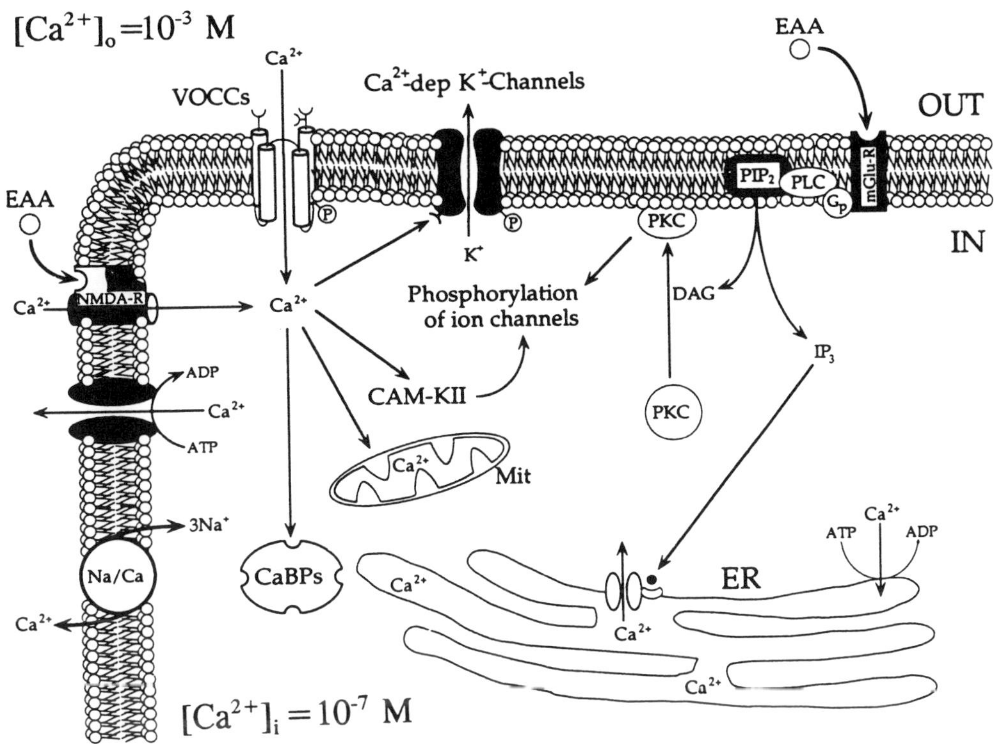

Fig. 1. Pathways leading to influx and intracellular release of [Ca^{2+}]$_i$. Main sources of Ca^{2+} influx are mediated by ligand-operated calcium channels and voltage-operated calcium channels. The ligand-sensitive channels are here exemplified on the NMDA channel being a major and direct source of Ca^{2+} influx on stimulation by glutamate. Secondary influx of Ca^{2+} through VOCCs results from depolarization of the membrane potential. Excitatory amino acids (EAA) not only affect NMDA (and others like AMPA channels), but also evoke more indirect release of Ca^{2+} from intracellular stores through metabotropic glutamate receptors (mGlu-R) and hydrolysis of the phospholipid PIP$_2$ and IP$_3$-sensitive receptors on the endoplasmic reticulum (ER). Activation of many other transmitter receptors, such as muscarinic receptors yield similar effects. Elevated Ca^{2+} affects membrane channels like the K$^+$ channels and kinases like calmodulin-dependent kinase II (CAM-KII) and protein kinase C (PKC) and has a direct influence on gene expression. Excess of [Ca^{2+}]$_i$ is bound to Ca^{2+}-binding proteins (CaBPs), is sequestered in intracellular stores like ER and mitochondria (Mit), or transported over the plasmalemma. (Reprinted with permission from ref. *177*.)

IP$_3$ *(10,11)*. Although both pathways converge on IP$_3$ formation and intracellular Ca^{2+} release, each pathway has different dynamics. There is evidence that the tyrosine kinase-linked receptor pathway and subsequent Ca^{2+} release is much slower, but yields an effect of longer duration than signal transduction relayed by G-protein-coupled receptors *(12)*.

2.1.3. Ca^{2+} Increase via Ryanodine Receptors

Besides intracellular Ca^{2+} release via IP$_3$ receptors on the endoplasmic reticulum, excitable cells such as neurons are also endowed with a second intracellular Ca^{2+} release channel protein: the ryanodine receptor (RyR) *(9,13,14)*. The ryanodine receptors have been identified in widespread areas of the brain in various vertebrate and mammalian species *(13,15,16)*. The presence of RyR is prominent in cerebellar Purkinje cells and hippocampal pyramidal neurons, notably in the cornu ammonis CA2 and CA3 areas. In many cell types, it is most likely that the RyR coexists with the IP_3 receptor between which there is a considerable molecular resemblance *(9)*. Both IP_3 and RyR appear to have different locations in nerve cells, which can vary from dominant positions in cell bodies, proximal or distal dendrites, spines or shafts. Consequently, Ca^{2+} release on activation of each receptor has a differential intracellular localization, which is accompanied by a different temporal profile. Recent investigations point to short-term phasic Ca^{2+} fluctuations after IP_3 stimulation and sustained long-term Ca^{2+} increases upon RyR stimulation by caffeine *(14)*. Moreover, the two Ca^{2+} channels can display complex patterns of interaction that can modulate intracellular Ca^{2+} release. As will be discussed later, the RyR may be of considerable importance in the sequence of events initiated by increased and sustained intracellular calcium involved in neuronal injury as a result of stroke. Calcium that enters the cytoplasm through voltage-sensitive plasma membrane calcium channels are probably the main signal for activation of the RyR *(9)*.

However, whereas the IP_3 receptor pathway can be activated by a large number of extracellular messengers as neurotransmitters, neuropeptides and neurotrophins, the precise mechanisms leading to RyR activation are far less clear. Calcium that has entered the cell through voltage-sensitive calcium channels was demonstrated in several cell types to induce a second intracellular Ca^{2+} release mediated by RyRs. This sensitivity of RyRs for Ca^{2+}, however, is not exclusive for the RyR since also the IP_3 receptor is dependent on the presence of Ca^{2+} for its initial activation by IP_3 *(9,17)*, high Ca^{2+} levels reduce IP_3 effects on its receptor *(9,18)*. Various other neuroactive and pharmacological compounds were shown to act as RyR receptor agonists in a number of neuronal and nonneuronal cell types. To date, however, Ca^{2+} remains the best candidate as the biological RyR agonist, possibly requiring some coupling mechanism between dihydropyridine receptors and RyR as was shown in skeletal muscle cells *(19)*.

2.2. Ca^{2+} Influx Mediated by Voltage-Gated Channels

Extracellular Ca^{2+} may enter neurons both by ligand-operated channels as already described above, and by voltage-activated or voltage-sensitive Ca^{2+} channels (VSCCs). Various types of VSCCs have now been identified in neurons, but their classification is subject to change as a result of burgeoning interest in structure and function of this class of channels. Miller in 1987 *(2)* and Tsien and colleagues in 1988 *(20)* distinguished three types of neuronal VSCCs: the dihydropyridine-sensitive L-channel and the dihydropyridine-insensitive N- and T-channels. T-channels respond to small depolarizations with transient, tiny Ca^{2+} currents, whereas stronger membrane depolarizations induce Ca^{2+} influx through N-channels. The L-type channel is a high-threshold or high-voltage activated calcium channel that responds to strong depolarizations and becomes very slowly inactivated *(21)*. More recently, the identification of a fourth type of neuronal VSCC was identified, that became known as the P-type channel *(22)*. P-channels, initially described for cerebellar Purkinje cells (hence its name) but also demonstrated now in cortex, hippocampus, and many other CNS regions, are dihydropyridine-insensitive channels that are apparently present in pre- and postsynaptic specializations *(22)*. Progress in the molecular biology of Ca^{2+} channels in recent years has considerably added to our knowledge of structure and function of the VSCC complex, whereas pharmacology and electrophysiology increased our insight in how these channels function in cellular activation and communication, which has been expertly surveyed *(4,23–26)*.

As already indicated, the four classes of VSCCs can be distinguished on their reactivity to high (the L-, N-, and P-channels) or low (the T channel) voltage changes, whereas each channel type displays a characteristic conductance level of 8 pS (T), 13 pS (N), 10-15 pS (P), and 25 pS (L) *(4,24)*. The specific sensitivity of the different classes of VSCCs to dihydropyridines or toxins like ω-conotoxin is of great significance for study of physiological effects *(24)* or cellular localizations of the different channels. VSCCs are composed of five subunits, the large α1 subunit that forms the ionic channel and an α2, β, γ, and δ peptide. The large α1 unit is the voltage-responding part of the complex and also forms the ionic channel. Molecular cloning technology of VSCCs has now demonstrated the existence of at least five α1 genes in the mammalian brain *(23,27)*. The unraveling of the molecular structure has now made it possible to further study the nature of VSCC channel selectivity to different membrane conditions and differences in Ca^{2+} permeability *(28)*.

A major question in this respect is which calcium channels are responsible for mechanisms of presynaptic transmitter release or postsynaptic activation *(23)*. This question is of key importance with regard to application of channel-specific Ca^{2+} antagonists and the functional consequences of such compounds and side-effects in therapeutic use. Such questions appear to address a matter of complex nature and a simple classification is not present. Studies on synaptosome preparations employing specific Ca^{2+} channel blockade by toxins allow the conclusion that multiple Ca^{2+} channels coexist in nerve terminals and participate in neurotransmitter release mechanisms *(29)*. Dependent on the brain area and species involved, N- and particularly P-channels and several noncharacterized Ca^{2+} channels, but not L-channels *(30)*, mediate the calcium flux necessary for exocytosis of transmitter release by presynaptic nerve endings. Such conclusions, however, highly depend on the neuronal system under study. In cholinergic terminals in the hippocampus, it was mainly the N-channel held responsible for acetylcholine release *(31)*. All studies, on the other hand, point to the L-channel in central neurons to be predominantly if not exclusively associated with postsynaptic actions, but care must be exercised with regard to the many exceptions of this too general assumption *(23)*.

2.3. Ca^{2+} *Binding to Proteins*

We have now briefly surveyed the various channels and receptor-mediated mechanisms that take part in the elevation of free intracellular calcium upon cellular activations. Elevated Ca^{2+} originates either from the large extracellular sink or from the intracellular storage pools *(1)*. Upon the temporary and often local increase of free intracellular calcium, the ions are able to bind to many cytosolic compounds and intracellular proteins, thereby triggering a large number of cellular actions. Several small molecules can bind Ca^{2+} including inositoltrisphosphate and probably also other forms of inositolphosphates. In view of the current survey, it is useful to distinguish between proteins that become activated after binding to Ca^{2+} and proteins that bind Ca^{2+}, thereby temporarily decreasing or buffering the local calcium concentration. Well known proteins that are activated by Ca^{2+} are protein kinase C and calmodulin, that after association with calcium, triggers its kinase CaM kinase II and other kinases *(32)*. In particular, the activation of protein kinase C is mentioned in the scope of this survey because of the effects attributed to activation of this kinase in neurotoxic mechanisms *(3,4)*.

Most Ca^{2+} binding proteins (CaBP) that are considered to exhibit a Ca^{2+} buffering action and proteins like calmodulin are provided with a motif specific for Ca^{2+} binding. This helix-loop-helix motif, also called the EF hand, contains a high-affinity Ca^{2+} binding site *(33)*. Calmodulin and calpain, but also the calcium buffering proteins parvalbumin, calretinin and calbindin-D28k have such EF-hand motifs and can absorb Ca^{2+} with high affinity but with different kinetics such as the slow binding by parvalbumin and the quick uptake by calbindin *(34)*.

2.3.1. Ca^{2+} Buffering Proteins

Parvalbumin, calbindin D28k, calretinin, and several other proteins are known to bind intracellular free Ca^{2+} when $[Ca^{2+}]_i$ sharply rises upon activation of the cell, and release Ca^{2+} when the $[Ca^{2+}]_i$ decreases *(35)*. Such a temporary binding by CaBP limits and modulates the Ca^{2+} rise and thus exerts a buffering function or plays a role in spatial distribution of the calcium signal *(35,36)*. By way of their Ca^{2+} buffering capacity, parvalbumin and calbindin were clearly demonstrated to curb the Ca^{2+} peaks, potently slow the rise in $[Ca^{2+}]_i$, yet having no effect on baseline Ca^{2+} levels *(36)*. Other functions of the calcium buffering proteins, however, are only poorly understood *(34)*. Their distribution in the nervous system is rather complex. Some types of nerve cells possess more than one CaBP, whereas others are devoid of such proteins *(37)*. On the other hand, the consistent coexistence of certain CaBP, with particular neurotransmitters such as parvalbumin with GABA inhibitory neurons allowed to use some CaBP, as selective neuronal markers. An interesting finding in this respect is further that the distribution of calbindin closely resembles that of VSCCs *(37)*.

Because of their particular capacity to bind surplus free intracellular calcium, CaBPs have drawn considerable attention as proteins that may play a role in neuroprotection in conditions in which intracellular calcium levels reach persistent pathogenic levels *(38,39)* as is possibly the case in aging *(40)*, ischemia, and neurodegenerative diseases *(41)*. As we will see later, there is growing evidence that changes in CaBP expression are intimately involved in such degenerative processes and may play a role in the intrinsic resistance and protective properties against neuronal dysfunction or cell death induced by overload of $[Ca^{2+}]_i$. In this regard, we demonstrated strong potentiation of the expression of CaBP parvalbumin, calbindin, and S-100 after nimodipine treatment during brain development. The potentiated developmental presence of these CaBPs coincided with a significant resistance against growth retardation induced by perinatal hypoxia in nimodipine-treated animals *(42)*.

2.4. Ca^{2+} and Gene Expression

As mentioned, the rise of $[Ca^{2+}]_i$ on cellular activation has a direct effect on various categories of proteins either functioning as calcium buffers, or becoming in an activated state, e.g., enzymatic active state. The different physiological effects triggered by the rise of $[Ca^{2+}]_i$ follow a different time-course. For example, transmitter release invoked by Ca^{2+} influx is in the millisecond range, whereas the activation of Ca^{2+}-dependent enzyme pathways can take up to seconds.

More long-term physiological effects, from minutes to hours, of elevated calcium are established when Ca^{2+} affects genetic expression yielding adaptive changes of the brain. There are probably several pathways that can mediate the intracellular calcium message to gene expression, as is the case in induction of transcription of immediate early genes. It is assumed that the increase of $[Ca^{2+}]_i$ is the message that translates activation of nicotinic receptors, NMDA receptors, and L-type calcium channels in immediate early gene responses *(43–45)*. In general, it is predicted that for Ca^{2+} signals to become physiologically effective, the rise in $[Ca^{2+}]_i$ should be local, fast, and large *(46)*. The spatially differentiated changes in Ca^{2+} concentration resulting from different influx pathways and the cascade of events leading to immediate early gene expression are thought to activate the transcription of different early genes. A major step in the genetic activation by Ca^{2+} is the phosphorylation of the cAMP response element binding protein (CREB); CREB phosphorylate is catalyzed by calcium-activated CaM kinase *(47)*. It has been argued that calcium that has entered the cells through different channels triggers distinct signalling pathways, thus yielding differentiated responses *(48,49)*. Entrance into the cell by NMDA channels or L-type calcium channels differentially influences gene transcription by way of different DNA-regulating elements *(49,50)*. More recent views argue for a role of increased $[Ca^{2+}]_i$ in transcription of so-called cell death genes. Notably, the persistent, prolonged Ca^{2+}-induced expression of immediate early genes like c-*fos* and c-*jun* has been associated with the process of delayed cell death *(51)*. Other more direct pathways are the activation of DNA degrading enzymes like endonucleases by sustained elevated nuclear calcium concentrations, whereas also the Ca^{2+}-dependent stimulation of proteases and phosphatases are considered as degrading pathways leading to nuclear breakdown *(52)*.

2.5. Ca^{2+} Extrusion from the Cytosol

Nerve cells possess a number of different mechanisms to restore the intracellular calcium concentration to baseline levels. In rest conditions the concentration of free $[Ca^{2+}]_i$ will be very low and intracellular calcium will be bound within subcellular organelles primarily into the endoplasmic reticulum (ER) and the mitochondria. Uptake of free Ca^{2+} by these organelles is a prerequisite to ensure sufficient Ca^{2+} reserves for required intracellular Ca^{2+} release in subsequent activations. Especially in presynaptic structures, sufficient Ca^{2+} storage in the ER is essential for transmitter release, since depletion of calcium from the ER prevents transmitter release *(53)*.

Shortly after a physiological cellular stimulation, the net charge with Ca^{2+} will be too large for intracellular binding to proteins alone, and additional transport mechanisms to balance temporary excess of Ca^{2+} become active. For transport of Ca^{2+} against large negative gradients of cations over the ER and plasma membranes, two mechanisms are present for intracellular sequestration in subcellular organelles or extrusion of Ca^{2+} over the plasmalemma. The ER and plasma membranes are endowed with Ca^{2+} pumps that act in a ATP-dependent fashion, whereas the plasmalemmal

Ca^{2+}-ATPase is modulated by the presence of calmodulin. In that sense, the plasmalemma Ca^{2+}-ATPase differs from the calmodulin-insensitive Ca^{2+}-ATPase in the membranes of the ER *(54)*.

However, sequestration of Ca^{2+} in the ER is a slow process that alone cannot bring down Ca^{2+} at the same rate at which Ca^{2+} enters the nerve cell after stimulation *(55)*. Additional transport of Ca^{2+} over membranes is achieved by a second mechanism that is based on ionic exchange of Ca^{2+} and Na$^+$. This Na$^+$/Ca^{2+} exchanger has a high transport capacity, but its exchange rate is strongly modulated by the local gradients of Na$^+$ and Ca^{2+} over the membranes. The Na$^+$/Ca^{2+} exchanger will obviously function dependent on the internal vs external Na$^+$ gradient, that is optimal when the external Na$^+$ concentration is restored after depolarizations. It was demonstrated that the magnitude of the Na$^+$ influx component participating in membrane potential shifts is a determinant factor in the efficacy of Ca^{2+} extrusion by the Na$^+$/Ca^{2+} exchanger *(56)*. Also, mitochondria were shown to be able to sequester Ca^{2+}, but storage in these organelles under physiologically normal conditions is limited by the low rate of Ca^{2+} uniporters as a result of a low affinity of the mitochondrial transporter for Ca^{2+} *(57)*. The Ca^{2+} uptake by mitochondria may increase with higher [Ca^{2+}]$_i$, that has been suggested to represent a mechanism for adapting neuronal metabolic activity to increased neuronal activity *(53)*. The limited mitochondrial storage of Ca^{2+}, however, can change dramatically when calcium concentrations in the neurons reach pathological levels. Excessive Ca^{2+} accumulation was demonstrated in conditions of severe Ca^{2+} overload such as after ischemia *(58)*.

3. CALCIUM AND CALCIUM HOMEOSTASIS IN CEREBROVASCULAR ENDOTHELIUM

The endothelial cells lining the lumen of the large and fine vessels of the cerebral circulation in the last decade have become a cell type of growing interest. The endothelium obviously is in a key position to influence functional integrity of the vascular domain as a whole. Notably in the larger but probably also in the small vessels of the brain, endothelial cells transfer signals to vascular responses, whereas at the same time they form the major cellular component of the blood-brain barrier. Blood vessels for their relaxation are dependent on the presence of an intact endothelium as was first demonstrated by Furchgott and Zawadski *(59)*. The endothelial cells are endowed with several enzyme systems of which nitric oxide synthase (NOS) in recent years was shown to be pivotal for the production of NO, which was previously known as endothelium-derived relaxing factor. NOS is one of the enzymes that acts in a calcium-dependent fashion *(60)*, as apparently many of the endothelial processes are, indicating the importance of intracellular calcium regulation of endothelial cells. Compared to our knowledge of neuronal or smooth muscle cell calcium homeostasis, data on the control of endothelial Ca^{2+} balance is limited.

Calcium concentrations in endothelial cells, e.g., from aorta origin increase upon cholinergic stimulation, which includes calcium fluxes mediated by both IP$_3$ and ryanodine receptors. These cells were demonstrated to be also endowed with intracellular Ca^{2+} storage mechanisms employing Ca^{2+}-ATPases and Ca^{2+} influx pathways from extracellular surroundings *(61)*. Other studies comparing endothelial cells from different vascular origins demonstrated that all these cells contained functional ryanodine receptors that may influence the increase of $[Ca^{2+}]_i$ in different ways *(62)*.

L-type calcium channels on brain endothelial cells have not been directly demonstrated, but various experimental studies indicate their presence in endothelial cells. For example, vascular preparations of aorta show that calcium-dependent contractions can be differentially modulated by the dihydropyridines nitrendipine and BAY K 8644 in the presence of endothelial lining of the vessels *(63)*. Similar studies by Vanhoutte *(60)* demonstrated prevention of calcium-triggered vasorelaxation by dihydropyridines and thus also provide evidence for voltage-dependent Ca^{2+} channels on endothelial cells. The studies mentioned above use peripheral arterial vessels as model of study. Reports on brain endothelial cells describe changes of $[Ca^{2+}]_i$ mediated by endothelin receptors and intracellular stores, but effects of dihydropyridines remain unclear. On the other hand, good evidence is available for stretch-activated cation channels that are permeable for Ca^{2+} with increasing channel opening by depolarizing currents. Bossu et al. *(64)* demonstrated in confluent endothelial cell cultures Ca^{2+} fluxes into the endothelium mediated by L-type channels with a relatively high conductance of 23 pS. These hitherto undetected channels, however, only became activated by highly depolarized potentials.

Our own studies on the profound impact of the L-type channel blocker nimodipine on aging-related microvascular degeneration clearly demonstrates the obvious role of calcium and DHP-sensitive channels in the maintenance of microvascular integrity. In these studies, we showed that chronic nimodipine application potently delays the age-related deposition of perivascular collagen and basement membrane thickening in cerebral capillaries *(65–67)*. Interestingly, the process of microvascular degradation was strongly accelerated in hypertensive stroke prone rat strains *(68)*.

4. ISCHEMIC STROKE

4.1. Introduction

Although the pathogenesis and tissue damage in stroke cannot be dissociated from a multitude of variables such as type of stroke, type and location of cerebral vessel involved, degree of vascular obstruction, brain region affected, and factors like age, stress parameters, and history of arterial hypertension, we will primarily center the present survey on ischemic stroke as a result of vascular occlusion due to athero-thrombosis, which is the principal cause of stroke.

Occlusion of a cerebral artery creates a condition for the neuronal environment that may result in acute, short-term, and long-term changes in cellular physiology, eventually leading to permanent cell dysfunction or cell death. The changes in the neuronal environment triggered by the ischemic event have been associated with numerous mechanisms that each in itself or in relation to each other challenge the viability of nerve cells and other cells of the CNS. Although this overview focuses on calcium homeostasis and the effect of a specific way of limiting calcium influx, it should be realized that the self-imposed limitation of interest bears the unmistakable risk of a partial view.

On the other hand, the evidence is abundant that derangement of intracellular calcium balance is essential in putative pathways leading to cell destruction, also indicated as the "final common pathway" of toxic cell death *(69)*. This description not only points to the role of Ca^{2+} in intracellular pathways of cell damage, but also to Ca^{2+} as common denominator in cytotoxic cell processes in general *(69,70)*, which both applies to cells of different origin as well as to the source of toxic damage and the nature of the lethal process *(1,69–71)* (*see* Chapter 1).

During and after ischemic insults, the alterations in the neuronal environment are multifold, but major factors leading to neuronal damage have been associated with the formation of free oxygen radicals (*see* Chapter 7), metabolic changes induced by hypoglycemia, acidosis and general derangement of intra- and extracellular ionic dysbalance, and excessive release of excitatory amino acids, notably glutamate and subsequent derangement of intracellular calcium regulation *(72–75)*.

4.2. Ischemic Stroke, Glutamate Release, and Intracellular Calcium

The sequence of events in the pathophysiology of ischemic stroke including release of extracellular glutamate, sustained induction of Ca^{2+} currents, and elevation of intracellular Ca^{2+}, and Ca^{2+}-induced derangement of cell maintenance culminating in cell damage has become well established *(73,75,76)*. Ischemic insults were demonstrated by way of measurements with the technique of microdialysis *(77,78)* to evoke a sharp 10-fold increase in the concentrations of extracellular glutamate and to a somewhat lesser degree of aspartate and other neurotransmitters *(77,79)* (*see* Chapters 5 and 6). The observed increase of glutamate appears to be the outcome of a number of regulating parameters.

Hypoglycemia or anoxia alone may not necessarily be responsible for increased release of glutamate from glutamatergic terminals. However, ischemia or combined anoxia and hypoglycemia strongly potentiates potassium-induced glutamate from terminals, whereas at the same time ischemia creates conditions that block the uptake of released glutamate by neurons and astroglia *(78)*. The ischemia-evoked release of glutamate requires activation of presynaptic N-type calcium channels in a calcium-dependent fashion, a mechanism that may explain the reduction of glutamate in cerebral ischemia by N-channel antagonists *(80)*. Berg-Johnson et al. *(81)* showed in brain slice exposed to ischemic conditions that the increase of extracellular glutamate was directly correlated with anoxic depolarizations with a concomitant ATP depletion, which only occurred when hypoxia was combined with glucose deprivation. This view is also supported by studies on the effect of hyperglycemia on tissue resistance against ischemia that point to a higher capacity of ATP production as a result of preischemic high glucose levels *(82)*. Besides, since glucose accessibility to the brain in general and to neurons in particular is highly influenced by adrenal hormones as glucocorticoids, it is obvious that the production of glutamate and its impact on neuronal viability will also be dependent on the current corticosteroid levels *(83)*. One of the mechanisms involved in this respect is that glucocorticoids enhance release of excitatory amino acids during ischemia *(84)* (*see* Chapter 9).

4.2.1. Glutamate and Calcium-Dependent Cell Damage

The cytotoxic and, in particular, neurotoxic action of sustained or high concentrations of glutamate and other excitotoxic amino acids dates back from the 1950s and

1960s *(85)*. Early studies with high doses of glutamate injected in the general circulation showed to result in acute neuronal death in the hypothalamus of mice and monkeys *(86,87)*, which proved to be a calcium-dependent cytotoxic process *(88,89)*. The observed excessive increase of extracellular glutamate during ischemia and the vulnerability of nerve cells to high glutamate concentrations then led to the hypothesis of glutamate-induced, calcium-dependent brain damage to explain neuro-degenerative mechanisms.

It has been this hypothesis based on the presumed sequence of biochemical events that has triggered an overwhelming research effort toward the cytotoxic mechanism of ischemic tissue damage and strategies to limit or prevent brain injury after occurrence of stroke. Ever since, the importance of the deranged calcium regulation in the molecular intracellular pathways leading to cell death has gained increasing momentum and is now thought to play a key function in many neurodegenerative syndromes and aging-related neuronal decline *(90)*. For further detail, *see* Chapters 3 and 5.

4.2.2. Calcium and Cell Death

However, evidence has accumulated in recent years that the neurodegenerative events initiated by the ischemia-induced rise of intracellular calcium are more complex than previously anticipated and subject to a number of interacting processes and variables *(91–93)*. The sustained elevation to pathologic levels of the intracellular calcium concentration is well known to give rise to a cascade of molecular and genetic changes leading to cell and membrane injury and eventually cell death. These biochemical pathways underlying the process of short-term and long-term or delayed cell death are subject to continuing developments. Interestingly, recent progress is now made in unraveling direct calcium-induced cell disintegration and necrotic cell death vs delayed cell death. Activation of distinct calcium-triggered pathways appears to be based on the way that calcium enters the cell *(49,50)*. In this respect, Ca^{2+} fluxes associated with early neurotoxic damage imply the involvement of the NMDA receptor channel *(94)*. This conclusion is consistent with the observed cell death after brief periods of exposure to high glutamate concentrations, that were followed by a degree of Ca^{2+} accumulation being proportional to the amount of neuronal death and that could only be effectively prevented by NMDA antagonists *(95)*. Additional and convincing evidence has now been presented that delayed cell death up to several days after transient ischemia is the effect of apoptotic mechanisms *(96)*. It was previously considered that persistent elevations of $[Ca^{2+}]_i$ after exposure of neurons to cytotoxic doses of glutamate were a prerequisite to explain the subsequent neurotoxic cell death *(97)*. More recent observations, however, point to the importance of a sharp initial rise of $[Ca^{2+}]_i$ after short toxic glutamate exposure, that is followed by a period of recovery to baseline $[Ca^{2+}]_i$ levels preceding a second delayed Ca^{2+} increase and cell damage *(98,99)* (Fig. 2).

Apparently, long-term high $[Ca^{2+}]_i$ does not necessarily have to precede cell degeneration. Cell death can be triggered in very early stages of sudden Ca^{2+} overload and includes cytotoxic pathways that require the activation of the NMDA channel. Unfortunately, such early but dramatic changes in intracellular calcium in stroke, which are so decisive for the further course of the neurodegenerative process, are unlikely to offer possibilities for postischemic therapeutic interventions.

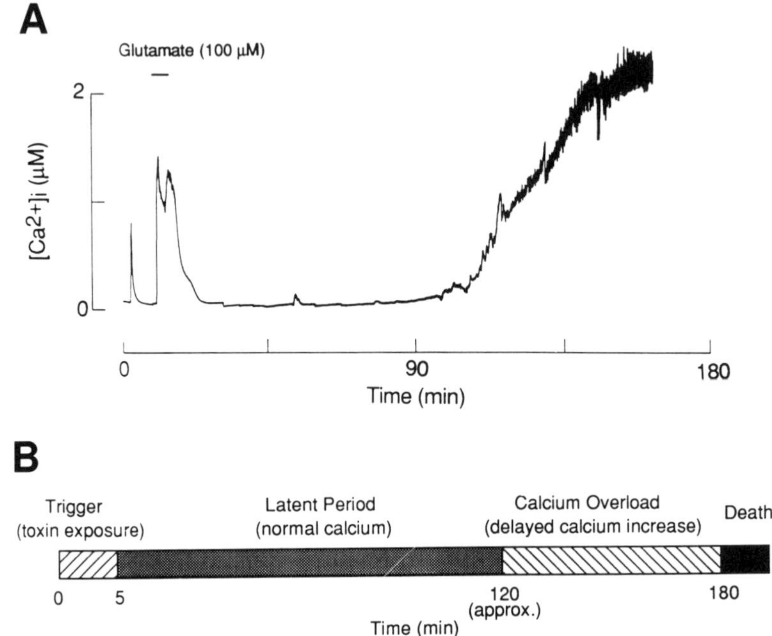

Fig. 2. Long-term effects of a short initial exposure to a toxic dose of glutamate to cultured hippocampal neurons. Monitoring of intracellular calcium concentrations **(A)** indicates a short but potent elevation of $[Ca^{2+}]_i$ as a direct response to the glutamate exposure, which is followed by a delayed secondary calcium increase that eventually leads to cell death by apoptotic mechanisms. This pattern of derangement of intracellular calcium homeostasis has been indicated by the time bar in B. (Reprinted with permission from ref. *98*.)

4.2.3. Glutamate Toxicity and Aging

It has to be considered that most observations on the behavior of $[Ca^{2+}]_i$ are performed in cultured nerve cells that lack the physiological environment of in vivo conditions. Besides, experimental studies on ischemia in cell culture are typically carried out on neonatal brain cells. In clinical practice, however, ischemic stroke generally is manifested at advanced ages, when nerve cells display considerable changes in homeostatic calcium regulation. The temporal profile of calcium transients after glutamate stimulation in young vs old neurons were elegantly demonstrated by Verkhratsky et al. *(40)*.

In short, in young neurons, Ca^{2+} levels sharply rise upon stimulation by K^+ or glutamate, and quickly return to baseline after withdrawal of the stimulus. Aged cells have a much higher baseline $[Ca^{2+}]_i$ and Ca^{2+} rises more slowly to lower peak levels, but the excess Ca^{2+} is very slowly removed from the cytoplasm (Fig. 3). There is no doubt that the age-related change of a worn calcium homeostasis will have a profound although unknown effect on the temporal calcium profile in ischemic conditions. It is anticipated, however, that during ischemia a much prolonged elevation after the initial stimulation will be present. On the other hand, a counteractive action may be expected from the fact that during aging the density of NMDA channels is decreased *(100)*,

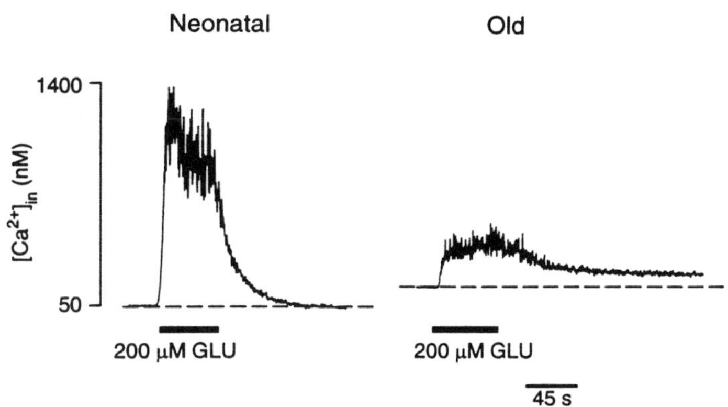

Fig. 3. The influence of the aging process is clearly indicated by the response of cultured neurons to stimulation by glutamate (and other agents). Cells from neonatal animals have a low basal level of $[Ca^{2+}]_i$ and respond to stimulation by a sharp rise of calcium elevation directly followed by reduction of the calcium concentration to the initial baseline level. In the dorsal root ganglion cells of aged animals, the basal level is much higher and the glutamate response characterized by lack of peak levels but extremely prolonged, the latter probably the result of deranged sequestering and calcium pumping mechanisms. (Reprinted with permission from ref. *40.*)

whereas the density of L-type calcium channels does not appear to be changed in senescence. This would imply a reduced vulnerability of aged nerve cells to glutamate toxicity invoked by calcium currents mediated through NMDA channels. This assumption finds support in a recent experiment in which we compared the neurotoxic effect of similar concentrations of NMDA infused in the magnocellular basal nucleus in young vs aged rat *(101)*. When comparing the neurotoxic effects of 40 nmol injection in the nucleus basalis, the neuronal damage was reduced from 49% in young animals to 36% in the aged rats of 26 mo *(101)*. This lower rate of cell death was proportional to the decreased presence of NMDA channels in the aging brain *(100)*.

4.3. Pharmacological Approaches in Stroke

It is obvious from the foregoing that the putative sequence of events after stroke that ultimately leads to brain tissue damage, at various points offers possibilities for pharmacological intervention to counteract the presumed cytotoxic mechanisms. Major efforts both preclinically and clinically have been directed toward resolving the thrombus as primary cause of the ischemic stroke, limiting calcium influx in the ischemic region either by way of NMDA and AMPA receptor blockade, or blockade of voltage-sensitive Ca^{2+} channels, and enhancement of the activity of free radical scavengers. Other approaches include suppression of glutamate release, inhibition of Ca^{2+}-dependent protease activity *(102)*, application of adenosine analogs or adenosine transport inhibitors *(103,104)* and serotonin 1A receptor agonists *(105,106)*, and blockade of intracellular ryanodine receptors *(107)*, lowering brain temperature during ischemia *(108)*, to mention only some of the reported strategies in ischemia research. The present paper specifically deals with nimodipine as a representative of the third category of putative anti-ischemic drugs directed towards limiting Ca^{2+} influx through the L-type Ca^{2+} channel.

5. NIMODIPINE IN EXPERIMENTAL HYPERTENSION, CEREBRAL ISCHEMIA, AND STROKE

5.1. Pharmacology

Among dihydropyridines (DHP) with L-type Ca^{2+} channel antagonistic activity, nimodipine has been identified as having both cerebrovasodilatory and neuronal effects at doses that have little or no effect on peripheral circulation *(109,110)* (Fig. 4). It is mainly these characteristics associated with high passage rate over the blood-brain barrier that defines the usefulness of nimodipine in stroke. Nimodipine binds the DHP-receptor site of the L-channel in a saturable, reversible, and stereospecific fashion *(111)*. The DHP-affinity is strongly increased by depolarization, whereby a preferential interaction of nimodipine with the inactivated state exists. Therefore, it is postulated that channels that open briefly may not permit sufficient time for pharmacologic action, whereas under conditions of prolonged stimulation of L-type channels, the antagonizing effects of nimodipine will become more prominent *(24,112)*.

5.2. Pharmacodynamics

5.2.1. Cerebrovasodilatory Effects

Concerning the vascular effects, nimodipine shows marked specificity for cerebral vessels *(113)*, dilating cerebral vessels at much lower concentrations than required for reduction of systemic arterial pressure. In vitro studies with human arteries have demonstrated that nimodipine is 10-fold more potent on cerebral than on temporal arteries *(114)*. In several animal models intracarotid administration of nimodipine increases cerebral blood flow, especially when cerebrovascular resistance is increased *(115,116)*. However, the intravenous administration of nimodipine results in variable effects *(117–122)*. The observed increases in cerebral blood flow are mostly accompanied by only minor decreases in arterial blood pressure *(113)*.

5.2.2. Neuronal Effects

The distribution of binding sites for nimodipine in the rat brain varies according to the brain region involved. Binding density is highest in the olfactory bulb and hippocampus, intermediate in the caudate nucleus and cerebral cortex, and lowest in the cerebellum *(123)*. Electrophysiological studies with mammalian neurons demonstrate that nimodipine in concentrations as low as 10 nM is able to block inward Ca^{2+} currents depending on the state of depolarization *(124)*. However, only at high concentrations (0.1–10 µM) has nimodipine been reported to block the release of some neurotransmitters and hormones from neuronal tissue *(109)*. Under normal metabolic circumstances, nimodipine at concentrations up to 1 µM, has also no effect on oxygen consumption by isolated brain slices *(109)*. In vivo studies on healthy animals demonstrate that nimodipine, although increasing blood flow, has no effect on cerebral metabolism, cortical ATP levels, or phosphocreatine levels *(109)*. In healthy volunteers, EEG studies show normalization of vigilance by administration of nimodipine in case of both high and low initial vigilance levels. Other investigations have shown higher scores in "alertness" and "memory function" after nimodipine treatment *(110)*. With regard to cognitive function, Disterhoft et al. *(90)* described the striking influence of nimodipine in potently reducing the age-related enhancement of after-hyperpolarization of pyrami-

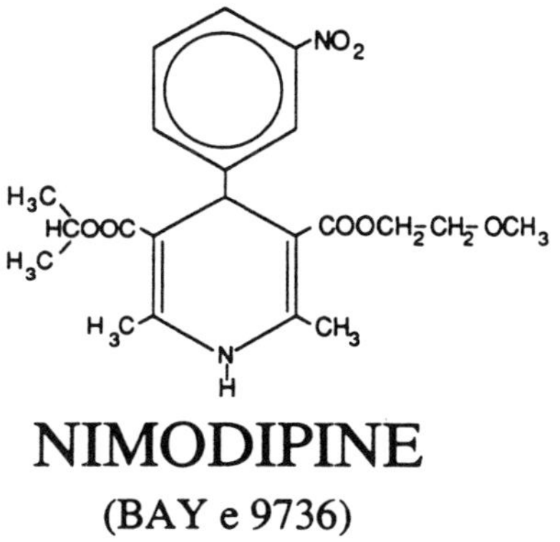

$$NIMODIPINE$$

NIMODIPINE

(BAY e 9736)

Fig. 4. Dihydropyridine structure of the L-type calcium antagonist nimodipine.

dal cells in the hippocampus. It is likely this effect that underlies improved cognitive performance in aging after application of compounds like nimodipine.

5.3. Pharmacokinetic Properties of Nimodipine

5.3.1. Absorption and Distribution

When nimodipine is administered to healthy humans by continuous iv infusion, steady-state plasma nimodipine concentrations are reached after 12–18 h *(113)*. After oral administration of nimodipine (tablets or capsules) in humans, maximum plasma levels are reached after 30–60 min, whereby a linearity exists between dose and plasma concentration *(125)*. Absorption of nimodipine from the gastrointestinal tract following oral administration is nearly complete, but bioavailability is only approx 10%, because of the first-passage hepatic metabolism *(113)*. Animal studies have shown that nimodipine is quickly distributed throughout all tissues and organs after iv or oral administration, and that approx 98% of unchanged nimodipine is bound to plasma proteins *(113,126)*. After administration, nimodipine is also found in the cerebrospinal fluid (CSF) although concentrations here are much lower than the mean plasma concentrations. This can be explained by the fact that in plasma only the 2% unbound fraction of nimodipine can reach equilibrium with the brain *(110,125)*.

5.3.2. Metabolism and Elimination

The major step in the metabolic pathway of nimodipine is hydrogenation of the dihydropyridine nucleus to form a pyridine analog. Next, hepatic demethylation occurs either before or after formation of the pyridine analog. Subsequent metabolism involves ester hydrolysis and hydroxylation of a methyl group *(113)*. The main metabolites of nimodipine have little or no pharmacological activity *(127)*. Following a single oral dose in healthy humans, 50% of the administered dose is excreted in the urine within 4 d, almost exclusively in the form of metabolites. An additional 32% is excreted in feces, likely due to biliary excretion *(128)*. In rats, after intravenous (as well as after oral administration of nimodipine), 20% of the dose is excreted in the urine and 80% in feces *(129)*. Plasma elimination half-life values in patients and healthy volunteers ranges from 0.9–1.5 h after iv administration, and from 1.7–5.6 h after oral administration. In patients with liver dysfunction, clearance of nimodipine is considerably impaired *(110)*.

5.4. Nimodipine in Experimental Stroke Models

As described in previous paragraphs, Ca^{2+} entry into brain cells and vascular smooth muscle plays an important role in the pathophysiology of brain ischemia. Nimodipine has therefore been widely tested for its potential in therapeutic intervention in preclinical trials using animal models of cerebral ischemia. We have to realize that no stroke model will ever be able to exactly reproduce the circumstances of clinical stroke in humans. However, in the last decade, some models have been developed that closely approximate certain aspects of the clinical situation. Because of their close resemblance to the cerebrovascular anatomy and physiology in higher species, and their low cost, rodents (notably rats and gerbils) are often used for these studies. Cerebral ischemia models can be divided into models of transient global ischemia, resulting in selective neuronal injury within vulnerable brain regions, and focal ischemia, typically giving rise to localized brain infarction *(130)*.

5.4.1. Nimodipine in Global Cerebral Ischemia Models

The first observation that nimodipine can prevent postischemic impairment of cerebral blood flow raised the interest to test nimodipine in global ischemia models. In the rat two-vessel occlusion model, whereby reversible bilateral common carotid artery occlusion (CCAO) is combined with systemic hypotension, resulting in marked reduction of forebrain blood flow and concomitant forebrain ischemia, nimodipine has proven to be an effective drug. Treatment before *and* after the start of ischemia resulted in significant neuroprotection in the CA1 area of the hippocampus *(131–133)*. In contrast to what was expected, the protective effects of pretreatment with nimodipine was not mediated by an increased postischemic cerebral blood flow (CBF) *(131,133)*. A nimodipine infusion for 24 h, starting 20 min after the ischemic period, resulted in a 50% reduction of mortality and revealed neuroprotection especially in the hippocampus, whereas no protection was seen in the striatum and the cortex *(132)*.

Nimodipine has also been tested in a rat model in which reversible bilateral CCAO is combined with bilateral vertebral artery occlusion. An advantage of this model is the ability to achieve a high grade forebrain ischemia in awake, freely moving rats. A nimodipine infusion started directly after an ischemic period prevented calcium-calmodulin binding, but did not result in significant histological protection *(134)*. Cerebroprotective effects of nimodipine have also been evaluated in gerbils subjected to transient bilateral CCAO. The absence of posterior communicating arteries in these animals is responsible for severe bilateral forebrain ischemia following bilateral CCAO. Transient occlusion lasting 5 min or less results in delayed cell death of the hippocampal CA1 pyramidal neurons *(130)*. Nimodipine applied 1 h before an ischemic period of 5 min was not able to avoid increased calcium levels in the CA1 region of the hippocampus, and had no significant effect on neuronal survival in that area *(135)*.

Delayed neuronal death is preceded by prolonged inhibition of protein synthesis. Nimodipine administered shortly after an ischemic period did not influence postischemic recovery of protein synthesis *(136)*. Another study with the same ischemia model, in which a therapeutic time window for nimodipine was investigated, showed also that treatment shortly before and after ischemia had the least protective effects, whereas late post-ischemia treatment (up to 24 h after ischemia) strongly prevented the delayed neuronal death of the sector CA1 pyramidal cells *(137)*.

In the late postischemic stages, no microcirculatory abnormalities are observed, which indicates that the mechanism of protection of nimodipine in this model is possibly the result of direct neuronal effects instead of cerebrovascular effects. In a study using a rabbit model of transient global cerebral ischemia, performed by intrathoracic artery occlusion combined with hypotension, the effects of local nimodipine application to the hippocampus was compared with the effects of systemic administration of nimodipine. Both treatments started before ischemia and continued after ischemia, did not affect extracellular concentrations of amino acids, but enhanced recovery and normalization of electroencephalographic activity and protected hippocampal neurons from early morphologic changes. Only nimodipine given systemically was able to attenuate the ischemia-induced drop of extracellular Ca^{2+} and completely prevented postischemic leakage of the blood–brain barrier *(138)*. From these results, it was concluded that the vasotropic action of nimodipine played a major role in the mechanism of neuronal protection. In a model of cardiac arrest of cats, no effect upon neuronal survival was found by postischemic nimodipine administration *(139)*.

5.4.2. Nimodipine in Focal Cerebral Ischemia Models

Nimodipine was extensively tested in rats with focal cerebral ischemia caused by occlusion of the middle cerebral artery (MCA). This model has proven to be a very useful approach to mimic the ischemic hemispherical infarction in humans and was therefore very suitable to test the therapeutic efficacy of drugs (such as Ca^{2+} antagonists) prior to human clinical trials. One of these models is the Tamura model *(140)*, which consists of reversible or irreversible proximal MCA occlusion, giving rise to infarction of both cortex and caudoputamen. Application of the Tamura model, however, requires craniotomy. A more recent method without the need for craniotomy is the intraluminal filament occlusion of the MCA *(141)*. In this model, a filament is introduced into the internal carotid artery by which the origin of the MCA can be reversibly occluded.

Nimodipine is not consistently effective in rats with MCA occlusion. Apart from the experimental conditions, the time of drug administration, size and severity of the infarct, dose of the drug, and duration of treatment influence the outcome. When nimodipine infusion was started before permanent MCA occlusion (according to the Tamura method) and continued until sacrifice, cerebral blood flow was increased in the parietal cortex compared to controls *(142)*. Also the volume of the ischemic damage was less in the cortex. However, the extent of the damage in the central core was not decreased by the drug administration. In the same study, it was found that nimodipine did not affect CBF in the caudate nucleus, which was associated with the lack of effect on ischemic injury in this region.

Pretreatment with nimodipine prior to MCA occlusion has been shown to prevent early loss of hippocampal CA1 parvalbumin immunorcactivity, indicating an early role of L-type channels in neuronal alterations *(143)*. Nimodipine given 30 min after permanent MCA occlusion showed also marked improvement on parameters of infarct size, and biochemical markers of infarction *(144)*. Curiously in another study in which nimodipine was also applied 30 min after MCA occlusion, it was reported that the drug treatment failed to augment CBF *(145)*. Investigators that started treating hypertensive rats with nimodipine before irreversible MCA occlusion, found that nimodipine improved circulation through ischemic tissue *(146)*. Also edema volume and cortical infarct volume was decreased by nimodipine. Based on further calculations, it was concluded that in the hypertensive rat, the improvement of CBF could not be the only factor responsible for the focal ischemic injury reduction. Also the modulating effect on calcium influx of neurons is most likely another protective role of nimodipine during permanent MCA occlusions.

Radiolabeled ligand studies with ^{3}H-nimodipine in rats with MCA occlusion show increased binding of nimodipine in severely ischemic regions and in later stages (after 4 h) also in regions with penumbral blood flow levels *(147)*. This can account for the results from an experiment in which the effects of nimodipine were recorded at different intervals after occlusion. After only 6 h, nimodipine treated rats started to show a reduced infarct compared to vehicle-treated rats *(148)*. These results provide further evidence that nimodipine prevents the progression of tissue necrosis in the penumbral zone of rats with permanent MCA occlusion, whereas the core region adjacent to the site of occlusion cannot be protected by the Ca^{2+} antagonist.

In a model of transient MCA occlusion by an intraluminal filament, nimodipine application directly after the ischemic period reduced abnormal Ca^{2+} accumulation in the MCA supply domain and anterior cerebral cortex, together with reduced neuronal damage in the MCA cortex, whereas neuronal injury in the striatum was unaltered *(149)*. Drug treatment during reversible MCA occlusion in cats and continued during reperfusion resulted in greater recovery of EEG *(150)*. During reperfusion, dramatic hyperperfusion was also seen compared to vehicle-treated animals that displayed normal perfusion levels. Next to these results, much lower increases in Ca^{2+} levels were found during ischemia and reperfusion, and less focal cortical damage was observed. In baboons, nimodipine infusion started before a period of 6 h MCA occlusion and continued for 4 d revealed a 100% survival, whereas in the vehicle-treated group, mortality was 33% *(151)*. There was also less clinical evidence for infarction 7 d after the ischemic period. Infarction areas observed after 14 d reperfusion tended to be smaller, although this was not significant. Nimodipine has also been tested in a rat model of photochemically induced focal cerebral thrombosis. Following an intravenous injection of a photosensitive dye, a light of a special wavelength is used that penetrates the skull, and induces platelet aggregation and thrombosis in the vessels directly under the dura mater. In this way, focal infarcts can be placed in any desired cortical location *(152)*. Nimodipine injection in this model 5 min after irradiation did not reduce the cortical infarct area examined 24 h after thrombosis.

5.4.3. Nimodipine in Stroke-Prone Spontaneously Hypertensive Rat Strains

A close reproduction of clinical stroke in humans associated with hypertension is a stroke in the stroke-prone strain of spontaneously hypertensive rats (SHR-SP). Both hemorrhages and infarcts occur spontaneously in >80% of the male animals *(153,154)*. Major sites with a high risk for stroke in these rats include the anteromedial and occipital cortex and basal ganglia *(130)*. A difficulty of studying the effects of neuroprotective agents in this model is the large variability of timing, location, and pathologic features

Table 1
Effects of Nimodipine Treatment of Stroke-Prone Spontaneously Hypertensive (SP) Rats Compared to Age-Matched Controls of the Wistar-Kyoto Strain (WK)

Experimental group	Age, wk	N	Body wt, g	Brain wt, g	Rel. brain wt (per 100 g body wt)	Number of rats with strokes
SP-ONSET	46	5	313 ±15.4	2.0 ± 0.20	0.68 ± 0.11	2
SP-PLAC	56	7	281 ± 9.3	2.5 ± 0.11	0.91 ± 0.06	7
SP-NIMO	56	7	355 ± 7.1	1.8 ± 0.02	0.51 ± 0.01	2
WK-ONSET	46	5	412 ± 3.6	2.1 ± 0.01	0.50 ± 0.002	0
WK-PLAC	56	8	412 ±10.2	2.0 ± 0.02	0.47 ± 0.01	0
WK-NIMO	56	8	405 ± 3.8	2.0 ± 0.03	0.49 ± 0.01	0

Of each rat strain, three groups were compared, the values of body and brain weight, ratio of brain to body weight, and animals with stroke were measured at onset of the experiment at the age of 46 wk and after a 10-wk treatment period (from the age of 46–56 wk). The major outcome was that nimodipine prevented loss of body wt and the occurrence of stroke in the SP animals. Nimodipine treatment characteristically prevented the high increase of brain weight in the SP rats, which resulted from the development of brain edema concomitant with stroke *(154)*.

of the strokes. On the other hand, this model is very suitable for testing prophylactic effects of calcium entry blockers on the occurrence of strokes and concomitant brain edema *(154)*. In aging SHR-SP rats treated chronically with nimodipine (applied in food 1000 ppm) for 10 wk, all animals survived the experimental period, while 2 out of 7 rats developed minor strokes, without any significant reduction of systolic blood pressure (Table 1).

In contrast, all placebo-treated animals developed severe strokes with a mortality of 50% *(154)*. Selective GABA-ergic and parvalbumin-positive neuronal death that occurred in placebo-treated SHR-SP rats was prevented by nimodipine treatment. Prophylactic effects of nimodipine were also seen in other studies *(155,156)*. Nimodipine postponed the time point when stroke started to develop and reduced mortality. Even in low doses (2 mg/kg/d), nimodipine delayed the onset of stroke *(155)*. During the occurrence of an acute stroke in these strain of rats, nimodipine improved the cerebral microcirculatory derangement *(157)*. Also, after onset of stroke, nimodipine treatment has shown beneficial effects in SHR-SP. Administration of the Ca^{2+} antagonist after symptoms of stroke, increased the survival time and attenuated the increase of the Ca^{2+} content of the brain *(158)*. Besides, fewer hypertensive cerebral lesions were reported and in some areas of tissue rarefaction an increase in number of functional capillaries was observed. Investigators, using the active avoidance task as a memory test, found ameliorating, probably secondary effects of nimodipine on the impaired learning ability in poststroke SHR-SP *(159)*.

In summary, in models of transient global ischemia, the efficacy of nimodipine depends on the type of animal model and the mode of treatment. Protection is especially prominent in the penumbral zone of the hippocampus, whereas there is generally no protection found in the striatum and cortex. In stroke models that evoke delayed neuronal death, nimodipine is most effective in the late postischemic stages when CBF has reached normal levels, indicating a direct cytoprotective effect. In focal cerebral ischemia caused by permanent MCA occlusion, both nimodipine pretreatment and post-treatment prevents progression of tissue necrosis in the penumbral zones of the cortex. Also, in models of transient MCA occlusion, nimodipine showed neuroprotection in cortical brain areas. In SHR-SP rats, nimodipine had both prophylactic and protective effects in case of stroke. Improvement of cerebral microcirculatory disturbance by nimodipine is responsible for these effects.

Disagreement exists about the protective mechanism of nimodipine in models of global and focal ischemia. There is evidence for protective effects through L-type channel blockade both in neuronal membranes and in the cerebrovascular domain. Most likely, both mechanisms are involved in neuroprotective effects of nimodipine in stroke. Unfortunately only few investigators have studied the temporal course of neuro-protection by nimodipine in the differential animal stroke models. Data about the effects of delayed treatment with nimodipine after onset of stroke, which resemble the clinical situation, are lacking.

6. CLINICAL RESULTS WITH NIMODIPINE IN ACUTE ISCHEMIC STROKE

6.1. Introduction

The promising results obtained in the early experimental studies on dihydropyridines in cerebrovascular and stroke model studies opened the door for application of these compounds and nimodipine in particular in clinical ischemic stroke trials. In the early period of nimodipine application, however, the L-type channel blockers were prima-rily considered as vasoactive compounds albeit with a preferential action on cerebral vasculature *(160)*. Although some reports already defined views on the relationship between neuronal calcium and cell death in stroke-affected brain regions *(161)*, the breakthrough in insight of neuronal damage mechanisms in ischemic conditions was yet to come *(75)*. A major characteristic of nimodipine was its capacity of transport over membranes and passage over the blood–brain barrier, in particular the entry of the cerebrovascular wall by passage over the endothelial lining of intraparenchymal vas-culature. This way, a compound like nimodipine was essentially capable of induc-ing-vasoactive responses in intracerebral vessels in the stroke affected brain. As has now become well recognized, a second way of action of dihydropyridines is their impact on maintenance or restoring Ca^{2+} levels in neuronal tissue. This charac-teristic feature of DHP as nimodipine is particularly important where Ca^{2+} balance is under threat as a result of ischemic events and leading to irreversible processes of neuronal injury or cell death.

6.2. Clinical Trials with Nimodipine in Ischemic Hemispheric Stroke

The first experience with nimodipine in clinical use was reported by Gelmers *(162)* in 1984. In this preliminary single-blind study in the Netherlands, 29 patients suffering from acute ischemic stroke were treated with 120 mg/d for a period of 28 d and com-pared to 31 controls. Treatment effects were followed for mortality and neurological scores, the latter assessed with an adapted Mathew Scale. The outcome of this study was promising and indicated a neurologic improvement in the 4-wk treatment period, which was significantly better for the nimodipine-treated patients compared to the placebo group. The study did not specifically mention the average time point of the start of the treatments after the onset of stroke.

Based on this preliminary study, Gelmers et al. *(163)* published a second double-blind multicenter trial of nimodipine in ischemic stroke. This trial included 93 patients in each group that were selected and treated in basically the same fashion as in the 1984 study. Treatments in this study were reported to have started within 24 h after the occurrence of stroke. The larger number of patients and improved methodology, however, permitted a more refined analysis of the outcome and identification of more parameters. In the nimodipine-treated group, the mortality rate during the 4-wk treatment and in a 6-mo follow-up period was found to be significantly lower than in controls. However, the better outcome in terms of reduced mortality was only valid for the male segment of the nimodipine-treated patients. There was also a significantly better improvement as assessed by the degree of neurologic impairment in the nimodipine group. Interestingly, the neurologic improvement by nimodipine in this study was most significant in patients defined by the Mathew score as showing a moderate-to-severe neurologic deficit at the start of the treatment program. Patients with either minor neurological damage or extremely poor neurological scores did not benefit from the nimodipine treatment.

The 1988 paper of Gelmers *(163)* was followed by a third report *(164)* in which the authors pooled the outcome of five nimodipine trials in the Netherlands, Spain, Italy, Germany, and Austria *(165,166)*. All trials were performed with basically similar protocols and methods of treatment assessment as previously indicated *(162,163)*. In this larger survey that involved 335 nimodipine-treated patients and 346 placebo-controls, again mortality and neurologic outcome at the end of treatment were recorded. The size of the trial allowed a further differentiation in treatment results and yielded the following major conclusions. Mortality during the 3- or 4-wk treatment period was 51% lower in the nimodipine group, whereas a significant increased improvement of neurologic outcome was also established for the entire nimodipine group. It was further calculated that the impact of nimodipine is only positive for patients with an initial neurologic score of moderate-to-severe. Furthermore, patients >65 yr old were more likely to benefit from the nimodipine treatment, which was a more specific outcome of the Spanish trial *(165)*. A clear and critical finding in the Italian study *(166)* was that a start of treatment within 12 h after the occurrence of stroke symptoms significantly increased the chance for a better outcome.

Not all clinical trials with nimodipine, however, came to the same moderately positive conclusions for the efficacy of treatment in stroke. Nimodipine (120 mg/d) given to a group of 24 patients (vs 28 placebo-controls) during 2 wk with Mathew scale scores of around 65 at the start of treatment did not provide any change in the normal development of neurologic score during the treatment period or the 4-mo follow-up *(167)*. It appears that the outcome of this rather limited trial confined to stroke patients with only a moderate neurologic deficit at the entry of the trial is in line with similar observations in the Gelmers reports on subgroup analysis based on severity of the neurologic dysfunction.

In the same year, an extensive British trial on 607 nimodipine and 608 placebo-control cases was published by the Trust Study Group *(168)*. Treatment in this trial was also based on 120 mg/d with nimodipine for 21 d started within 48 h of stroke onset, but the end-point of the impact of treatment was assessment of functional recovery after 6 mo by use of the Barthel index and an arbitrary Activities of Daily Living Scale. The outcome of this study was negative in the sense that no beneficial effect of nimodipine could be recorded after a follow-up of 24 wk, neither in functional recovery nor mortality. In fact, the nimodipine group tended to show a somewhat delayed recovery speed. Subgroup analysis of start of treatment within 24 h also showed no positive influence of the dihydropiridine.

A second major, carefully designed multicenter clinical trial on nimodipine in acute ischemic stroke carried out in the United States provided the following outcome on efficacy and also safety of nimodipine treatment *(169)*. The trial was based on a placebo group of 264 patients and three groups of 265, 268, and 267 stroke patients receiving 60, 120, or 240 mg/d nimodipine, respectively, during 21 d. Follow-up check was carried out after 3 and 6 mo. Neurological function was determined using the Toronto and motor strength scales. In spite of the large number of cases, there were no differences in mortality rates between the placebo and the three nimodipine groups. Also clinical outcome values for all patients that started treatments within a 48 h limit after stroke onset, did not reveal any differences either established by the Toronto and motor score, or other scores including the Mathew scores used by the Gelmers trials. Further subgroup analysis demonstrated the importance of the interval after stroke. In particular, the subgroup receiving 120 mg/d nimodipine that started within 18 h after stroke symptoms had a better neurologic outcome expressed as a significantly lower rate of neurologic worsening. The critical importance of starting point of treatment in this American study was further emphasized by the positive correlation between nimodipine treatment efficacy when started within 18 h from onset and a negative computed tomographic scan. This latter finding may prove to be of crucial value since it suggests that a positive outcome in clinical efforts to limit brain damage is reasonably only to be expected in the time period before irreversible infarction has developed.

A more recently reported clinical study *(170)* was carried out in Finland on 176 nimodipine treated (120 mg/d for 21 d) and 174 placebo-controls. A subgroup stratification was included as to address analysis parameters that proved to be of potential importance in treatment efficacy as became apparent from previous studies such as age (16–59 yr and 60–69 yr), start of treatment (within 24 h and 24–48 h), and initial severity of stroke (mild and severe). The results of this study show some contrast to the earlier findings of Gelmers et al. *(162–164)*. Mortality in the nimodipine group was higher than in the controls, but this difference was not significant. Comparison of the neurologic outcome of all subgroups mentioned with regard to age, severity of stroke, or start of treatment in this study showed no differences between nimodipine and placebo treatments.

Based on the positive neuroprotective results yielded with intravenous nimodipine infusions during stroke models in primates *(151)*, several clinical trials were undertaken with intravenous application of the drug. In a Canadian study, 84 patients entered the trial within 48 h after stroke and received 2 mg/h intravenous nimodipine for 10 d, followed by oral (180 mg/d) nimodipine for 6 mo. The drug-treated patients had a mildly better neurologic score at the end of the treatment period, which was, however, not statistically significant. A European study *(171)* comprised groups of stroke patients that either received intravenous nimodipine of 1 mg/h or 2 mg/h for 5 d after which treatment was continued with oral nimodipine application of 120 mg/d for 21 d. The trial, however, was prematurely terminated because of safety reasons, whereas the placebo groups scored better than the intravenous drug-treated groups. Analysis of diastolic blood pressure in the various groups indicated a relatively fast reduction of postischemic blood pressure that positively correlated with an unfavorable neurologic outcome.

The overall conclusions of all clinical data that became available on nimodipine application in ischemic hemispheric stroke were recently summarized by Kakarieka et al. *(173)* and Mohr et al. *(174)*. As was also indicated in the survey above, *(163)* older patient groups with moderate to severe neurologic deficit and subjected to treatment within 12 h may benefit from the calcium antagonist. In particular, the time interval between start of treatment after stroke, the initial improvement of residual blood flow, and maintenance of blood pressure and the restoration of blood flow have come into focus apart from the neuroprotection of the jeopardized neuronal tissue in the ischemic penumbra *(173)*. In this respect, it was demonstrated by a large-scale MCA occlusion study in monkeys that occlusion of more than 8 h generally resulted in extensive and probably irreversible cellular necrosis *(175)*.

In this line of evidence, an interesting and recent clinical nimodipine study was initiated by Limburg and colleagues who in 1994 started a double-blind, placebo-controlled ischemic stroke trial aimed at 1750 patients (M. Limburg, personal communication). This "very early nimodipine use in stroke" (VENUS) trial carried out in the Netherlands is characterized by the start of treatment within 6 h after onset of stroke. The protocol requires participation of general practitioners, who in the Netherlands are usually the first physicians to see a patient. Patients receive 30-mg tablets four times daily for 10 d, after which the condition of patients is assessed by the Rankin Handicap Scale to be repeated after a follow-up period of 3 mo. It may be anticipated that the outcome of this study will, at least, provide more conclusive data on the essential issue of early intervention.

6.3. Summary and Conclusions

Our knowledge on stroke, the pathophysiologic mechanisms leading to stroke-induced brain damage and possibilities for therapy have dramatically increased during the last two decades. Experimental research in in vivo animal models as well as in cell cultures have greatly added to our insight in the cellular and molecular processes involved in neuronal injury and neuronal cell death in ischemic conditions. However, the preclinical scientific approach was often directed towards the unraveling of cellular and molecular steps in the cascade of events leading to cell death, and much less towards the design of a realistic therapeutic application. It may be expected that the much better understanding of the pathophysiology of ischemic stroke will yield better and more adequate therapies or combinations of therapies. In this respect, it appears that the strong urge for stroke treatments has given way to somewhat premature clinical trials. Nonetheless, the experience obtained with clinical and preclinical applications of compounds like nimodipine and other calcium antagonists have greatly contributed to the current views on the crucial role of intracellular calcium homeostasis in vascular and neuronal tissue in health and disease. Such knowledge not only applies to conditions with acute threat to brain tissue as in stroke, but also in relation to chronic-progressive decline of neural function and aging-associated neurodegeneration as in dementia and Alzheimer's disease *(68,176,177)*.

An important more recent finding of experimental study of ischemic brain damage is the impact of the initial early phase of glutamate exposure of the postsynaptic neurons. There is now evidence that when NMDA channel-mediated calcium elevations reach toxic threshold levels, degenerative processes are triggered that probably set in motion the train of programmed cell death. The question then arises whether the severe ischemic event once started and the subsequent neurodegenerative processes can still be influenced in such a way that cell death on a large scale can be prevented. The outlook is not overtly optimistic but it has been argued that very early removal of the primary cause of atherothrombosis by thrombolytic treatment combined with secondary neuroprotective therapy may prove a fruitful strategy in stroke treatment *(178)*.

7. FUTURE OUTLOOK

From the tangibly modest clinical benefits of various stroke treatments, it may be expected that future strategies will follow different approaches of this devastating disease in the aging population, notably of developed nations. In view of the powerful effects of several anti-ischemic drugs when applied before the application of stroke or the protective effects in stroke-prone rat models, a prophylactic approach with Ca^{2+} antagonists might be anticipated. Short- or long-term prophylactic application of neuroprotective drugs to high-risk populations has been suggested as a reasonable way to go when safe compounds for lengthy administration are available *(179)*.

The question remains of what possibilities are left when the neuronal injury has become a fact, which will be the reality in most cases in which treatment cannot be started early enough for successful intervention, which is probably the case in acute insults as in stroke. A major new effort in CNS research is starting to surface since results of experimental study point to the challenging possibilities offered by the activity of growth factors. This more recently discovered class of peptides with receptors responding to them on various groups of nerve cells, are potentially promising tools for activation of remaining neuronal systems. It has long been known that both axonal and dendritic neuritic processes possess an impressive capacity of outgrowth in denervated brain regions, a process indicated as neuritic sprouting. Recent studies indicated that regenerative phenomena may considerably be promoted by growth factors *(180,181)*.

REFERENCES

1 Clapham, D. E., Calcium signaling, *Cell,* **80** (1995) 259–268.
2 Miller, R. J., Multiple calcium channels and neuronal function, *Science,* **235** (1987) 46–52.
3 Nicotera, P., Bellomo, G., and Orrenius, S., Calcium-mediated mechanisms in chemically induced cell death, *Annu. Rev. Pharmacol. Toxicol.,* **32** (1992) 449–470.
4 Bertolino, M. and Llinás, R. R., The central role of voltage–activated and receptor–operated calcium channels in neuronal cells, *Annu. Rev. Pharmacol. Toxicol.,* **32** (1992) 399–421.
5 MacDermott, A. B., Mayer, M. L., Westbrook, G. L., Smith, S. J., and Barker, J. L., NMDA-receptor activation increases cytoplasmic calcium concentration in cultured spinal cord neurones, *Nature,* **321** (1986) 519–522.
6 Nowak, L., Bregestovski, P., Ascher, P., Herbet, A., and Prochiantz, A., Magnesium gates glutamate-activated channels in mouse central neurones, *Nature,* **307** (1984) 462–465.

7 Schneggenburger, R., Tempia, F., and Konnerth, A., Glutamate- and AMPA-mediated calcium influx through glutamate receptor channels in medial septal neurons, *Neuropharmacology,* **32** (1993) 1221–1228.

8 Rogers, M. and Dani, J. A., Comparison of quantitative calcium flux through NMDA, ATP and ACh receptor channels, *Biophys. J.,* **68** (1995) 501–506.

9 Berridge, M. J., Inositol trisphosphate and calcium signalling, *Nature,* **361** (1993) 315–325.

10 Kim, U. H., Fink, Jr., D., Kim, H. S., Park, D. J., Contreras, M. L., Guroff, G., and Rhee, S. G., Nerve growth factor stimulates phosphorylation of phospholipase C-γ in PC12 cells, *J. Biol. Chem.,* **266** (1991) 1359–1362.

11 Maness, L. M., Kastin, A. J., Weber, J. T., Banks, W. A., Beckman, B. S., and Zadina, J. E., The neurotrophins and their receptors: structure, function and neuropathology, *Neurosci. Biobehav. Rev.,* **18** (1994) 143–159.

12 Renard, D. C., Bolton, M. M., Rhee, S. G., Margolis, B. L., Zilberstein, A., Schlessinger, J., and Thomas, A. P., Modified kinetics of platelet-derived growth factor-induced Ca^{2+} increases in NIH-3T3 cells overexpressing phospholipase $C\gamma_1$, *Biochem. J.,* **281** (1992) 775–784.

13 Walton, P. D., Airey, J. A., Sutko, J. L., Beck, C. F., Mignery, G. A., Südhof, T. C., Deerinck, T. J., and Ellisman, M. H., Ryanodine and inositol trisphosphate receptors coexist in avian cerebellar Purkinje neurons, *J. Cell Biol.,* **113** (1991) 1145–1157.

14 Seymour-Laurent, K. J. and Barish, M. E., Inositol 1,4,5-trisphosphate and ryanodine receptor distribution and patterns of acetylcholine- and caffeine-induced calcium release in cultured mouse hippocampal neurons, *J. Neurosci.,* **15** (1995) 2592–2608.

15 Sharp, A. H., McPherson, P. S., Dawson, T. M., Aoki, C., Campbell, K. P., and Snyder, S. H., Differential immunohistochemical localization of inositol 1,4,5-trisphosphate and ryanodine-sensitive Ca^{2+} release channels in rat brain, *J. Neurosci.,* **13** (1993) 3051–3063.

16 Nakanishi, S., Kuwajima, G., and Mikoshiba, K., Immunohistochemical localization of ryanodine receptors in mouse central nervous system, *Neurosci. Res.,* **15** (1992) 130–142.

17 Swann, K., Inositol trisphosphate and calcium signalling, *Biochem. J.,* **287** (1992) 79–84.

18 Mignery, G. A., Johnston, P. A., and Südhof, T. C., Mechanism of Ca^{2+} inhibition of InsP3 binding to the cerebellar InsP3 receptor, *J. Biol. Chem.,* **267** (1993) 7450–7455.

19 Ehrlich, B. E., Functional properties of intracellular calcium-release channels, *Curr. Opin. Neurobiol.,* **5** (1995) 304–309.

20 Tsien, R. W., Lipscombe, D., Madison, D. V., Bley, K. R., and Fox, A. P., Multiple types of neuronal calcium channels and their selective modulation, *TINS,* **11** (1988) 431–438.

21 Nowycki, M. C., Fox, A. P., and Tsien, R. W., Three types of neuronal calcium channels with different calcium agonist sensitivity, *Nature,* **316** (1985) 440–443.

22 Llinás, R., Sugimori, M., Hillman, D. E., and Cherksey, B., Distribution and functional significance of the P-type, voltage-dependent Ca^{2+} channels in the mammalian central nervous system, *TINS,* **15** (1992) 351–355.

23 McCleskey, E. W., Calcium channels: cellular roles and molecular mechanisms, *Curr. Opin. Neurobiol.,* **4** (1994) 304–312.

24 Triggle, D. J., Molecular pharmacology of voltage-gated calcium channels, *Ann. NY Acad. Sci.,* **747** (1994) 267–281.

25 Snutch, T. P. and Reiner, P. B., Ca^{2+} channels: diversity of form and function, *Curr. Opin. Neurobiol.,* **2** (1992) 247–253.

26 Catterall, W. A., Structure and function of voltage-sensitive ion channels, *Science,* **242** (1988) 50–61.

27 Sather, W. A., Tanabe, T., Zhang, J. F., and Tsien, R. W., Biophysical and pharmacological characterization of a class A calcium channel, *Ann. NY Acad. Sci.,* **747** (1994) 294–301.

28 Yang, J., Ellinor, P. T., Sather, W. A., Zhang, J. F., and Tsien, R. W., Molecular determinants of Ca^{2+} selectivity and ion permeation in L-type Ca^{2+} channels, *Nature,* **366** (1993) 158–161.

29 Dunlap, K., Luebke, J. I., and Turner, T. J., Exocytotic Ca^{2+} channels in mammalian central neurons, *TINS,* **18** (1995) 89–98.

30 Takahashi, T. and Momiyama, A., Different types of calcium channels mediate central synaptic transmission, *Nature,* **366** (1993) 156–158.

31 Clos, M. V., Garcia Sanz, A., Sabria, J., Pastor, C., and Badia, A., Differential contribution of L- and N-type calcium channels on rat hippocampal acetylcholine release, *Neurosci. Lett.,* **182** (1994) 125–128.

32 Nishizuka, Y., Protein kinase C and lipid signaling for sustained cellular responses, *FASEB J.,* **9** (1995) 484–496.

33 Kretsinger, R. H., Structure and evolution of calcium-modulated proteins, *CRC Crit. Rev. Biochem.,* **8** (1989) 119–174.

34 Andressen, C., Blümcke, I., and Celio, C. R., Calcium-binding proteins: selective markers for nerve cells, *Cell Tissue Res.,* **271** (1993) 181–208.

35 Baimbridge, K. G., Celio, M. R., and Rogers, J. H., Calcium-binding proteins in the nervous system, *TINS,* **15** (1992) 303–308.

36 Chard, P. S., Bleakman, D., Christakos, S., Fullmer, C. S., and Miller, R. J., Calcium buffering properties of calbindin D28k and parvalbumin in rat sensory neurones, *J. Physiol. Lond.,* **472** (1993) 341–357.

37 Celio, M. R., Calbindin D-28k and parvalbumin in the rat nervous system, *Neuroscience,* **35** (1990) 375–475.

38 Mattson, M. P., Protective roles for calcium binding proteins, *Neurosci. Facts,* **2** (1991) 2,3.

39 Goodman, J. H., Wasterlain, C. G., Massaweh, W. F., Dean, E., Sollas, A. L., and Sloviter, R. S., Calbindin-D28k immunoreactivity and selective vulnerability to ischemia in the dentate gyrus of the developing rat, *Brain Res.,* **26** (1993) 309–314.

40 Verkhrastsky, A., Shmigol, A., Kirischuk, S., Pronchuk, N., and Kostyuk, P., Age-dependent changes in calcium currents and calcium homeostasis in mammalian neurons, *Ann. NY Acad. Sci.,* **747** (1994) 365–381.

41 Khachaturian, Z. S., Calcium hypothesis of Alzheimer's disease and brain aging, *Ann. NY Acad. Sci.,* **747** (1994) 1–11.

42 Luiten, P. G. M., Buwalda, B., Traber, J., and Nyakas, C., Induction of enhanced postnatal expression of immunoreactive calbindin-D28k in rat forebrain by the calcium antagonist nimodipine, *Dev. Brain Res.,* **79** (1994) 10–18.

43 Greenberg, M. E., Ziff, E. B., and Greene, L. A., Stimulation of neuronal acetylcholine receptors induces rapid gene transcription, *Science,* **234** (1986) 80–83.

44 Murphy, T. H., Worley, P. F., and Baraban, J. A., L-type voltage channels mediate synaptic activation of immediate early genes, *Neuron,* **7** (1991) 625–635.

45 Cole, A. J., Saffen, D. W., Baraban, J. M., and Worley, P. F., Rapid increase and an immediate early gene messenger RNA in hippocampal neurons by synaptic NMDA receptor activation, *Nature,* **340** (1989) 474–476.

46 Augustine, G. J. and Neher, E., Neuronal Ca^{2+} signalling takes the local route, *Curr. Opin. Neurobiol.,* **2** (1992) 302–307.

47 Sheng, M., Thompson, M. A., and Greenberg, M. E., CREB: a Ca^{2+} regulated transcription factor phosphorylated by calmodulin-dependent kinases, *Science,* **252** (1991) 1427–1430.

48 Gallin, W. J. and Greenberg, M. E., Calcium regulation of gene expression in neurons: the mode of entry matters, *Curr. Opin. Neurobiol.,* **5** (1995) 367–374.

49 Ghosh, A. and Greenberg, M. E., Calcium signaling in neurons: molecular mechanisms and cellular consequences, *Science,* **268** (1995) 239–247.

50 Bading, H., Ginty, D. D., and Greenberg, M. E., Regulation of gene expression in hippocampal neurons by distinct calcium signaling pathways, *Science,* **260** (1993) 181–186.

51 Schreiber, S. S. and Baudry, M., Selective neuronal vulnerability in the hippocampus—a role for gene expression? *TINS,* **18** (1995) 446–451.

52 Nicotera, P., Zhivotovsky, B., and Orrenius, S., Nuclear calcium transport and the role of calcium in apoptosis, *Cell Calcium,* **16** (1994) 279–288.

53 Blaustein, M. P., Calcium transport and buffering in neurons, *TINS,* **11** (1988) 438–443.

54 Carafoli, E., Intracellular Ca^{2+} homeostasis, *Annu. Rev. Biochem.,* **56** (1987) 395–433.

55 Nachshen, D. A., The early time course of potassium-stimulated calcium uptake in presynaptic nerve terminals isolated from rat brain, *J. Physiol. Lond.,* **361** (1985) 251–268.

56 Kiedrowski, L., Brooker, G., Costa, E., and Wroblewski, J. T., Glutamate impairs neuronal calcium extrusion while reducing sodium gradient, *Neuron,* **12** (1994) 295–300.

57 Pozzan, T.,Rizzuto, R., Volpe, P., and Meldolesi, J., Molecular and cellular physiology of intracellular calcium stores, *Physiol. Rev.,* **74** (1994) 595–636.

58 Rasgado-Flores, H. and Blaustein, M. P., Na/Ca exchange in barnacle muscle cells has a stoichiometry of 3 Na^+/1 Ca^{2+}, *Am. J. Physiol.,* **252** (1987) C499–C504.

59 Furchgott, R. F. and Zawadski, J. V., The obligatory role of endothelial cells in the relaxation of arterial smooth muscle by acetylcholine, *Nature,* **288** (1989) 373–376.

60 Vanhoutte, P. M., Vascular endothelium and Ca^{2+} antagonists, *J. Cardiovasc. Pharmacol.,* **12** (1988) S21–S28.

61 Wang, X., Lau, F., Li, L., Yoshikawa, A., and van Breemen, C., Acetylcholine-sensitive intracellular Ca^{2+} store in fresh endothelial cells and evidence for ryanodine receptors, *Circ. Res.,* **77** (1995) 37–42.

62 Ziegelstein, R. C., Spurgeon, H. A., Pili, R., Passaniti, A., Cheng, L., Corda, S., Lakatta, E. G., and Capogrossi, M. C., A functional ryanodine-sensitive Ca^{2+} store is present in vascular endothelial cells, *Circ. Res.,* **74** (1994) 151–156.

63 Van Overloop, B., Stoclet, J. C., and Gairard, A., Age, endothelium and aortic reactivity to calcium in Wistar rat, *Life Sci.,* **53** (1993) 821–831.

64 Bossu, J. L., Elhamdani, A., and Feltz, A., Voltage-dependent calcium entry in confluent capillary endothelial cells, *FEBS Lett.,* **299** (1992) 239–242.

65 De Jong, G. I., de Weerd, H., Schuurman, T., Traber, J., and Luiten, P. G. M., Microvascular changes in aged rat forebrain. Effects of chronic nimodipine treatment, *Neurobiol. Aging,* **11** (1990) 381–389.

66 De Jong, G. I., Horvath, E., and Luiten, P. G. M., Effects of early onset of nimodipine treatment on microvascular integrity in the aging rat brain, *Stroke,* **21** (1990) IV113–116.

67 Luiten, P. G. M., de Jong, G. I., Mulder, A. B., Horvath, E., Schuurman, T., and Traber, J., Ultrastructural changes in microvascular morphology in the senescent rat brain: effects of long-term treatment with the calcium antagonist nimodipine. In J. Traber and W. H. Gispen (eds.), *Nimodipine and Central Nervous System Function: New Vistas*, Schattauer Verlag, Stuttgart, 1989, pp. 239–256.

68 Luiten, P. G. M., de Jong, G. I., and Schuurman, T., Cerebrovascular, neuronal and behavioral effects of long-term Ca^{2+} channel blockade in aging normotensive and hypertensive rat strains, *Ann. NY Acad. Sci.,* **747** (1994) 431–451.

69 Schanne, F. A. X., Kane, A. B., Young, E. E., and Farber, J. L., Calcium dependence of toxic cell death: a final common pathway, *Science,* **206** (1979) 700–702.

70 Farber, J. L., The role of calcium in lethal injury, *Chem. Res. Toxicol.,* **3** (1990) 503–508.

71 Orrenius, S., McCabe, M. J., Jr., and Nicotera, P., Ca^{2+}-dependent mechanisms of cytotoxicity and programmed cell death, *Toxicol. Lett.,* **64/65** (1992) 357–364.

72 Siesjo, B. K. and Bengtsson, F., Calcium fluxes, calcium antagonists, and calcium-related pathology in brain ischemia, hypoglycemia, and spreading depression: a unifying hypothesis, *J. Cereb. Blood Flow Metab.,* **9** (1989) 127–140.

73 Pulsinelli, W., Pathophysiology of acute ischaemic stroke, *Lancet,* **339** (1992) 533–536.

74 Swann, J. D., Smith, M. W., Phelps, P. C., Maki, A., Berezesky, I. K., and Trump, B. F. Oxidative injury induces influx-dependent changes in intracellular calcium homeostasis, *Toxicol. Pathol.,* **19** (1991) 128–137.

75 Siesjo, B. K., Pathophysiology and treatment of focal cerebral ischemia, *J. Neurosurg.*, **77** (1992) 337–354.

76 Choi, D. W., Glutamate neurotoxicity and diseases of the nervous system, *Neuron,* **1** (1988) 623–634.

77 Benveniste, H., Drejer, J., Schousboe, A., and Diemer, N. H., Elevation of the extracellular concentrations of glutamate and aspartate in rat hippocampus during transient cerebral ischemia monitored by intracerebral microdialysis, *J. Neurochem.*, **43** (1984) 1369–1374.

78 Drejer, J., Benveniste, H., Diemer, N. H., and Schousboe, A., Cellular origin of ischemia-induced glutamate release from brain tissue in vivo and in vitro, *J. Neurochem.*, **45** (1985) 145–151.

79 Obrenovitch, T. P. and Richards, D. A., Extracellular neurotransmitter changes in cerebral ischemia, *Cerebrovasc. Brain Metab. Rev.*, **7** (1995) 1–54.

80 Takizawa, S., Matsushima, K., Fujita, H., Nanri, K., Ogawa, S., and Shinohara, Y., A selective N-type calcium channel antagonist reduces extracellular glutamate release and infarct volume in focal cerebral ischemia, *J. Cereb. Blood Flow Metab.*, **15** (1995) 611–618.

81 Berg-Johnson, J., Grondahl, T. O., Langmoen, I. A., Haugstad, T. S. and Hegstad, E., Changes in amino acid release and membrane potential during cerebral hypoxia and glucose deprivation, *Neurol. Res.*, **17** (1995) 201–208.

82 Ekholm, A., Katsura, K., and Siesjo, B. K., Coupling of energy failure and dissipative K+ flux during ischemia: role of preischemic plasma glucose concentration, *J. Cereb. Blood Flow Metab.*, **13** (1993) 193–200.

83 Sapolsky, R. M., Potential behavioral modification of glucocorticoid damage to the hippocampus, *Behav. Brain Res.*, **30** (1993) 175–182.

84 Chou, Y. C., Lin, W. J., and Sapolsky, R. M., Glucocorticoids increase extracellular [3H]D-aspartate overflow in hippocampal cultures during cyanide-induced ischemia, *Brain Res.*, **654** (1994) 8–14.

85 Westbrook, G. L., Glutamate receptors and excitotoxicity. In S. G. Waxman (ed.) *Molecular and Cellular Approaches to the Treatment of Neurological Disease,* Raven Press, New York, 1993, pp. 35–50.

86 Olney, J. W., Brain lesions, obesity, and other disturbances in mice treated with monosodium glutamate, *Science,* **164** (1969) 719–721.

87 Olney, J. W. and Sharpe, L. G., Brain lesions in an infant Rhesus monkey treated with monosodium glutamate, *Science,* **166** (1969) 386–388.

88 Garthwaite, G. and Garthwaite, J., Neurotoxicity of excitatory amino acid receptor agonists in rat cerebellar slices: dependence on calcium concentration, *Neurosci. Lett.*, **66** (1986) 193–198.

89 Choi, D. W., Glutamate neurotoxicity in cortical cell culture is calcium dependent, *Neurosci. Lett.*, **58** (1985) 293–297.

90 Disterhoft, J. F., Gispen, W. H., Traber, J., and Khachaturian, Z. S. (eds.) Calcium Hypothesis of Aging and Dementia, *Ann. NY Acad. Sci.*, *747* (1994) 382–406.

91 Choi, D. W., Calcium: still center-stage in hypoxic-ischemic neuronal death, *TINS,* **18** (1995) 58–60.

92 Choi, D. W., Cerebral hypoxia: some new approaches and unanswered questions, *J. Neurosci.*, **10** (1990) 2493–2501.

93 Mattson, M. P., Murrain, M., Guthrie, P. B., and Kater, S. B., Fibroblast growth factor and glutamate: opposing roles in the generation of hippocampal neuroarchitecture, *J. Neurosci.*, **9** (1989) 3728–3740.

94 Tymianski, M. T., Charlton, M. P., Carlen, P. L., and Tator, C. H., Source specificity of early calcium neurotoxicity in cultured embryonic spinal neurons, *J. Neurosci.*, **13** (1993) 2085–2104.

95 Hartley, D. M., Kurth, M. C., Bjerkness, L., Weiss, J. H., and Choi, D. W., Glutamate receptor induced 45Ca accumulation in cortical cell culture correlates with subsequent neuronal degeneration, *J. Neurosci.*, **13** (1993) 1993–2000.

96 Nitatori, T., Sato, N., Waguri, S., Karasawa, Y., Araki, H., Shibanai, K., Kominami, E., and Uchiyama, Y., Delayed neuronal death in the CA1 pyramidal cell layer of the gerbil hippocamous following transient ischemia is apoptosis, *J. Neurosci.*, **15** (1995) 1001–1011.

97 Manev, H., Favaron, M., Guidotti, A., and Costa, E., Delayed increase of Ca^{2+} influx elicited by glutamate: role in neuronal death, *Mol. Pharmacol.*, **36** (1989) 106–112.

98 Randall, R. D. and Thayer, S. A., Glutamate-induced calcium transient triggers delayed calcium overload and neurotoxicity in rat hippocampal neurons, *J. Neurosci.*, **12** (1992) 1882–1895.

99 Dubinsky, J. M., Intracellular calcium levels during the period of delayed excitotoxicity, *J. Neurosci.*, **13** (1993) 623–631.

100 Tamaru, M., Yoneda, Y., Ogita, K., Shimizu, J., and Nagata, Y., Age-related decreases of the N-methyl-d-aspartate receptor complex in the rat cerebral cortex and hippocampus, *Brain Res.*, **542** (1991) 83–90.

101 Luiten, P. G. M., Douma, B. R. K., Van der Zee, E. A., and Nyakas, C., Neuroprotection against NMDA induced cell death in rat nucleus basalis by Ca^{2+} antagonist nimodipine; influence of aging and developmental drug treatment, *Neurodegeneration,* **4** (1995) 307–314.

102 Lee, K. S., Frank, S., Vanderklish, P., Arai, A., and Lynch, G., Inhibition of proteolysis protects hippocampal neurons from ischemia, *Proc. Natl. Acad. Sci. USA,* **88** (1991) 7233–7237.

103 Rudolphi, K. A., Schubert, P., Parkinson, F. E., and Fredholm, B. B., Adenosine and brain ischemia, *Cerebrovasc. Brain Metab. Rev.,* **4** (1992) 346–369.

104 Héron, A., Lekieffre, D., Le Peillet, E., Labennes, F., Seylaz, J., Plotkine, M., and Boulu, R. G., Effects of an A adenosine receptor agonist on the neurochemical, behavioral and histological consequences of ischemia, *Brain Res.,* **64** (1994) 217–224.

105 Prehn, J. H. M., Welsch, M., Backhauss, C., Nuglisch, J., Ausmeier, F., Karkoutly, C., and Krieglstein, J., Effects of serotonergic drugs in experimental brain ischemia: evidence for a protective role of serotonin in cerebral ischemia, *Brain Res.,* **630** (1993) 10–20.

106 Bielenberg, G. W. and Burkhardt, M. 5-Hydroxytryptamine1A agonists. A new therapeutic principle for stroke treatment, *Stroke,* **2(Suppl. IV)** (1990) IV161–IV163.

107 Zhang, L., Andou, Y., Masuda, S., Mitani, A., and Kataoka, K., Dantrolene protects against ischemic, delayed neuronal death in gerbil brain, *Neurosci. Lett.,* **158** (1993) 105–108.

108 Bruno, V. M., Goldberg, M. P., Dugan, L. L., Giffard, R. G., and Choi, D. W., Neuroprotective effect of hypothermia in cortical cultures exposed to oxygen-glucose deprivation or excitatory amino acids, *J. Neurochem.,* **63** (1994) 1398–1406.

109 Scriabine, A., Schuurman, T., and Traber J., Pharmacological basis for the use of nimodipine in central nervous system disorders, *FASEB J.,* **3** (1989) 1799–1806.

110 Wadworth, A. N. and McTavish, D., Nimodipine: a review of its pharmacological properties, and therapeutic efficacy in cerebral disorders, *Drugs & Aging,* **2** (1992) 262–286.

111 Janis, R. A., Silver, P. J., and Triggle, D. J., Drug action and cellular calcium regulation, *Adv. Drug Res.,* **16** (1987) 309–591.

112 Triggle, D. J., Ferrante, J., Kwon, Y. W., Skattebøl, A., Hawthorn, M., Langs, D. A., and Bangalore, R., 1,4-Dyhydropyridine-sensitive neuronal Ca^{2+} channels: pharmacology and regulation. In J. Traber J and W. H. Gispen (eds.) *Nimodipine and central nervous system function: New Vistas,* Schattauer, Stuttgart, 1989, pp. 3–16.

113 Langley, M. S. and Sorkin, E. M., Nimodipine: a review of its pharmacolodynamic and pharmacokinetic properties, and therapeutic potential in cerebrovascular disease, *Drugs,* **37** (1989) 669–699.

114 Jansen, I., Tfeld-Hansen, P., and Edvinsson, L., Comparison of the calcium entry blockers nimodipine and flunarizine on human cerebral and temporal arteries: role in cerebrovascular disorders, *Eur. J. Clin. Pharmacol.,* **40** (1991) 7–15.

115 Harper, A. M., Craigen, L., and Kazda, S., Effect of the calcium antagonist, nimodipine, on cerebral blood flow and metabolism in the primate, *J. Cereb. Blood Flow Metab.,* (1981) 349–356.

116 Kanda, K. and Flaim, S. F., Effects of nimodipine on cerebral blood flow in conscious rat, *J. Pharmacol. Exp. Ther.,* **236** (1986) 41–47.

117 Grabowski, M. and Johansson, B. B., Nifedipine and nimodipine: effect on blood pressure and regional cerebral blood flow in conscious normotensive and hypertensive rats, *J. Cardiovasc. Pharmacol.,* **7** (1985) 1127–1133.

118 Harris, R. J., Branston, N. M., Symon, L., M. Bayhan, M., and Watson, A., The effects of a calcium antagonist, nimodipine, upon physiological responses of the cerebral vasculature and its possible influence upon focal cerebral ischemia, *Stroke,* **13** (1982) 759–766.

119 Kazda, S., and Towart, R., Nimodipine, a new calcium antagonist with a preferential cerebrovascular action, *Acta Neurochirur.,* **63** (1982) 259–265.

120 McCalden, T. A., Nath, R. G., and Thiele, K., The effects of a calcium antagonist (nimodipine) on basal cerebral blood flow and reactivity to various agonists, *Stroke,* **15** (1984) 527–530.

121 Mohamed, A. A., McCulloch, T., Mendelow, A. D., Teasdale, G. M., and Harper, A. M., Effect of the calcium antagonist nimodipine on local cerebral blood flow: relationship to arterial pressure, *J. Cereb. Blood Flow Metab.,* **4** (1984) 206–211.

122 Talley, P. W., Sundt, T. M., Jr., and Anderson, R. E., Improvement of cortical perfusion, intracellular pH, and electrocorticography by nimodipine during transient focal cerebral ischemia, *Neurosurgery,* **24** (1989) 80–87.

123 Skattebøl, A. and Triggle, D. J., Regional distribution of calcium channel ligand (1,4–dihydropyridine) binding sites and 45Ca uptake processes in brain, *Biochem. Pharmacol.,* **36** (1987) 443–461.

124 McCarthy, R. T., Nimodipine block of L-type calcium channels in dorsal root ganglion cells. In J. Traber and W. H. Gispen (eds.) *Nimodipine and Central Nervous System Function: New Vistas,* Schattauer Verlag, Stuttgart, 1989, pp. 3–16.

125 Rämsch, K. D., Ahr, G., Tettenborn, D., and Auer, L. M., An overview on pharmacokinetics of nimodipine in healthy volunteers and in patients with subarachnoid hemorrhage, *Neurochirurgia,* **28** (1985a) 74–78.

126 Rämsch, K. D., Graefe, K. H., and Sommer, J., Pharmacokinetics and metabolism of nimodipine. In Betz et al. (eds.) Nimodipine: pharmacological and clinical properties, Schattauer Verlag, Stuttgart, 1985b, pp. 147–161.

127 Towart, R., Wehinger, E., Meyer, H., and Kazda, S. ,The effects of nimodipine, its optical isomers and metabolites on isolated vascular smooth muscle, *Arzneim.-Forsch.,* **32(I)** (1982) 338–346.

128 Kamath, B., Lettieri, J., Krol, G., Raemusch, K., and Yeh, S., Pharmacokinetics and metabolism of radiolabeled nimodipine, *Pharmacol. Res.,* **4** (1987) S80.

129 Maruhn, D., Siefert, H. M., Weber, H., Rämsch K., and Suwelack, D., Pharmacokinetics and metabolism of nimodipine. I. Communication: absorption, concentration in plasma and excretion after single administration of [^{14}C]nimodipine in rat, dog and monkey, *Arzneim.-Forsch.,* **35** (1985) 1781–1786.

130 Ginsberg, M. and Busto, R., Rodent models of cerebral ischemia, *Stroke,* **20** (1989) 1627–1642.

131 Rami, A. and Krieglstein, J., Neuronal protective effects of calcium antagonists in cerebral ischemia, *Life Sci.,* **55** (1994) 2105–2113.

132 Rod, M. R. and Auer, R. N., Combination therapy with nimodipine and dizocilpine in a rat model of transient forebrain ischemia, *Stroke,* **23** (1992) 725–732.

133 Welsch, M., Nuglisch, J., and Krieglstein, J., Neuroprotective effect of nimodipine is not mediated by increased cerebral blood flow after transient forebrain ischemia in rats, *Stroke,* **2** (1990) IV105–107.

134 Grotta, J. C., Picone, C. M., Dedman, J. R., Rhoades, H. M., Strong, R. A., Earls, R. M., and Yao, L. P., Neuronal protection correlates with prevention of calcium-calmodulin binding in rats, *Stroke,* **21**, (1990) III28–31.

135 Nakamura, K., Hatakeyama, T., Furuta, S., and Sakaki, S., The role of early Ca^{2+} influx in the pathogenesis of delayed neuronal death after brief forebrain ischemia in gerbils, *Brain Res.,* **613** (1993) 181–192.

136 Xie, Y., Seo, K., Ishimaru, K., and Hossmann, K. A., Effect of calcium antagonists on postischemic protein synthesis in gerbil brain, *Stroke,* **23** (1992) 87–92.

137 Mossakowski, M. J. and Gadamski, R., Nimodipine prevents delayed neuronal death of sector CA pyramidal cells in short-term forebrain ischemia in Mongolian gerbils, *Stroke,* **21** (1990) IV120–122.

138 Lazarewicz, J. W., Pluta, R., Puka, M., and Salinska, E., Diverse mechanisms of neuronal protection in experimental rabbit brain ischemia, *Stroke,* **21** (1990) IV108–10.

139 Tateishi, A., Fleischer, J. E., Drummond, J. C., Scheller, M. S., Zornow, M. H., Grafe, M. R., and Shapiro, H. M., Nimodipine does not improve neurologic outcome after 14 minutes of cardiac arrest in cats, *Stroke,* **20** (1989) 1044–1050.

140 Tamura, A., Graham, D. I., McCulloch, J., and Teasdale, G. M., Focal cerebral ischemia in the rat: 1. Description of technique and early neuropathological consequences following middle cerebral artery occlusion, *J. Cereb. Blood Flow Metabol.,* **1** (1981) 53–60.

141 Nagasawa, H. and Kogure, K., Correlation between cerebral blood flow and histologic changes in a new rat model of middle cerebral artery occlusion, *Stroke,* **20** (1989) 1037–1043.

142 Teasdale, G., Mendelow, A. D., Graham, D. I., Harper, A. M., and McCulloch, J., Efficiacy of nimodipine in cerebral ischemia or hemorrhage, *Stroke,* **2** (1990) IV123–125.

143 Benyó, Z., De Jong, G. I., and Luiten, P. G. M., Nimodipine prevents early loss of hippocampal CA parvalbumin immunoreactivity after focal cerebral ischemia in the rat, *Brain Res. Bul.,* **36** 1995) 569–572.

144 Sauter, A., Rudin, M., Wiederhold, K. H., and Hof, R. P., Cerebrovascular, biochemical and cytoprotective effects of isradipine in laboratory animals, *Am. J. Med.,* **86** (1989) 134–146.

145 Dirnagl, U., Jacewicz, M., and Pulsinelli, W., Nimodipine posttreatment does not increase blood flow in rats with focal cortical ischemia, *Stroke,* **2** (1990) 1357–1361.

146 Jacewicz, M., Brint, S., Tanabe, J., Wang, X. J., and Pulsinelli, W. A., The effect of nimodipine pretreatment on focal cerebral ischemia in hypertensive rat. In A. Scriabine, G. M. Teasdale, D. Tettenborn, and W. Young (eds) *Nimodipine: Pharmacological and Clinical Results in Cerebral Ischemia,* Springer–Verlag, Berlin, 1990, pp. 33–42.

147 Hogan, M., Gjedde, A., and Hakim, A. M., Nimodipine binding in focal cerebral ischemia, *Stroke,* **21** (1990) IV78–80.

148 Bielenberg, G. W., Burniol, M., Rosen, R., and Klaus, W., Effects of nimodipine on infarct size and cerebral acidosis after middle cerebral artery occlusion in the rat, *Stroke,* **21** (1990) IV90–92.

149 Hara, H., Nagasawa, H., and Kogure, K., Nimodipine prevents postischemic brain damage in the early phase of focal cerebral ischemia, *Stroke,* **2** (1990) IV102–104.

150 Greenberg, J. H., Uematsu, D., Araki, N., Hickley, W. F., and Reivich, M., Cytosolic free calcium during focal cerebral ischemia and the effects of nimodipine on calcium and histologic damage, *Stroke,* **2** (1990) IV72–77.

151 Hadley, M. N., Zabramski, J. M., Spetzler, R. F., Rigamonti, D., Fifield, M. S., and Johnson, P. C., The efficacy of intravenous nimodipine in the treatment of focal cerebral ischemia in a primate model, *Neurosurgery,* **25** (1989) 63–70.

152 Watson, B. D., Dietrich, W. D., Busto, R., Wachtel, M. S., and Ginsberg, M. D., Induction of reproducible brain infarction by photochemically initiated thrombosis, *Ann. Neurol.,* **17** (1985) 497–504.

153 Okamoto, K., Yamori, Y., and R Nagaoka, A., Establishment of the stroke-prone spontaneously hypertensive rat (SHR), *Circ. Res.,* **34,35** (1974) I143–153.

154 Luiten, P. G. M., de Jong, G. I., Van der Zee, E. A., Braaksma, M., Maes, F. W., Schuurman, T., and Nyakas, C., Neuroprotection by chronic nimodipine treatment in aging hypertensive stroke-prone rats, *Drugs Dev.,* **2** (1993) 183–191.

155 Liu, L. S., Zhao, Y., Lei, Y. P., Wang, W., Zhang, X. E., and Jin, L., Calcium antagonists in prevention of hypertension and stroke in stroke-prone spontaneously hypertensive rats, *Chin. Med. J. Engl.,* **102** (1989) 106–113.

156 Ueda, M., Masui, M., Kawakami, M., Matsunaga, K., Gemba, T., Ninomiya, M., Nakano, A., Torii, M., Adachi, I., and Ito, H., Pharmacological studies on a new dihydro-thienopyridine calcium antagonist. 4th communication: prophylactic and therapeutic effects of S-(+)-metyl-4,7-dihydro-3-isobutyl-6-methyl-4-(3-nitrophenyl)thieno[2,3-b] pyridine-5-carboxylate in stroke-prone spontaneously hypertensive rats, *Arzneim.-Forsch.,* **43** (1993) 1291–1303.

157 Zai, Q. H. and Duan, C. G., Cerebral microcirculatory disturbances of stroke prone spontaneously hypertensive rats with acute stroke and mechanism of treatment, *Chin. Med. J. Engl.,* **106** (1993) 518–521.

158 Lenz, P., Luckhaus, G., Stasch, J. P., and Kazda, S., Therapy of diseased stroke-prone spontaneously hypertensive rats with nimodipine, *Stroke,* **2** (1990) IV111,112.

159 Kinoshita, K., Yamamura, M., Matsuoka, Y., Ameliorating effect of clentiazem (TA-3090), a new Ca antagonist, on the impaired learning ability of poststroke SHRSP, *Pharmacol. Biochem. Behav.,* 50 (1995) 509–515.

160 White, R. P., Cunningham, M. P., and Robertson, J. T., Effect of the calcium antagonist nimodipine on contractile responses of isolated canine basilar arteries induced by serotonin, prostaglandin F2α, thrombin, and whole blood, *Neurosurgery,* **10** (1982) 344–348.

161 Siesjo, B. K., Cell damage in the brain: a speculative synthesis, *J. Cereb. Blood Flow Metab.,* **1** (1981) 155–185.

162 Gelmers, H. J., The effects of nimodipine on the clinical course of patients with acute ischemic stroke, *Acta Neurol. Scand.,* **69** (1984) 232–239.

163 Gelmers, H. J., Gorter, K., de Weerdt, C. J., and Wiezer, H. J. A., A controlled trial of nimodipine in acute ischemic stroke, *N. Engl. J. Med.,* **318** (1988) 203–207.

164 Gelmers, H. J. and Hennerici, M., Effects of nimodipine on acute ischemic stroke. Pooled results from five randomized trials, *Stroke,* **2** (1990) IV8–84.

165 Martinez-Vila, E., Guillen, F., Villanueva, A., Matias-Guiu, J., Bigorra, J., Gil, P., Carbonell, A., and Martinez-Lage, J., Placebo-controlled trial of nimodipine in the treatment of acute ischemic infarction, *Stroke,* **2** (1990) 1023–1028.

166 Paci, A., Ottaviano, P., Trenta, A., Iannone, G., De Santis, L., Lancia, G., Moschini, E., Carosi, M., Aigoni, S., and Caresia, L., Nimodipine in acute ischemic stroke: a double-blind controlled study, *Acta Neurol. Scand.,* **80** (1989) 282–286.

167 Bogousslavsky, J., Regli, F., Zumstein, V., and Köbberling, W., Double-blind study of nimodipine in non-severe stroke, *Eur. Neurol.,* **30** (1990) 23–26.

168 Trust Study Group, Randomised, double-blind, placebo-controlled trial of nimodipine in acute stroke, *Lancet,* **336** (1990) 1205–1209.

169 The American Nimodipine Study Group, Clinical trial of nimodipine in acute ischemic stroke, *Stroke,* **23** (1992) 3–8.

170 Kaste, M., Fogelholm, R., Erilä, T., Palomäki, H., Murros, K., Rissanen, A., and Sarna, S., A randomized, double-blind, placebo-controlled trial of nimodipine in acute ischemic hemispheric stroke, *Stroke,* **25** (1994) 1348–1353.

171 Wahlgren, N.G., MacMahon, D. G., De Keyser, J., Indredavik, B., and Ryman, T., Intravenous nimodipine west european stroke trial (INWEST) of nimodipine in the treatment of acute ischaemic stroke. *Cerebrovasc. Dis.,* **4** (1994) 204–210.

172 Norris, J. W., LeBrun, L. H., and Anderson, B. A., The Canwin Study Group, Intravenous nimodipine in acute ischaemic stroke, *Cerebrovasc. Dis.,* **4** (1994) 194–196.

173 Kakarieka, A., Schakel, E. H., and Fritze, J., Clinical experience with nimodipine in cerebral ischemia, *J. Neural Transm.,* **43(Suppl)** (1994) 13–21.

174 Mohr, J. P., Orgogozo, J. M., Harrison, M. J. G., Hennerici, M., Wahlgren, N.G., Gelmers, J. H., Martinez-Vila, E., Dycka, J., and Tettenborn, D., Meta-analysis of oral nimodipine trials in acute ischemic stroke, *Cerebrovasc. Dis.,* **3** (1994) 197–203.

175 DeGirolami, U., Crowell, R. M., and Marcoux, F. W., Selective necrosis and total necrosis in focal cerebral ischemia. Neuropathologic observations on experimental middle cerebral artery occlusion in the Macaque monkey, *J. Neuropathol. Exp. Neurol.,* **43** (1984) 57–71.

176 Mattson, M. P., Rydel, R. E., and Smith-Swintosky, V. L., Altered calcium signaling and neuronal injury: stroke and Alzheimer's disease as examples, *Ann. NY Acad. Sci.,* **679** (1993) 1–21.

177 Disterhoft, J. F., Moyer, J. R., Jr., and Thompson, L. T., The calcium rationale in aging and Alzheimer's disease, *Ann. NY Acad. Sci.,* **747** (1994) 382–406.

178 Fieschi, C., *Acute ischemic stroke therapy. Abstr. 6th Int. Symp. Pharmacol. Control Calcium Potass. Homeost,* Florence, 1994, pp. 42.

179 Fisher, M., Jones, S., and Sacco, R. L. Prophylactic neuroprotection for cerebral ischemia, *Stroke,* **25** (1994) 1075–1080.

180 Barger, S. W. and Mattson, M. P. Excitatory amino acids and growth factors: biological and molecular interactions regulating neuronal survival and plasticity. In: *CNS Neurotransmitters and Neuromodulators,* vol. 2, CRC, 1995, pp. 273–294.

181 Alderson, R. F., Wiegand, S. J., Anderson, K. D., Cai, N., Cho, J., Lindsay, R. M., and Altar, C. A., Neurotrophin-4/5 maintains the cholinergic phenotype of axotomized septal neurons, Eur. *J. Neurosci.,* **8** (1996) 282–290.

5

Glutamate Neurotoxicity and Stroke

Tihomir P. Obrenovitch

1. INTRODUCTION

Glutamate is the principal excitatory neurotransmitter in the mammalian central nervous system (CNS), subserving postsynaptic excitatory responses in many cortical efferent systems and in intrahippocampal pathways. When glutamatergic terminals are depolarized, vesicular glutamate is released into the synaptic cleft in a Ca^{2+}-dependent manner *(1)*. There, glutamate interacts with several distinct families of receptors located principally on postsynaptic neurons. *N*-methyl-D-aspartate (NMDA), α-amino-3-hydroxy-5-methyl-4-isoxazole-propionic acid (AMPA), and kainate receptors are termed *ionotropic* because they open cation channels upon their activation by an agonist *(2,3)*. Excitatory postsynaptic potentials are commonly composed of both AMPA and NMDA receptor-mediated components. In contrast, metabotropic receptors are linked to second messenger systems. Depending on the cell type and receptor subtype, metabotropic receptors might mediate inositol phosphate metabolism, release of arachidonic acid, or changes in cyclic adenosine monophosphate (cAMP) levels *(4)*. Glutamate is removed from the synaptic cleft and extracellular space by Na^+-dependent, high-affinity monocarboxylic acid carriers located in both neurons and glia *(1)*. Efficient uptake of glutamate contributes to the termination of its actions on synaptic receptors since there is no extracellular enzyme to breakdown glutamate *(5)*.

Neuronal death subsequent to excessive glutamate-mediated excitation, often referred to excitotoxicity, stands out as a critical factor common to a variety of neurological disorders, including stroke *(6,7)*. Excessive inward currents of Ca^{2+} and Na^+ through glutamate-gated ion channels, possibly supplemented by release of Ca^{2+} from intracellular stores subsequent to metabotropic receptor activation, leads to intracellular Ca^{2+} overload, which is a common mediator of cell death *(8–10)*. As glutamate antagonists are neuroprotective in stroke models, lesions of excitatory pathways reduce hippocampal cell death, and extracellular glutamate concentrations markedly increase during ischemia, it is generally assumed that the trigger of ischemia-induced excitotoxicity is high extracellular glutamate *(11–13)* (Fig. 1). This chapter has three objectives:

1. To illustrate that the current concept of ischemia-induced excitotoxicity (Fig. 1), centered on high extracellular glutamate, conflicts with key experimental evidence.
2. To show that this oversimplified theory may lead to inappropriate strategies for the protection of neurons against ischemia.

From: Clinical Pharmacology of Cerebral Ischemia *Edited by:* G. J. Ter Horst and J. Korf *Humana Press Inc., Totowa, NJ*

101

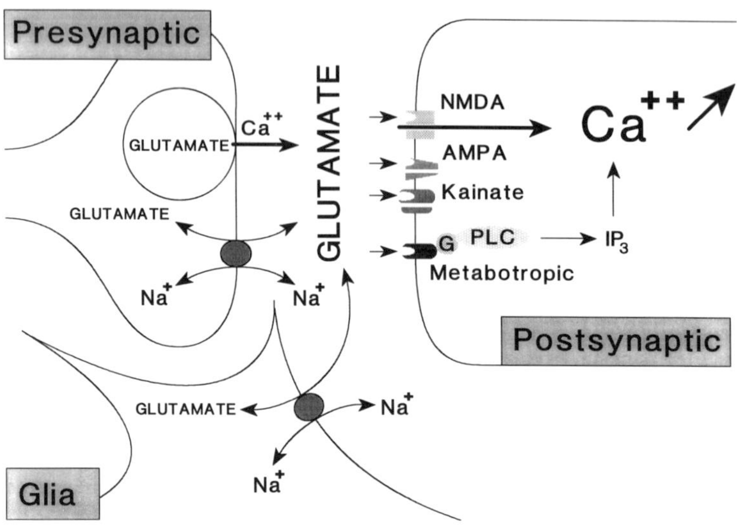

Fig. 1. Schematic illustration of the current hypothesis of ischemia-induced excitotoxicity, centered on high extracellular glutamate subsequent to excessive presynaptic release and deficient/reversed glutamate uptake. G, G-protein; PLC, phospholipase C; IP$_3$, inositol-1,4,5-triphosphate.

3. To propose alternative mechanisms, involving glutamatergic neurotransmission, which may be detrimental to neuronal survival in ischemia.

Some of the arguments put forward below have been developed in more detail in recent reviews *(14–17)*.

2. NEUROTRANSMITTER EFFLUX IN CEREBRAL ISCHEMIA

2.1. Excessive Efflux During Ischemia Is Not Specific to Excitatory Neurotransmitters

Intracerebral microdialysis has made it possible to examine changes in the composition of the brain extracellular fluid (ECF) in living experimental models *(18)*. This method has rapidly become popular in stroke research because the ECF is the neuronal microenvironment in which neurotransmitters and drugs are available to synaptic receptors, and through which nonsynaptic cell-cell chemical signalling takes place. Since Benveniste and coworkers first reported in 1984 *(19)* that extracellular concentrations of the excitatory amino acids glutamate and aspartate increased markedly in the rat hippocampus during ischemia, numerous studies have confirmed and characterized this phenomenon in different brain regions, with various models of focal and global ischemia, and also in humans (Fig. 2) *(15,20)*. This efflux of excitatory transmitters is constantly referred to in stroke pharmacology, but ischemia provokes similar changes in inhibitory and modulatory compounds that are potentially protective *(15,21)*.

Changes in GABA and taurine are often overlooked despite relevant findings: The relative increase of extracellular GABA is more pronounced than that of glutamate *(22–24)*; ischemic efflux of excitatory and inhibitory amino acids are synchronous

(23,24); and K⁺-evoked release of GABA and taurine remained unchanged up to 5 days after 20 min of forebrain ischemia, whereas those of glutamate and aspartate were reduced to 35–40% of control *(25)*. The latter finding suggests that either the presynaptic release of glutamate is more vulnerable to ischemic damage than that of inhibitory amino acids, or a selective vulnerability of glutamatergic neurons *(25,26)*.

The efflux of adenosine is also relevant because this inhibitory neuromodulator protects against cerebral ischemia *(27)*, reduces excitotoxicity both in vitro and in vivo *(28–30)*, and is released at a higher blood flow threshold than glutamate *(31)*. As adenosine inhibits excitatory amino acid release by acting on presynaptic A_1 receptors, several studies have examined whether the protective effect of adenosine correlates with reduced glutamate release. Conclusive evidence for such a link is still lacking, however, and adenosine may be neuroprotective via other mechanisms *(15,27)*.

2.2. The Large Glutamate Efflux Produced by Severe Transient Ischemia Is Rapidly Cleared During Reperfusion

Glutamate released into the extracellular space during ischemia is rapidly cleared during reperfusion, and to basal levels. This is another important and well-established feature *(15)* that is seldom mentioned when the excitotoxic hypothesis of ischemia-induced neuronal loss is discussed. By incorporating an electrode within the microdialysis probe to record electroencephalogram (EEG) and extracellular direct current (DC) potential at the dialysate sampling site *(32)*, we showed that ECF glutamate began to recover as DC potential started to normalize, i.e., with cellular repolarization (Fig. 2) *(33)*. Even after 20 min of forebrain ischemia with sustained depolarization, which produces marked histologic damage *(34)*, both glutamate and aspartate returned to normal levels within 10–15 min *(24)*. Ischemia-induced glutamate release is also reversible with repeated transient ischemic insults, and there is no evidence of a cumulative effect *(35,36)*. These data clearly indicate that glutamate uptake, deficient or even reversed during ischemia, recovers rapidly as transmembrane ionic gradients are restored. Postischemic recovery of presynaptic glutamate uptake was also demonstrated in striatal synaptosomes *(37)* and this mechanism was apparently preserved 4 d after ischemia in the rat hippocampal CA1 *(38)*. Rapid return to baseline during reperfusion was also observed with other neurotransmitters *(15)*.

The rapid clearance of excitatory amino acids from the extracellular space during reperfusion conflicts with findings suggesting that postischemic events related to glutamatergic systems may contribute to delayed neuronal death subsequent to transient ischemia. For example, lesions of the CA1 afferent immediately after 20-min forebrain ischemia offered neuroprotection *(39)*, as did postischemic administration of glutamate receptor antagonists *(40–43)* or voltage-gated Na⁺-channel modulators *(44,45)* in some studies.

2.3. There Is No Sustained Increase of Extracellular Glutamate in the Penumbra

The concept of penumbra in focal ischemia was introduced to define regions with blood flow below that needed to sustain electrical activity (i.e., neuronal function), but above that required to maintain cellular ionic gradients, and leads in time to irreversible cellular damage *(16,46)*. Determination of glutamate efflux in the penumbra is important because it is especially responsive to protection with glutamate receptor antagonists *(14,47)*. It is a difficult task, however, because the penumbra is a narrow, unstable rim whose location cannot be accurately predicted in individual experiments, and implantation of a microdialysis probe is likely to alter its fragile state *(16)*.

To circumvent this methodological problem, we have modified the rat four-vessel occlusion model of cerebral ischemia *(48)* to produce sustained and controlled periods of penumbral ischemia in the forebrain (i.e., electrical silence without anoxic depolarization). This approach is similar to that previously described to create a consistent "metabolic penumbral zone" *(49)*. The severity of ischemia was continuously regulated using EEG and DC potential as feedback parameters, recorded with an electrode within the microdialysis probe used to monitor neurotransmitter changes *(32)*. During 30 min of penumbral ischemia, extracellular concentrations of excitatory amino acids in the striatum rose only slightly. Massive overflow of neuroactive compounds only occurred with sustained anoxic depolarization (Fig. 2) *(24)*. These findings strongly suggest that excitotoxic processes in the penumbra are not related to sustained increases in the extracellular level of excitatory amino acids, initiated within this region.

With focal ischemia, however, excessive glutamate could conceivably leak from the ischemic core, damaging nearby neurons, resulting in further efflux of glutamate, and thus extending tissue damage according to a self-propagating event *(9)*. However, careful examination of glutamate toxicity in rat cerebellar slices has invalidated this hypothesis. As long as the transmembrane Na^+ and K^+ gradients were maintained, which is the case in the penumbra *(16,46)*, 100 μM glutamate in the incubation medium was only toxic to the outermost regions of the slice *(51)*.

Another cause of transient glutamate efflux in the penumbra is recurrent spreading depression (SD), which propagates in regions adjacent to the ischemic core and contributes to the extent of tissue damage (*see* Section 4.2.). By slightly increasing the K^+ concentration in the medium perfused through the microdialysis probe to avoid buffering the transient increase in extracellular K^+ associated with propagating SD, we have demonstrated a large, synchronous, transient increase in extracellular glutamate *(52,53)*, probably of presynaptic vesicular origin *(17)*. It is very likely that in the penumbra, where energy supply is compromised, the clearance of such transient glutamate release is slower because high-affinity glutamate uptake is tightly coupled to transmembrane Na^+/K^+ ionic gradients *(1)*, and low blood flow reduces the capacity of the brain tissue to restore normal ionic gradients *(54,55)*. Nevertheless, as the concept of the ischemic penumbra implies absence of anoxic depolarization in this region, there should be no sustained increase in extracellular glutamate in this condition. The contrary would indicate progression towards infarction.

2.4. Ischemia-Induced Histological Damage: Is It Linked to High Extracellular Glutamate?

Between 10 and 20 min of a uniform severe reduction of cerebral blood flow in rats (two-vessel occlusion combined with systemic hypotension) produced similar increases in extracellular glutamate in the dorsal hippocampus, anterior thalamus, somatosensory cortex, and dorsolateral striatum, i.e., vulnerable and nonvulnerable brain regions *(56)*. Similarly, extracellular glutamate levels were increased in both CA1 and CA3 fields of the gerbil hippocampus during 5 min of ischemia, which resulted in selective

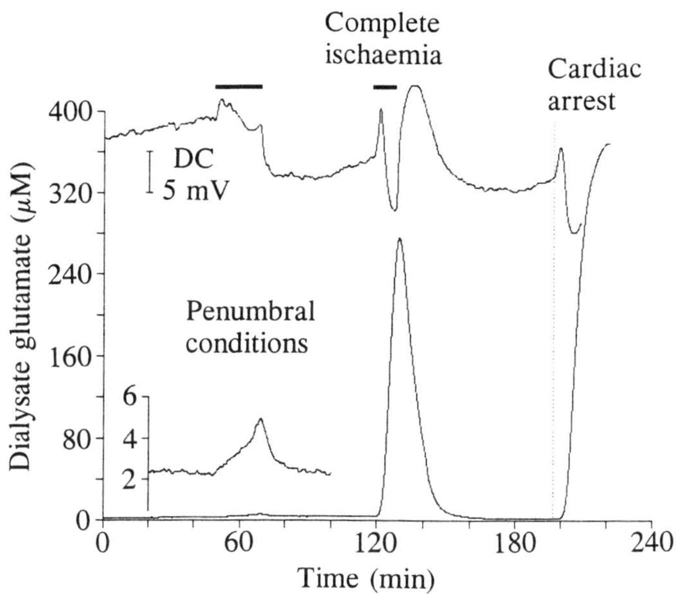

Fig. 2. Changes in dialysate glutamate during cerebral ischemia of graded severity (bottom traces). Microdialysis probe, incorporating an electrode for the recording of EEG and DC potential (upper trace), were implanted in the striatum of anesthetized rats. Dialysate glutamate concentration was continuously monitored by enzyme-fluorescence *(50)*. Two to three hours after implantation, animals were subjected to the following procedure: 20 min of penumbral conditions (i.e., EEG silence without anoxic depolarization) produced by controlled four-vessel occlusion *(24)*; 40 min of recirculation; 5 min of complete forebrain ischemia; >60 min recirculation before cardiac arrest. Note that marked glutamate efflux only occurred during complete ischemia with anoxic depolarization. Data are from a single representative experiment. The bottom left insert shows the changes in dialysate glutamate during penumbral conditions; its scale is 20-fold smaller than that used for representing glutamate changes throughout the procedure. Figure reproduced from Obrenovitch (1995) *(17)* with permission of Springer Verlag (Wien).

delayed neuronal death only of the pyramidal neurons; CA3 neurons were spared *(57)*. Therefore, regional susceptibility to ischemic damage does not appear to be directly linked to glutamate efflux.

A significant correlation was found between the magnitude of glutamate efflux and the volume of ischemic damage in rat brain following middle cerebral artery occlusion (MCAO) *(58,59)*. However, this does not imply that poor neuropathologic outcome was a consequence of a more elevated ECF glutamate during ischemia because there was also a significant correlation between tissue damage and efflux of other neuroactive amino acids. Therefore, it appears that neurotransmitter efflux is a generalized consequence of ischemia, increasing with the severity of the insult *(15)*. This efflux is presumably not limited to neurotransmitters because similar changes were observed with compounds which have a high intra-/extracellular concentration ratio but no clear involvement in neurotransmission. A good example of such a compound is *N*-acetylaspartate (NAA) *(60)*.

Numerous experimental and pharmacological procedures have been carried out to test whether neuroprotection correlates with reduced glutamate efflux during ischemia. These include lesions of excitatory pathways, hypothermia, fasting, and a wide range of drugs. Overall, these experiments have provided conflicting data and failed to produce convincing evidence *(15)*. Testing the hypothesis that high extracellular glutamate correlates with ischemic injury is clearly beset by the difficulty of separating cause from effect.

2.5. Does High Extracellular Glycine Contribute to Excitotoxicity?

It is established that the NMDA receptor has a distinct binding site for glycine, and that occupation of this site by glycine is a requirement for NMDA/glutamate receptor activation *(61–63)*. Submicromolar concentrations of glycine markedly potentiate NMDA responses in cultured neurons and other in vitro preparations *(64)*. At higher concentrations, glycine may also act as an agonist at the NMDA recognition site *(65)*. As extracellular glycine levels were found to remain elevated in the rat hippocampus for up to 80 min after transient global ischemia in some studies *(56,66–68)*, it was speculated that high ECF glycine may contribute to ischemia-induced NMDA/glutamate receptor activation and, therefore, to excitotoxicity. Several factors however make this hypothesis questionable *(15)*.

Firstly, the persistent increase in ECF glycine during reperfusion was moderate *(56,66)* and further work provided conflicting evidence: No sustained increase of ECF glycine was found in the striatum of rats subjected to 15–20 min of forebrain ischemia *(24,69)*; there was no evidence of persistent changes in ECF glycine in the hippocampus of gerbils subjected to three episodes of 2-min forebrain ischemia, despite marked neuronal damage 4 d later *(36)*; and extracellular glycine did not change during 30 min of penumbral conditions *(24)*.

Secondly, the functional importance of the potentiation of NMDA responses by glycine in vivo remains unclear *(62,63,70,71)*. This potentiation was detectable with glycine concentrations as low as 10 nM in cultured neurons *(64)* and its EC$_{50}$ varied from 0.2–0.7 µM with various in vitro preparations. These low concentrations contrast with the much higher glycine levels present in the ECF (5–12 µM) *(56)*. The hypothesis that glycine levels may be kept lower close to synaptic receptors is debatable *(63)*. Glycine uptake systems are present in synaptosomes as well as in neurons and glial cells but they may not be capable of reducing ECF glycine concentrations below a few µM since their Km is 10–100 µM. Even the cloned neuron-specific glycine transporter expressed at high levels in regions with a high density of NMDA receptors had a Km >100 µM *(72)*. Studies initiated to verify whether or not the NMDA receptor glycine site was saturated in vivo have provided conflicting data *(15,62,63)*. Recently, we have shown that very high concentrations of glycine (>1 mM) were necessary to evoke a detectable depolarization in the rat striatum, and that glycine did not potentiate NMDA-evoked responses (Fig. 3) *(71)*. The possibility remains that an endogenous glycine antagonist such as kynurenate may keep the NMDA receptors in a glycine-sensitive state *(73)*.

A third element conflicts with the concept that high extracellular glycine is neuro-toxic: Glycine activates strychnine-sensitive glycine-operated Cl$^-$ channels resulting in hyperpolarization and inhibition *(74)*. The inhibitory action of glycine is important

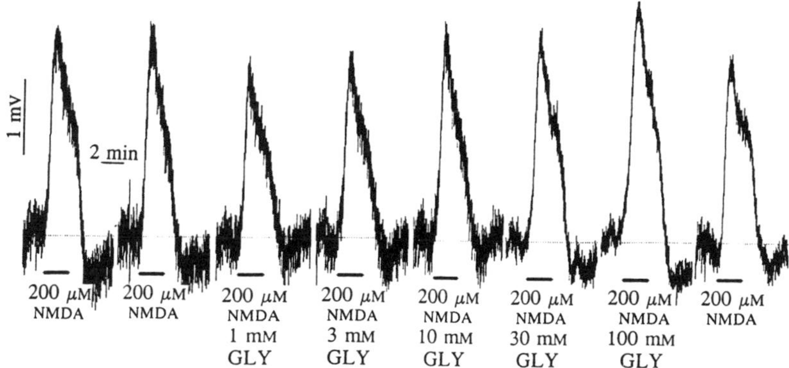

Fig. 3. Effect of increased extracellular glycine on NMDA-evoked responses in the rat striatum. Step increased concentrations of glycine (1–100 m*M*) were coapplied with 200 μ*M* NMDA for 2 min through a microdialysis electrode, which made it possible to monitor changes in the DC potential precisely at the site of drug application *(77)*. Only >10 m*M* glycine significantly potentiated NMDA-evoked responses, by approx 10%. These representative data from a single experiment were taken from Hardy et al. (1995) *(71)*.

in the spinal cord and brain stem, but generally considered of minor significance in upper brain regions. Nevertheless, glycine-induced inhibitory responses, antagonized by strychnine, were observed in supraspinal areas *(75)*, and glycine by itself has modest anticonvulsant properties in a number of seizure models, including against NMDA-induced seizures *(76)*.

Therefore, the potential excitatory and deleterious action of high extracellular glycine remains unproven, and the increase in dialysate glycine observed in ischemia could equally be considered as inhibitory and potentially protective *(78,79)*. Nevertheless, as occupation of the glycine site coupled to the NMDA receptor is required for NMDA/glutamate receptor activation, this site remains an attractive target for potential neuroprotective agents *(80)*.

To summarize this section, the widely accepted concept that glutamate efflux is critical to neuronal survival in ischemia (Fig. 1) conflicts with important elements: There is no obvious excitatory/inhibitory imbalance at the extracellular level; in transient ischemia, glutamate accumulation is cleared within minutes of recirculation, a feature that is not compatible with postischemic pharmacological protection and delayed neuronal death; the penumbra is especially responsive to protection with glutamate receptor antagonists, but sustained, high extracellular glutamate does not occur in this region; and regional vulnerability to ischemic damage is presumably not linked to glutamate efflux.

3. KINETICS AND ORIGINS
OF GLUTAMATE EFFLUX IN SEVERE ISCHEMIA:
IMPLICATIONS FOR NEUROPROTECTION

3.1. Possible Origins of Glutamate Efflux in Severe Ischemia

Under resting conditions, the level of glutamate in the cytoplasm of brain cells is around 10,000-fold higher than that in the extracellular space, a gradient maintained by high-affinity acidic amino acid carriers present in both presynaptic and glial plasma membranes *(1)*. High-affinity glutamate uptake requires the simultaneous presence of external Na^+ and internal K^+, and its efficacy is dependent on the Na^+/K^+ gradient across the plasma membrane *(81)*. In glial cells isolated from the retina of salamander, raising external K^+ to 10 mM was sufficient to reverse glutamate uptake; this effect was activated by intracellular glutamate and Na^+, and increased by membrane depolarization *(82)*. Intracellular acidosis during ischemia may also contribute to inhibition/reversal of glutamate uptake *(83–85)*. The reversal of glutamate uptake has two important characteristics: It reflects a change in the driving forces (i.e., ionic gradients) and not an alteration of the carrier protein; and the resulting efflux of glutamate is of cytoplasmic origin (metabolic pool of glutamate) and therefore Ca^{2+}-independent. As cerebral ischemia implies deficiency of the Na^+/K^+ ATPase *(86)* leading to disruption of the transmembrane ionic gradients with anoxic depolarization *(87)*, moderate insults are likely to produce an imbalance between glutamate efflux and uptake *(88)*, whereas more severe insults may reverse glutamate uptake processes *(13)*.

Another potential source of glutamate efflux in ischemia is exocytosis of neurotransmitter glutamate, i.e., Ca^{2+}-dependent release from presynaptic vesicles *(1,89)*. However, all exocytosis processes share a feature that is critical within the context of ischemia—energy dependency. Vesicular release in response to a stimulus is tightly regulated, mainly at the fusion step, and this step involves ATP hydrolysis *(90)*. As ATP rapidly decreases during ischemia *(91,92)*, exocytosis of glutamate is likely to be limited in this condition.

Cellular swelling subsequent to ischemia *(93,94)* may also contribute to increasing extracellular glutamate concentration because hypotonic swelling of cultured astrocytes inhibited glutamate uptake and increased its release *(95,96)*. Swelling-induced glutamate release, which did not appear to involve reversal of the Na^+-dependent uptake carrier, was inhibited by a number of anion transport inhibitors *(95)*, including L-644,711, which improved outcome in an experimental brain trauma/hypoxia model *(97)*. Finally, in addition to the potential mechanisms outlined above, cellular lysis may also be responsible for glutamate leakage and high extracellular glutamate levels in ischemia.

3.2. Time Course of Glutamate Efflux in the Ischemic Core

We have studied the time course of changes in extracellular glutamate and their Ca^{2+}-dependency in the rat striatum during focal ischemia produced by MCAO, using microdialysis coupled to enzymatic flow analysis of the dialysate for optimum time resolution *(98)*. When the probe was perfused with control artificial CSF, ischemia produced a biphasic increase in extracellular glutamate that started from the onset of ischemia. During the first phase, lasting around 10 min, dialysate glutamate increased to around 35 μM (vs baseline of approx 5.8 μM). Then, dialysate glutamate increased progressively to its maximum (80–85 μM), reached after around 1 h of ischemia, where it remained for at least 3 h *(98)*. A similar pattern of changes was obtained during complete ischemia produced by cardiac arrest, with a probe implanted in the rat cerebral cortex and dialysate glutamate recorded by enzyme-amperometry (Fig. 4) *(99)* or measured each minute by high performance liquid chromatography (HPLC) *(100)*. Biphasic changes in extracellular glutamate during severe ischemia indicate the involvement of several mechanisms and multiple subcellular origins.

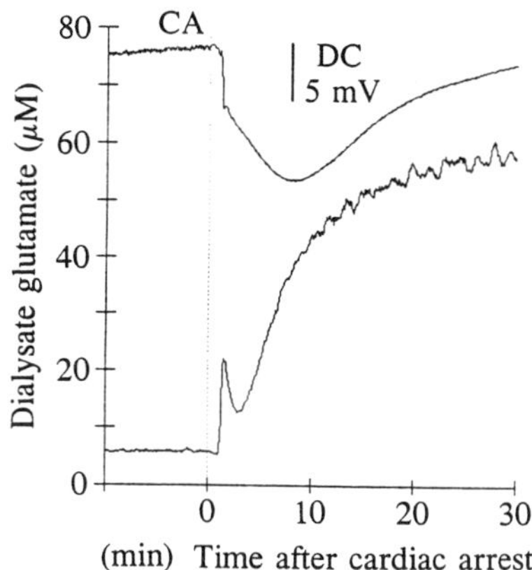

Fig. 4. Biphasic changes in dialysate glutamate (bottom trace) and corresponding DC potential (upper trace) during complete ischemia produced by cardiac arrest (CA). Microdialysis probes were implanted in the cortex of anesthetized rats, and glutamate recorded by enzyme-amperometry. In this experiment, the exocytotic phase (1st peak) of glutamate release was especially marked; note that it was closely associated with anoxic depolarization. Data are from Zilkha et al. (1995) *(99)*, reprinted with permission of the authors and Elsevier Science-NL.

3.3. Magnitude of Exocytotic Glutamate Release in the Ischemic Core

The initial component of the extracellular glutamate kinetic was no longer detectable when Ca^{2+} was omitted from the perfusion medium *(98)*, suggesting that it originated from vesicular glutamate. This implies that residual ATP was maintained at a level sufficient to sustain exocytotic release for several minutes after ischemia onset, probably at the expense of phosphocreatine whose depletion precedes that of ATP *(92)*. Experiments with cardiac arrest (Fig. 4) showed that exocytotic release during ischemia was closely associated with anoxic depolarization, an event that combines massive Ca^{2+} influx *(87)* with residual ATP levels of around 1/3 normal *(101)*.

The findings outlined in Sections 3.2. and 3.3. strongly suggest that the magnitude of Ca^{2+}-dependent glutamate release during ischemia (Fig. 4, 1st peak) is minor in comparison with that of the total efflux of glutamate. A number of previous studies agree with this characteristic *(15)*: Exocytosis could not be sustained for more than a few minutes because magnetic resonance spectroscopy (MRS) showed total ATP depletion within 10 min of complete ischemia *(102)*; a large part of glutamate released by synaptosomes during cyanide-induced anoxia was independent of Ca^{2+} *(103)*; and glutamate was released from cultured astrocytes exposed to hypoxic-hypoglycemic conditions and the magnitude of this release was larger than that from neurons, even though astrocytes do not have any presynaptic terminals *(104)*.

3.4. Does the Reversal of Glutamate Uptake Contribute to Efflux of Metabolic Glutamate In Vivo?

In order to test whether the second phase of glutamate release in ischemia was due to reversal of its uptake, we have studied the effect of selective blockade of high-affinity glutamate transporters, using L-*trans*-pyrrolidine-2,4-dicarboxylate (L-*trans*-PDC), a new selective inhibitor of these transporters *(105)*. As the reversal of glutamate uptake results from alterations of transmembrane ionic gradients, and not from a change in the carrier protein, L-*trans*-PDC inhibits the reversed as well as normal glutamate uptake *(5)*. The L-*trans*-PDC was applied through the microdialysis probe, starting 10 min before cardiac arrest and continued throughout the postmortem recording period. Although 2.5 m*M* L-*trans*-PDC markedly increased basal levels of dialysate glutamate, the time course of postmortem glutamate changes was barely altered *(17)*. The maximum rate of release and the increase in dialysate glutamate appeared slightly exacerbated during the first exocytotic phase, but the kinetics of the second, Ca^{2+}-independent phase were essentially unchanged. In contrast to data obtained with rat cultured astrocytes *(84)* and hippocampal slices *(106)*, these findings obtained in vivo do not support the reversal of glutamate uptake as a major contributor to increased extracellular glutamate during complete ischemia. The primary mechanism of glutamate efflux in the ischemic brain remains to be determined.

3.5. Implications for Neuroprotection

In addition to contradicting important data (*see* Section 2.), the common concept of ischemia-induced excitotoxicity illustrated in Fig. 1 may lead to inappropriate therapeutic strategies. Indeed, if we accept that high extracellular glutamate is the trigger for neuronal loss in ischemia, then rational targets for drug development should be the reduction of ischemia-induced glutamate efflux through the two mechanisms that are the most often advanced: presynaptic, vesicular release, and reversed electrogenic uptake (Fig. 1) *(13)*.

3.5.1. Inhibition of Presynaptic, Vesicular Glutamate Release

The ischemic penumbra is not exposed to sustained high extracellular levels of glutamate (*see* Section 2.3.). Transient glutamate release, presumably of vesicular origin, only occurs with propagating SD, and high extracellular glutamate may not contribute to SD initiation or propagation (*see* Section 4.2.). In the ischemic core, most of the released glutamate is of metabolic origin, probably from both neurons and glia (*see* Section 3.3.). By themselves, these findings question the validity of therapeutic strategies aimed at preventing or reducing excessive release of neurotransmitter glutamate in ischemia, although the possibility that glutamate changes at the synaptic level may be small, but pathologically important and cannot yet be totally disclaimed. In addition, selective inhibition of presynaptic vesicular release may be difficult because the synaptic machinery for release of neurotransmitter glutamate is not unique to glutamatergic synapses *(89)*, nor to synaptic transmission in general, but common to a wide range of secretory and membrane-fusion processes *(90)*.

The reduction of ischemia-induced glutamate efflux by the antiepileptic drug lamotrigine and its analog (BW1003C87 and BW619C89) is often considered as causative of neuroprotection *(44,79)*, but this prevalent interpretation conflicts with key findings *(15,107)*:

1. The action of these drugs on high extracellular levels of glutamate during ischemia is not specific. Extracellular aspartate, GABA, glycine, and taurine are reduced to a similar extent *(44,79)*. Presumably, the nonspecific reduction in neurotransmitter efflux is due to these compounds reducing the severity of ischemia (*see* Section 2.1.) *(107,108)*.

2. BW1003C87 reduced hippocampal CA1 lesions even when administered up to 2 h after forebrain ischemia, i.e., long after complete return of extracellular glutamate to normal levels *(33,35,44)*.
3. BW1003C87 reduced extracellular glutamate levels in both cortex and caudate during MCAO in rats, but only the cortex was protected *(79)*. Similarly, the reduction in microdialysate glutamate concentration by lamotrigine did not relate to cerebroprotection in the striatum *(109)*.

The cerebroprotective actions of lamotrigine and structural analogs are more likely a direct consequence of use-dependent down-modulation of voltage-gated Na^+ channels, and this mechanism of action should be recognized as such *(107)*. Again, reduction of ischemia-induced neurotransmitter efflux by putative neuroprotective agents does not imply that the former is causative.

The beneficial effects of riluzole in rodent models of ischemia are also attributed to an antiglutamate action because of unusual actions on glutamatergic transmission *(110,111)*. Riluzole-inhibited responses evoked by excitatory amino acids without binding to ionotropic glutamate receptors *(112)*. It also reduced spontaneous glutamate release and increased the size of its potassium-releasable pool *(113)*. However, as with the lamotrigine-related compounds, riluzole may be neuroprotective because it is a highly specific blocker of Na^+ channels in their inactivated state *(107)*.

3.5.2. Inhibition of Reversed Glutamate Uptake

In addition to the disputed contribution of reversed uptake to glutamate efflux in stroke (*see* Section 3.4.), interfering with glutamate transporters to reduce this form of release may be a problematic strategy. Firstly, competitive inhibitors of glutamate carriers, such as L-*trans*-PDC, may also evoke glutamate release by heteroexchange *(105,114)*. Secondly, as the reversal of electrogenic uptake mechanisms results from alterations of transmembrane ionic gradients and not from a change in the carrier protein, pharmacological inhibition of reversed glutamate uptake in severely ischemic regions would imply reduced efficacy of normal uptake in normal and penumbral areas.

The elements reviewed in Sections 2. and 3. also challenge the validity of ischemic models that rely on application of glutamate or its agonists to in vitro preparations *(14,115)* or to the brain of laboratory animals *(116)*.

4. ALTERNATIVE NEUROTOXIC MECHANISMS INVOLVING GLUTAMATERGIC SYSTEMS

Although the analysis (*see* Sections 2. and 3.) disputes the hypothesis that high extracellular glutamate is the triggering event leading to neuronal death in ischemia, one cannot deny the enormous body of pharmacological data indicating that glutamate antagonists are neuroprotective in a number of stroke models *(14)*. This apparent contradiction suggests a broader definition of excitotoxicity, whereas other abnormalities involving glutamatergic systems, in addition to the direct actions of increased extracellular levels of glutamate on its receptors, are neurotoxic (Fig. 5). A few selected possibilities are examined in this chapter.

4.1. Alterations of Glutamate Ionotropic Receptors

The exceptional diversity of glutamate-operated ion channels, the complexity of their modulation, and the probable involvement of second messenger systems, provide a

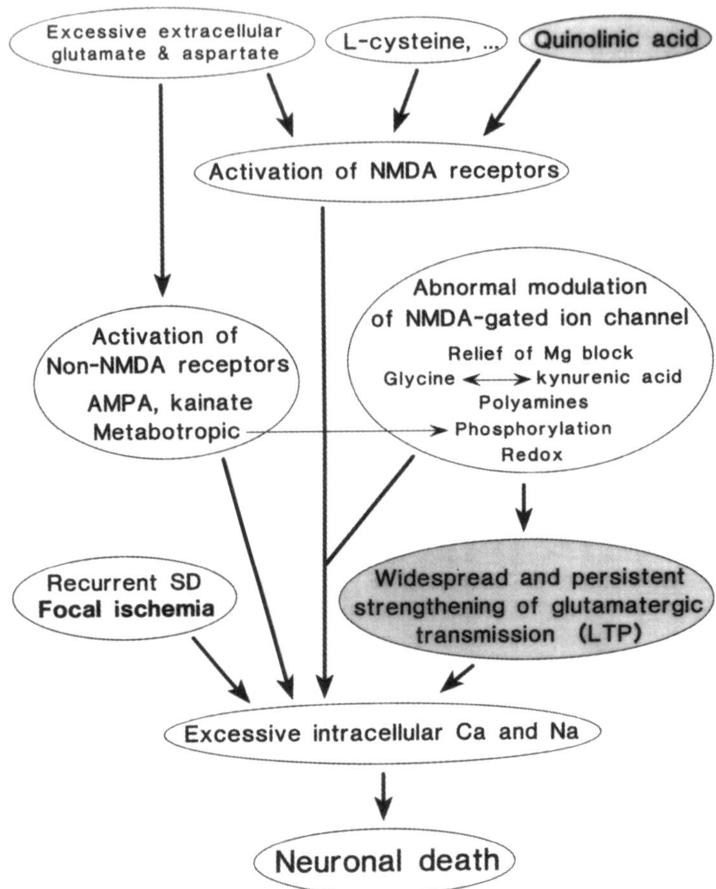

Fig. 5. Diagram illustrating multiple processes involving glutamatergic systems, initiated during ischemia or delayed (shaded captions), potentially capable of inducing neuronal death. SD, spreading depression. Reproduced from Obrenovitch and Richards (1995) *(15)*, with permission of Lippincott-Raven Publishers.

number of opportunities for detrimental mechanisms to occur. Among the glutamate ionotropic receptor family, this section focuses on the NMDA-gated channel because it opens at the lowest glutamate concentrations, is highly permeable to Ca^{2+}, and the complexity of its modulation is unique *(117)*. In addition to increased glutamate levels at the NMDA/glutamate receptor site, excessive and potentially deleterious cation permeability may result from different types of alterations: abnormal allosteric modulation, i.e., response to ligand binding (Mg^{2+}, Zn^{2+}, glycine, polyamines); chemical modification, namely redox and phosphorylation; and indirect modulation by activation of other glutamate-gated ion channels. The hypothesis that high extracellular glycine may contribute to excitotoxicity has been discussed previously in Section 2.5.

4.1.1. Relief of the Voltage-Dependent Mg^{2+}-Block of NMDA-Gated Channels

A defect of the Mg^{2+} block of NMDA-gated channels *(118)*, leading to persistent Ca^{2+} and Na^+ influx, stands out as a potential excitotoxic mechanism *(119)*. As the maintenance of cellular membrane potential is energy-dependent, this anomaly is certainly relevant to illnesses associated with inadequate energy supply (i.e., ischemia, anoxia, hypoglycemia, and so on) but it also provides a link between excitotoxicity and defective mitochondrial energy metabolism *(120)*. The data outlined below clearly suggest that the relief of the voltage-dependent Mg^{2+}-block is a key contributor to the transition of glutamate from neurotransmitter to neurotoxin.

With cultured cerebellar granule cells, the toxicity of glutamate is markedly increased by omitting glucose from the incubation medium, anoxia, inclusion of inhibitors of oxidative phosphorylation or Na^+/K^+-ATPase, or removal of Mg^{2+} *(121,122)*. This indicates that ATP production and functioning Na^+/K^+-ATPases are essential for generating a resting potential sufficient to maintain the voltage-dependent Mg^{2+} block of the NMDA receptor channel.

The gradient of K^+ across the cellular membrane is the principal determinant of its resting potential, and hyperpolarization by opening ATP-sensitive K+ channels abolished excitotoxicity and cyanide neurotoxicity in cultured neurons *(123,124)*. In accord with these results, the sensitivity of cortical cultures to hypoxia was markedly increased when the K^+ concentration in the serum-free medium was raised from 5.4 to 25 m*M* (i.e., K^+-induced chronic membrane depolarization), and hypoxia-induced cytotoxicity in these conditions was attenuated by NMDA receptor blockade *(125)*.

4.1.2. Polyamine Potentiation of NMDA/Glutamate-Evoked Responses in Ischemia

At low micromolar concentrations, the polyamines, spermine and spermidine, act at specific domains of the NMDA receptor ionophore complex to potentiate the effects of glutamate and glycine *(126)*, and polyamines are released by NMDA receptor activation *(127)*. Therefore, changes in endogenous polyamines during ischemia could contribute to neuronal damage by exacerbating NMDA receptor stimulation *(128)*. This is an interesting hypothesis but the multiple actions of polyamines on ion channels make it very difficult to test *(15)*. Inhibition of polyamine synthesis with difluoromethylornithine (DFMO) reduced the neurotoxic effect of NMDA applied to the striatum *(129)* and the volume of cortical infarct after MCAO in rats *(130)*. In contrast, treatment with polyamines protected hippocampal CA1 and striatal neurons from delayed degeneration after global ischemia in the gerbil *(131)*, and prevented glutamate neurotoxicity in the rat retina *(132)*. In addition, a large body of evidence supports the concept that polyamine responses after neural trauma may be critical components of a protective process in the injured neurons *(131)*. Whether or not polyamines contribute to ischemia-induced excitotoxicity, the polyamine site(s) remains a relevant target to develop NMDA receptor antagonists with a better therapeutic index *(133)*.

4.1.3. Redox Changes

The function of the NMDA receptor-ionophore complex can be modulated by redox changes, presumably by reversible oxidation and reduction of thiol groups located on the extracellular domains of the NMDA receptor complex *(134)*. In a number of preparations, NMDA-evoked responses are potentiated by exposure to disulfide-reducing agents such as dithiothreitol (DTT), whereas thiol oxidizing agents reduce NMDA receptor activation and NMDA neurotoxicity. It is noteworthy that fully oxidized NMDA receptors remain functional (i.e., NMDA-evoked currents are not completely blocked), suggesting that physiological redox modulation may occur to control the overactivity of NMDA receptors induced by a reductive process, without affecting their basal responses *(134)*.

Several endogenous redox compounds can contribute to maintain the redox equilibrium of NMDA receptors in vivo, and ischemia may favor reduction (i.e., excessive cation permeability). Measurements of NADH fluorescence consistently showed that brain tissue redox is in a reduced state during ischemia *(135–137)*. The concentration of L-cysteine, an excitotoxic, thiol amino acid, increases during ischemia both in brain tissue and extracellular fluid *(138,139)*, and L-cysteine is released by depolarization *(140,141)*. The oxidizing agent methoxatin, which inhibited NMDA-evoked responses in cortical neurons by acting on redox sites *(142)*, reduced the infarct volume in a rat model of hypoxic ischemic injury *(143)*. On the other hand, oxidation of NMDA receptor redox sites may provide the molecular basis for the neuroprotective effects of nitric oxide (NO) *(144)*, although NO still demonstrated inhibitory properties on NMDA receptors when their sulfhydryl groups were alkylated by *N*-ethylmaleimide, suggesting the presence of a supplementary, metal ion redox site *(145)*.

The redox site could constitute a relevant target for selectively preventing the deleterious consequence of enhanced NMDA receptor function because oxidation reduced NMDA-evoked responses without preventing the induction and expression of physiological long-term potentiation *(146)*. This suggests that this strategy may be used to block the toxic action of excessive NMDA receptor activation without detrimental effects on cognitive and memory processes mediated by this receptor (*see* Section 4.3.).

4.1.4. Protein Kinase C Activation

Activation of intracellular protein kinase C (PKC) enhances NMDA-evoked responses, apparently by increasing the probability of channel openings and reducing the voltage-dependent Mg^{2+} block of NMDA receptor-channels *(117)*. The modulation of NMDA receptor-channels by PKC provides a positive feedback loop that may potentially be excitotoxic, whereas activated PKC would result in larger and longer glutamate-mediated excitatory responses, causing more Ca^{2+} influx and further PKC activation *(147)*. Whether this occurs in stroke remains unclear (15). It is worth noting that gangliosides, which are potent inhibitors of PKC, protected cultured neurons against glutamate toxicity *(148)* and had some beneficial effects against cerebral ischemia *(149)* (*see*, however, ref. *150*).

Emphasis has been placed on abnormal modulation of the NMDA receptor ionophore complex in this section, but the role of endogenous excitotoxins such as quinolinic acid, L-cysteine, and acidic sulphur-containing amino acids cannot be ruled out (Fig. 5) *(15)*.

4.2. Recurrent Spreading Depression in Focal Ischemia

Spreading depression (SD), propagating transient suppression of electric activity associated with membrane depolarization, is well documented in focal ischemia *(14,16)*, and spontaneous cortical SD was reported in patients with severe head injury *(151)*. Spreading depression is deleterious to the outcome of focal ischemia, probably because it produces marked disruption in ionic homeostasis, acidosis, increased energy demand, and neurotransmitter efflux *(152)* in regions in which residual blood supply can only sustain basal ionic homeostasis *(14,16)*.

Two factors strongly suggest that the beneficial effect of NMDA receptor blockade in focal ischemia may result from SD inhibition: The penumbra is selectively protected by NMDA receptor antagonists, even when the drug is administered after ischemia onset *(14)*; and NMDA receptor activation is a prerequisite for SD, since both elicitation and propagation of SD are exclusively sensitive to competitive and noncompetitive antagonists of NMDA receptor-channel complex *(153,154)*. In the context of this article, it is interesting to mention that, although K^+-induced SD strictly requires NMDA receptor activation, neither its initiation nor its propagation requires high extracellular glutamate *(155,156)*. This clearly illustrates that *the sensitivity of an experimental or pathological phenomenon to glutamate receptor antagonists does not necessarily imply involvement of high extracellular glutamate in its genesis.*

4.3. Persistent Enhancement of Synaptic Efficacy

Delayed neuronal death subsequent to transient ischemia may result from a pathological extension of long-term potentiation (LTP), i.e., a form of synaptic plasticity exhibited by glutamic acid ion channel receptors *(157)*. In the hippocampus, a high-frequency stimulation of excitatory synapses leads to a persistent increase of the potency of AMPA receptors. The induction of this form of LTP (LTP_A), which is that occurring in physiological conditions, depends on the activation of NMDA receptor during the stimulation. Other forms of LTP can be induced, for example when Mg^{2+} concentrations are reduced to increase NMDA currents and/or AMPA receptors blocked *(158,159)*, when NMDA receptors are potentiated (LTP_N). Like LTP_A, LTP_N requires NMDA receptor activation for its induction.

Under normal conditions, LTP is input-specific, i.e., synapses are potentiated only if they are active when the postsynaptic membrane is sufficiently depolarized to relieve the Mg^{2+} block of the NMDA-operated channel *(157)*. However, during ischemia, either anoxic depolarization or recurrent SD, superimposed upon other abnormalities (*see* Section 4.1.), may induce widespread or generalized LTP, which might be hazardous to neurons as some of them receive thousands of excitatory synapses. This hypothetical mechanism that implies persistent changes in glutamatergic synaptic transmission, is compatible with the protective action of glutamate receptor antagonists administered after the insult (*see* Section 2.2.). It is also supported by the following findings: Exposure of rat hippocampal slices to a short anoxic-aglycemic episodes generate a form of LTP that is induced and expressed exclusively by NMDA receptors (i.e., analogous to LTP_N) *(160)*; NMDA receptor-mediated excitatory responses were enhanced in hippocampal slices prepared from gerbils that had been subjected to 5 min of ischemia *(161)*; and Ca^{2+} uptake evoked by electric stimulation was also enhanced postischemically *(162)*.

It is interesting to note that, in contrast to physiological LTP (LTP_A), both anoxia-induced LTP and LTP_N (induced by electrical stimulation under low Mg^{2+} and AMPA receptor block) were mediated by the NMDA receptor redox site since their induction was prevented by reduction of these sites *(146,163)*. This suggests that it may be possible to selectively block anoxia-induced LTP with drugs acting on the NMDA receptor redox sites *(135)* (*see* Section 4.1.3.).

5. CONCLUDING REMARKS

It has been established that excessive flux of Ca^{2+} (and presumably Na^+) through glutamate-operated ion channels leads to intracellular Ca^{2+} loading and neuronal death, but the origins of these malfunctions are unclear. The oversimplified concept that glutamate becomes neurotoxic because of high extracellular levels resulting from excessive presynaptic release or altered uptake (Fig. 1) conflicts with key findings. Furthermore, recent evidence seriously questions the dogma that high extracellular glutamate is excitotoxic in vivo. Extraordinarily high concentrations of glutamate must be applied to the brain to produce significant depolarizations *(77)*; marked increases of endogenous extracellular glutamate by drug-induced inhibition of glutamate uptake, to levels above those observed when anoxic depolarization occurs, only produces minor depolarizations *(164)*, and these are not due to activation of glutamate ionotropic receptors *(165)*; and accumulation of glutamate in the extracellular space subsequent to uptake inhibition was not sufficient for inducing neuronal damage *(166)*.

It has become important and timely to broaden the notion of excitotoxicity by including alternative hypotheses such as alterations of postsynaptic glutamate ionotropic receptors and metabotropic receptors, and derangements of glutamatergic systems leading to synchronized receptor activation (e.g., spreading depression) or persistent increase in synaptic efficiency (anoxia-induced LTP) (Fig. 5). This extended notion of excitotoxicity presumably applies to other neurological disorders, especially epilepsy *(167)*.

Finally, this analysis makes clear that a number of conclusions based on data obtained using cultured neurons do not apply to in vivo situations. Questions related to the pathophysiology and pharmacology of stroke must be addressed in the living, functional brain.

ACKNOWLEDGMENTS

The authors' work on cerebral ischemia and neuroprotection was supported by The Wellcome Trust, The Medical Council, The Brain Research Trust, The Mihara Trust (Tokyo), Pfizer Central Research (UK), and Johnson & Johnson (UK). We are grateful to Douglas A. Richards (Department of Pharmacology, University of Birmingham, Birmingham) and Jutta Urenjak (Discovery Biology, Pfizer Central Research, Sandwich, UK) for their constructive comments.

REFERENCES

1 Nicholls, D. G. and Attwell, D., The release and uptake of excitatory amino acids, *Trends Neurosci.,* **11** (1990) 462–468.
2 Mori, H. and Mishina, M., Structure and function of the NMDA receptor channel, *Neuropharmacology,* **34** (1995) 1219–1237.
3 Bettler, B. and Mulle, C., AMPA and kainate receptors, *Neuropharmacology,* **34** (1995) 123–139.
4 Pin, J.-P. and Duvoisin, R., The metabotropic glutamate receptors: structure and functions, *Neuropharmacology,* **34** (1995) 1–26.
5 Eliasof, S. and Werblin, F., Characterization of the glutamate transporter in retinal cones of the tiger salamander, *J. Neurosci.,* **13** (1993) 402–411.

6 Lipton, S. A. and Rosenberg, P. A., Mechanisms of disease: excitatory amino acids as a final common pathway for neurologic disorders, *N. Eng. J. Med.,* **330** (1994) 613–622.

7 Muir, K. W. and Lees, K. R., Clinical experience with excitatory amino acid antagonist drugs, *Stroke,* **26** (1995) 503–513.

8 Garthwaite, G., Hajós, F., and Garthwaite, J., Ionic requirements for neurotoxic effects of excitatory amino acid analogues in rat cerebellar slices, *Neuroscience,* **18** (1986) 437–447.

9 Choi, D. W., Calcium-mediated neurotoxicity: relationship to specific channel types and role in ischaemic damage, *Trends Neurosci.,* **11** (1988) 465–469.

10 Randall, R. D. and Thayer, S. A., Glutamate-induced calcium transient triggers delayed calcium overload and neurotoxicity in rat hippocampal neurons, *J. Neurosci.,* **12** (1992) 1882–1895.

11 Meldrum, B. S., Protection against ischaemic neuronal damage by drugs acting on excitatory neurotransmission, *Cereb. Brain Metabol. Rev.,* **2** (1990) 27–57.

12 Choi, D. W. and Rothman, S. M., The role of glutamate neurotoxicity in hypoxic-ischemic neuronal death, *Ann. Rev. Neurosci.,* **13** (1990) 171–177.

13 Szatkowski, M. and Attwell, D., Triggering and execution of neuronal death in brain ischaemia: two phases of glutamate release by different mechanisms, *Trends Neurosci.,* **17** (1994) 359–365.

14 Hossmann, K.-A., Glutamate-mediated injury in focal cerebral ischemia: the excitotoxin hypothesis revised, *Brain Pathol.,* **4** (1994) 23–36.

15 Obrenovitch, T. P. and Richards, D. A., Extracellular neurotransmitter changes in cerebral ischaemia, *Cerebrovasc. Brain Metab. Rev.,* **7** (1995) 1–54.

16 Obrenovitch, T. P., The ischaemic penumbra: twenty years on, *Cerebrovasc. Brain Metab. Rev.,* **7** (1995) 297–323.

17 Obrenovitch, T. P., Origins of glutamate release in ischaemia, *Acta Neurochir. Suppl. (Wien),* **66** (1995) 50-55.

18 Benveniste, H. and Hüttemeier, P. C., Microdialysis—theory and application, *Prog. Neurobiol.,* **35** (1990) 195–215.

19 Benveniste, H., Drejer, J., Schousboe, A., and Diemer, N. H., Elevation of the extracellular concentrations of glutamate and aspartate in rat hippocampus during transient cerebral ischemia monitored by intracerebral microdialysis, *J. Neurochem.,* **43** (1984) 1369–1374.

20 Bullock, R., Zauner, A., Woodward, J., and Young, H. F., Massive persistent release of excitatory amino acids following human occlusive stroke, *Stroke,* **26** (1995) 2187–2189.

21 Kanthan, R., Shuaib, A., Griebel, R., and Miyashita, H., Intracerebral human microdialysis: in vivo study of an acute focal ischemic model of the human brain, *Stroke,* **26** (1995) 870-873.

22 Shimada, N., Graf, R., Rosner, G., and Heiss, W.-D., Differences in ischemia-induced accumulation of amino acids in the cat cortex, *Stroke,* **21** (1990) 1445–1451.

23 Matsumoto, K., Graf, R., Rosner, G., Taguchi, J., and Heiss, W.-D., Elevation of neuroactive substances in the cortex of cats during prolonged focal ischemia, *J. Cereb. Blood Flow Metab.,* **13** (1993) 586–594.

24 Obrenovitch, T. P., Urenjak, J., Richards, D. A., Ueda, Y., Curzon, G., and Symon, L., Extracellular neuroactive amino acids in the rat brain striatum during moderate and severe transient ischemia, *J. Neurochem.,* **61** (1993) 178–186.

25 Matsumoto, K., Ueda, S., Hashimoto, T., and Kuriyama, K., Ischemic neuronal injury in the rat hippocampus following transient forebrain ischemia: evaluation using in vivo microdialysis, *Brain Res.,* **543** (1991) 236–242.

26 Johansen, F. F., Christensen, T., Jensen, M. S., Valente, E., Jensen, C. V., Nathan, T., Lambert, J. D. C., and Diemer, N. H., Inhibition of postischemic rat hippocampus: GABA receptors, GABA release, and inhibitory postsynaptic potentials, *Exp. Brain Res.,* **84** (1991) 529–537.

27 Rudolphi, K. A., Schubert, P., Parkinson, F. E., and Fredholm, B. B., Neuroprotective role of adenosine in cerebral ischemia, *Trends Pharmacol. Sci.,* **13** (1992) 439–445.

28 Arvin, B., Neville, L. F., and Roberts, P. J., 2–chloroadenosine prevents kainic acid-induced lesions of rat striatum, *Br. J. Pharmacol.,* **92(Suppl)** (1987) 540P.

29 Pan, J., Neville, L. F., Arvin, B., and Roberts, P. J., Adenosine analogs suppress K$^+$- and kainate-induced synaptosomal release of endogenous glutamate, *Br. J. Pharmacol.,* **95(Suppl)** (1988) 889P.

30 Connick, J. H. and Stone, T. W., Quinolinic acid neurotoxicity: protection by intracerebral phenyl-isopropyladenosine (PIA) and potentiation by hypotension, *Neurosci. Lett.,* **101** (1989) 191–196.

31 Matsumoto, K., Graf, R., Rosner, G., Shimada, N., and Heiss W.-D., Flow thresholds for extracellular purine catabolite elevation in cat focal ischemia, *Brain Res.,* **579** (1992) 309–314.

32 Obrenovitch, T. P., Richards, D. A., Sarna, G. S., and Symon, L., Combined intracerebral microdialysis and electrophysiological recording: methodology and applications, *J. Neurosci. Meth.,* **47** (1993) 139–145.

33 Ueda, Y., Obrenovitch, T. P., Lok, S.-Y., Sarna, G. S., and Symon, L., Efflux of glutamate produced by short ischemia of varied severity in the rat striatum, *Stroke,* **23** (1992) 253–259.

34 Pulsinelli, W. A., Brierley, J. B., and Plum, F., Temporal profile of neuronal damage in a model of transient forebrain ischemia, *Ann. Neurol.,* **11** (1982) 491–499.

35 Ueda, Y., Obrenovitch, T. P., Lok, S.-Y., Sarna, G. S., and Symon, L., Changes in extracellular glutamate concentration produced in the rat striatum by repeated ischemia, *Stroke,* **23** (1992) 1125–1131.

36 Nakata, N., Kato, H., and Kogure, K., Effects of repeated cerebral ischemia on extracellular amino acid concentrations measured with intracerebral microdialysis in the gerbil hippocampus, *Stroke,* **24** (1993) 458–464.

37 Silverstein, F. S., Buchanan, K., and Johnston, M. V., Hypoxia-ischemia causes severe but reversible depression of striatal synaptosomal ^{3}H-glutamate uptake, *Ann. Neurol.,* **18** (1985) 122.

38 Johansen, F. F., Jørgensen, M. B., von Lubitz, D. K. J. E., and Diemer, N. H., Selective dendrite damage in hippocampal CA1 stratum radiatum with unchanged axon ultrastructure and glutamate uptake after transient cerebral ischaemia in the rat, *Brain Res.,* **291** (1984) 373–377.

39 Johansen, F. F., Jørgensen, M. B., and Diemer, N. H., Ischaemia induced delayed neuronal death in the CA11 hippocampus is dependent on intact glutamatergic innvervation. In T. P. Hicks, D. Lodge, and H. McLennan (eds.) *Excitatory Amino Acid Transmission Neurology and Neurobiology,* Liss, New York, 1987, pp. 245–248.

40 Gill, R., Foster, A. C., and Woodruff, G. N., MK-801 is neuroprotective in gerbils when administered during the post-ischaemic period, *Neuroscience,* **25** (1988) 847–855.

41 Nellgard, B. and Wieloch, T., Postischemic blockade of AMPA but not NMDA receptors mitigates neuronal damage in the rat brain following transient severe cerebral ischemia, *J. Cereb. Blood Flow Metab.,* **12** (1992) 2–11.

42 Sheardown, M. J., Nielsen, E. O., Hansen, A. J., Jacobsen, P., and Honoré, T., 2,3-dihydro-6-nitro-7-sulfamyl-benzo(F)quinoxaline: a neuroprotectant for cerebral ischemia, *Science,* **247** (1990) 571–574.

43 Swan, J. H. and Meldrum, B. S., Protection by NMDA antagonists against selective cell loss following transient ischaemia, *J. Cereb. Blood Flow Metab.* **10** (1990) 343–351.

44 Lekieffre, D. and Meldrum, B. S., The pyrimidine-derivative, BW1003C87, protects CA1 and striatal neurons following transient severe forebrain ischaemia in rats. A microdialysis and histological study, *Neuroscience,* **56** (1993) 93–99.

45 Pratt, J., Rataud, J., Bardot, F., Roux, M., Blanchard, J.-C., Laduron, P. M., and Stutzmann, J.-M., Neuroprotective actions of riluzole in rodent models of global and focal cerebral ischaemia, *Neurosci. Lett.,* **140** (1992) 225–230.

46 Astrup, J., Siesjö, B. K., and Symon, L., Thresholds in cerebral ischemia—the ischemic penumbra, *Stroke,* **12** (1981) 723–725.

47 McCulloch, J., Ozyurt, E., Park, C. K., Nehls, D. G., Teasdale, G. M., and Graham, D. I., Glutamate receptor antagonists in experimental focal cerebral ischaemia, *Acta Neurochir. Suppl. (Wien),* **57** (1993) 73–79.

48 Pulsinelli, W. A. and Brierley, J. B., A new model of bilateral hemispheric ischemia in the unanesthetized rat, *Stroke,* **10** (1979) 267–272.

49 Busto, R. and Ginsberg, M. D., Graded focal ischemia in the rat by unilateral carotid artery occlusion and elevated intracranial pressure: hemodynamic and biochemical characterization, *Stroke,* **16** (1985) 466–476.

50 Obrenovitch, T. P., Sarna, G. S., Millan, M. H., Lok, S.-Y., Kawauchi, M., Ueda, Y., and Symon, L., Intracerebral dialysis with on-line enzyme fluorometric detection: a novel method to investigate the changes in the extracellular concentration of glutamic acid. In J. Krieglstein and H. Oberpichler (eds.) *Pharmacology of Cerebral Ischemia 1990,* Wissenschattliche Verlagsgesellschaft, Stuttgart, 1990, pp. 23–31.

51 Garthwaite, G., Williams, G. D., and Garthwaite, J., Glutamate toxicity: an experimental and theoretical analysis, *Eur. J. Neurosci.,* **4** (1992) 353–360.

52 Obrenovitch, T. P., Zilkha, E., and Urenjak, J., Intracerebral microdialysis: electrophysiological evidence of a critical pitfall, *J. Neurochem.,* **64** (1995) 1884–1887.

53 Obrenovitch, T. P. and Zilkha, E., Changes in extracellular glutamate concentration associated with propagating cortical spreading depression. In J. Olesen and M. A. Moskowitz (eds.) *Experimental Headache Models in Animal and Man,* Raven, New York, 1995, pp. 113–117.

54 Back, T., Zhao, W., and Ginsberg, M. D., Three-dimensional image-analysis of brain glucose metabolism/blood flow uncoupling and its electrophysiological correlates in the acute ischemic penumbra following middle cerebral artery occlusion, *J. Cereb. Blood Flow Metab.,* **15** (1995) 566–577.

55 Branston, N. M., Strong, A. J., and Symon, L., Kinetics of resolution of transient increases in extracellular potassium activity: relationships to regional blood flow in primate cerebral cortex, *Neurol. Res.,* **4** (1982) 1–19.

56 Globus, M. Y.-T., Busto, R., Martinez, E., Valdes, I., Dietrich, W. D., and Ginsberg, M. D., Comparative effect of transient global ischemia on extracellular levels of glutamate, glycine, and nonvulnerable brain regions in the rat, *J. Neurochem.,* **57** (1991) 470-478.

57 Mitani, A., Andou, Y., and Kataoka, K., Selective vulnerability of hippocampal CA1 neurons cannot be explained in terms of an increase in glutamate concentrations during ischemia in the gerbil: brain microdialysis study, *Neuroscience,* **48** (1992) 307–313.

58 Butcher, S. P., Bullock, R., Graham, D. I., and McCulloch, J., Correlation between amino acid release and neuropathologic outcome in rat brain following middle cerebral artery occlusion, *Stroke,* **21** (1990) 1727–1733.

59 Takagi, K., Ginsberg, M. D., Globus, M. Y.-T., Dietrich, W. D., Martinez, E., Kraydieh, S., and Busto, R., Changes in amino acid neurotransmitters and cerebral blood flow in the ischemic penumbral region following middle cerebral artery occlusion in the rat: correlation with histopathology, *J. Cereb. Blood Flow Metab.,* **13** (1993) 575–585.

60 Urenjak, J., Obrenovitch, T. P., Richards, D. A., Williams, S. R., and Symon, L., Is *N*-acetyl-aspartate involved in neurotransmission? A microdialysis study. In H. Rollema, B. Westerink and W. J. Drijfhout (eds.) *Monitoring Molecules in Neuroscience,* University Center for Pharmacy, Groningen, 1991, pp. 355–357.

61 Wafford, K. A., Kathoria, M., Bain, C. J., Marshall, G., Le Boudellès, B., Kemp, J. A., and Whiting, P. J., Identification of amino acids in the *N*-methyl-D-aspartate receptor NR1 subunit that contribute to the glycine binding site, *Mol. Pharmacol.,* **47** (1995) 374–380.

62 Thomson, A. M., Glycine modulation of the NMDA receptor/channel complex, *Trends Neurosci.,* **12** (1989) 349–353.

63 Wood, P. L., The co-agonist concept: is the NMDA-associated glycine receptor saturated *in vivo*? *Life Sci.,* **57** (1995) 301–310.

64 Johnson, J. W. and Ascher, P., Glycine potentiates the NMDA response in cultured mouse brain neurons, *Nature,* **325** (1987) 529–531.

65 Pace, J. R., Martin, B. M., Paul, S. M., and Rogawski, M. A., High concentrations of neutral amino acids activate NMDA receptor currents in rat hippocampal neurons, *Neurosci. Lett.,* **141** (1992) 97–100.

66 Globus, M. Y.-T., Ginsberg, M. D., and Busto, R., Excitotoxic index—a biochemical marker of selective vulnerability, *Neurosci. Lett.,* **127** (1991) 39–42.

67 Baker, A. J., Zornow, A. J., Scheller, M. S., Yaksh, T. L., Skilling, S. R., Smullin, D. H., Laarson, A. A., and Kuczenski, R., Changes in extracellular concentrations of glutamate, aspartate, glycine, dopamine, serotonin, and dopamine metabolites after transient global ischaemia in the rabbit brain, *J. Neurochem.,* **57** (1991) 1370-1379.

68 Cantor, S. L., Zornow, M. H., Miller, L. P., and Yaksh, T. L., The effect of cyclohexyl-adenosine on the periischemic increases of hippocampal glutamate and glycine in the rabbit, *J. Neurochem.,* **59** (1992) 1884–1892.

69 Lin, B., Globus, M. Y.-T., Dietrich, W. D., Busto, R., Martinez, E., and Ginsberg, M. D., Differing neurochemical and morphological sequelae of global ischemia: comparison of single- and multiple-insult paradigms, *J. Neurochem.,* **59** (1992) 2213–2223.

70 Wang, L.-Y. and MacDonald, J. F., Modulation by magnesium of the affinity of NMDA receptors for glycine in murine hippocampal neurones, *J. Physiol. (Lond.),* **486** (1995) 83–95.

71 Hardy, A. M., Urenjak, J., and Obrenovitch, T. P., High extracellular glycine does not potentiate depolarizations evoked by application of *N*-methyl-D-aspartate to the rat striatum *in vivo*, *Br. J. Pharmacol.,* **116(Suppl)** (1995) 112P.

72 Smith, K. E., Borden, L. A., Hargig, P. R., Branchek, T., and Weinsbank, R. L., Cloning and expression of a glycine transporter reveal colocalization with NMDA receptors, *Neuron,* **8** (1992) 927–935.

73 Swartz, K. J., During, M. J., Freese, A., and Beal, M. F., Cerebral synthesis and release of kynurenic acid: an endogenous antagonist of excitatory amino acid receptors, *J. Neurosci.,* **10** (1990) 2965–2973.

74 Bormann, J., Hamill, O. P., and Sakmann, B., Mechanisms of anion permeation through channels gated by glycine and gamma-aminobutyric acid in mouse cultured spinal neurones, *J. Physiol. (Lond.),* **385** (1987) 243–286.

75 Shirasaki, T., Klee, M. R., Nakaye, T., and Akaike, N., Differential blockade of bicuculline and strychnine on GABA- and glycine-induced responses in dissociated rat hippocampal pyramidal cells, *Brain Res.,* **561** (1991) 77–83.

76 Larson, A. A. and Beitz, A. J., Glycine potentiates strychnine-induced convulsions: role of NMDA receptors, *J. Neurosci.,* **8** (1988) 3822–3826.

77 Obrenovitch, T. P., Urenjak, J., and Zilkha, E., Intracerebral microdialysis combined with recording of extracellular field potential: a novel method for investigation of depolarizing drugs *in vivo*, *Br. J. Pharmacol.,* **113** (1994) 1295–1302.

78 Nilsson, G. E. and Lutz, P. L., Release of inhibitory neurotransmitters in response to anoxia in turtle brain, *Am. J. Physiol.,* **261** (1991) R32–R37.

79 Graham, S. H., Chen, J., Sharp, F. R., and Simon, R. P., Limiting ischemic injury by inhibition of excitatory amino acid release, *J. Cereb. Blood Flow Metab.,* **13** (1993) 88–97.

80 Kemp, J. A. and Leeson, P. D., The glycine site of the NMDA receptor—five years on, *Trends Pharmacol. Sci.,* **24** (1993) 20-25.

81 Kanner, B. I. and Bendahan, A., Binding order of substrates to the sodium and potassium ion coupled L-glutamic acid transporter from rat brain, *Biochemistry,* **21** (1982) 6327–6330.

82 Szatkowski, M., Barbour, B., and Attwell, D., Non-vesicular release of glutamate from glial cells by reversed electrogenic glutamate uptake, *Nature,* **348** (1990) 443–446.

83 Bouvier, M., Szatkowski, M., Amato, A., and Attwell, D., The glial cell glutamate uptake carrier countertransports pH-changing anions, *Nature,* **360** (1992) 471–474.

84 Gemba, T., Oshima, T., and Ninomiya, M., Glutamate efflux via the reversal of the sodium-dependent glutamate transporter caused by glycolytic inhibition in rat cultured astrocytes, *Neuroscience,* **63** (1994) 789–795.

85 Swanson, R. A., Farrell, K., and Simon, R. P., Acidosis causes failure of astrocyte glutamate uptake during hypoxia, *J. Cereb. Blood Flow Metab.* **15** (1995) 417–424.

86 Erecinska, M. and Silver, I. A., Ions and energy in mammalian brain, *Prog. Neurobiol.,* **43** (1994) 37–71.

87 Hansen, A. J., Effects of anoxia on ion distribution in the brain, *Physiol. Rev.,* **65** (1985) 101–148.

88 Bradford, H. F., Young, A. M. J., and Crowder, J. M., Continuous leakage from brain cells is balanced by compensatory high-affinity reuptake transport, *Neurosci. Lett.,* **81** (1987) 296–302.

89 McMahon, H. T., Foran, P., Dolly, J. O., Verhage, M., Wiegant, V. M., and Nicholls, D. G., Tetanus toxin and botulinum toxins A and B inhibit glutamate, gamna-aminobutyric acid, aspartate, and metenkephalin release from synaptosomes. Clues to the locus of action, *J. Biol. Chem.,* **267** (1992) 21,338–21,343.

90 Söllner, T. and Rothman, J. E., Neurotransmission: harnessing fusion machinery at the synapse, *Trends Neurosci.,* **17** (1994) 344–349.

91 Lowry, O. H., Passonneau, J. V., Hasselberger, F. X., and Schultz, D., Effects of ischemia on known substrates and cofactors of the glycolytic pathway in brain, *J. Biol. Chem.,* **239** (1964) 18–30.

92 Obrenovitch, T. P., Garofalo, O., Harris, R. J., Bordi, L., Ono, M., Momma, F., Bachelard, H. S., and Symon, L., Brain tissue concentration of ATP, phosphocreatine, lactate, and tissue pH in relation to reduced cerebral blood flow following experimental acute middle cerebral artery occlusion, *J. Cereb. Blood Flow Metab.,* **8** (1988) 866–874.

93 Kempski, O., Zimmer, M., Neu, A., von Rosen, F., Jansen, M., and Baethmann, A., Control of glial cell volume in anoxia: in vitro studies on ischemic cell swelling, *Stroke,* **18** (1987) 623–628.

94 Korf, J. and Postema, F., Rapid shrinkage of rat striatal extracellular space after local kainate application and ischemia as recorded by impedance, *J. Neurosci. Res.,* **19** (1988) 504–510.

95 Kimelberg, H. K., Goderie, S. K., Higman, S., Pang, S., and Waniewski, R. A., Swelling-induced release of glutamate, aspartate, and taurine from astrocyte cultures, *J. Neurosci.,* **10** (1990) 1583–1591.

96 Kimelberg, H. K., Rutledge, E., Goderie, S., and Charniga, C., Astrocytic swelling due to hypotonic or high K^+ medium causes inhibition of glutamate and aspartate uptake and increases their release, *J. Cereb. Blood Flow Metab.,* **15** (1995) 409–416.

97 Kimelberg, H. K., Rose, J. W., Barron, K. D., Waniewski, R. A., and Cragoe, E. J., Astrocytic swelling in traumatic-hypoxic brain injury. Beneficial effects of an inhibitor of anion exchange transport and glutamate uptake in glial cells, *Mol. Chem. Neuropathol.,* **1** (1989) 1–31.

98 Wahl, F., Obrenovitch, T. P., Hardy, A. M., Plotkine, M., Boulu, R., and Symon, L., Extracellular glutamate during focal cerebral ischaemia in rats: time course and calcium-dependency, *J. Neurochem.,* **63** (1994) 1003–1111.

99 Zilkha, E., Obrenovitch, T. P., Koshy, A., Kusakabe, H., and Bennetto, H. P., Extracellular glutamate: on-line monitoring using microdialysis coupled to enzyme-amperometric analysis, *J. Neurosci. Meth.,* **60** (1995) 1–9.

100 Fabricius, M., Jensen, L. H., and Lauritzen, M., Microdialysis of interstitial amino acids during spreading depression and anoxic depolarization in rat neocortex, *Brain Res.,* **612** (1993) 61–69.

101 Obrenovitch, T. P., Scheller, D., Matsumoto, T., Tegtmeier, F., Höller, M., and Symon, L., Rapid redistribution of hydrogen ions is associated with depolarization and repolarization subsequent to cerebral ischaemia-reperfusion, *J. Neurophysiol.*, **64** (1990) 1125–1133.

102 Shimizu, H., Graham, S. H., Chang, L.-H., Mintorovich, J., James, T. L., Faden, A. I., and Weinstein, P. R., Relationship between extracellular neurotransmitter amino acids and energy metabolism during cerebral ischemia in rats monitored by microdialysis and in vivo magnetic resonance spectroscopy, *Brain Res.*, **605** (1993) 33–42.

103 Sanchez-Prieto, J. and Gonzalez, P., Occurrence of a large Ca^{2+}-independent release of glutamate during anoxia in isolated nerve terminals, *J. Neurochem.*, **50** (1988) 1322–1324.

104 Ogata, T., Nakamura, Y., Shibata, T., and Kataoka, K., Release of excitatory amino acids from cultured hippocampal astrocytes induced by a hypoxic-hypoglycemic stimulation, *J. Neurochem.*, **58** (1992) 1957–1959.

105 Griffiths, R., Dunlop, J., Gorman, A., Senior, J., and Grieve, A., L-*trans*-pyrrolidine-2,4-dicarboxylate and *cis*-1-aminocyclobutane behave as transportable, competitive inhibitors of the high-affinity glutamate transporters, *Biochem. Pharmacol.*, **47** (1994) 267–274.

106 Madle, J. E. and Burgesser, K., Adenosine triphosphate depletion reverses sodium-dependent, neuronal uptake of glutamate in rat hippocampal slices, *J. Neurosci.*, **13** (1993) 4429–4444.

107 Urenjak, J. and Obrenovitch, T. P., Pharmacological modulation of voltage-gated Na^+ channels: a rational and effective strategy against ischemic brain damage, *Pharmacol. Rev.*, **48** (1996) 21–67.

108 Taylor, C. P., Burke, S. P., and Weber, M. L., Hippocampal slices: glutamate overflow and cellular damage from ischemia are reduced by sodium-channel blockade, *J. Neurosci. Meth.*, **59** (1995) 121–128.

109 Graham, S. H., Chen, J., Lan, J., Leach, M. J., and Simon, R. P., Neuroprotective effects of a use-dependent blocker of voltage dependent sodium channels, BW619C89, in rat middle cerebral artery occlusion, *J. Pharmacol. Exp. Ther.*, **269** (1994) 854–859.

110 Bardot, F., Roux, M., Blanchard, J. C., Laduron, P. M., and Stutzmann, J. M., Neuroprotective actions of riluzole in rodent models of global and focal cerebral ischaemia, *Neurosci. Lett.*, **140** (1992) 225–230.

111 Wahl, F., Allix, M., Plotkine, M., and Boulu, R. G., Effect of riluzole of focal cerebral ischemia in rats, *Eur. J. Pharmacol.*, **230** (1993) 209–214.

112 Benavides, J., Camelin, J. C., Mitrani, N., Flamand, F., Uzan, A., Legrand, J. J., Gueremy, C., and Le Fur, G., 2-Amino 6-trifluoromethoxy benzothiazole, a possible antagonist of the excitatory amino acid neurotransmission. II. Biochemical properties, *Neuropharmacology*, **24** (1985) 1085–1092.

113 Cheramy, A., Barbeito, L., Godeheu, P. and Glowinski, J., Riluzole inhibits the release of glutamate in the caudate nucleus of the cat in vivo, *Neurosci. Lett.*, **147** (1992) 209–212.

114 Waldmeier, P. C., Wicki, P., and Feldtrauer, J.-J., Release of endogenous glutamate from rat cortical slices in presence of the glutamate uptake inhibitor L-trans-pyrrolidine-2,4-dicarboxylic acid, *Naunyn Schmiedeberg's Arch. Pharmacol.*, **348** (1993) 478–485.

115 Tamura A., Animal models used in cerebral ischemia and stroke research. In G. J. Ter Horst and J. Korf (eds.) *Clinical Pharmacology of Cerebral Ischemia*, Humana, Totowa, NJ (1996), this volume.

116 Fujisawa, H., Dawson, D., Browne, S. E., MacKay, K. B., Bullock, R., and McCulloch, J., Pharmacological modification of glutamate neurotoxicity *in vivo*, *Brain Res.*, **629** (1993) 73–78.

117 McBain, C. J. and Mayer, M. L., *N*-methyl-D-aspartic acid receptor structure and function, *Physiol. Rev.*, **74** (1994) 723–760.

118 Ruppersberg, J. P., Kitzing, E., and Schoepfer, R., The mechanism of magnesium block of NMDA receptors, *Seminars Neurosci.*, **6** (1994) 87–96.

119 Turski, L. and Turski, W. A., Towards an understanding of the role of glutamate in neurodegenerative disorders: energy metabolism and neuropathology, *Experientia,* **49** (1993) 1064–1072.

120 Beal, M. F., Hyman, B. T., and Koroshetz, W., Do defects in mitochondrial energy metabolism underlie the pathology of neurodegenerative diseases? *Trends Neurosci.,* **16** (1993) 125–131.

121 Henneberry, R. C., Novelli, A., Cox, J. A., and Lysko, P. G., Neurotoxicity of *N*-methyl-D-aspartate receptor in energy-compromised neurons. An hypothesis for cell death in aging and disease, *Ann. NY Acad. Sci.,* **568** (1989) 225–233.

122 Brines, M. L. and Robbins, R. J. Inhibition of $\alpha 2/\alpha 3$ sodium pump isoforms potentiates glutamate neurotoxicity, *Brain Res.,* **591** (1992) 94–102.

123 Abele, A. E. and Miller, R. J., Potassium channel activators abolish excitotoxicity in cultured hippocampal pyramidal neurons, *Neurosci. Lett.,* **115** (1990) 195–200.

124 Patel, M. N., Yim, G. K. W., and Isom, G. E., Potentiation of cyanide neurotoxicity by blockade of ATP-sensitive potassium channels, *Brain Res.,* **593** (1992) 114–116.

125 Uchiyama-Tsuyuki, Y., Araki, H., and Otomo, S., Depolarization and hypoxia-induced cell damage in serum-free cultures of the rat cortex, and related extra cellular glutamate changes, *Eur. J. Pharmacol.,* **293** (1995) 245–250.

126 Williams, K., Romano, C., Dichter, M. A., and Molinoff, P. B., Modulation of the NMDA receptor by polyamines, *Life Sci.,* **48** (1991) 469–498.

127 Fage, D., Voltz, C., Scatton, B., and Carter, C., Selective release of spermine and spermidine from the rat striatum by *N*-methyl-D-aspartate receptor activation in vivo, *J. Neurochem.,* **58** (1992) 2170-2175.

128 Carter, C. J., Lloyd, K. G., Zivkovic, B., and Scatton, B., Ifenprodil and SL 820715 as cerebral anti-ischaemic agents. III. Evidence for antagonist effects at the polyamine site within the NMDA receptor complex, *J. Pharmacol. Exp. Ther.,* **253** (1990) 475–481.

129 Porcella, A., Fage, D., Voltz, C., Bourdiol, F., Benavides, J., Scatton, B., and Carter, C., Implication of the polyamines in the neurotoxic effects of *N*-methyl-D-aspartate, *Neurol. Res.,* **14** (1992) 181–183.

130 Muszynski, C. A., Robertson, C. S., Goodman, J. C., and Henley, C. M., DFMO reduces cortical infarct volume after middle cerebral artery occlusion in the rat, *J. Cereb. Blood Flow Metab.,* **13** (1993) 1033–1037.

131 Gilad, G. M. and Gilad, V. H., Polyamines can protect against ischemia-induced nerve cell death in gerbil forebrain, *Exp. Neurol.,* **111** (1991) 349–355.

132 Gilad, G. M. and Gilad, V. H., Treatment with polyamines can prevent monosodium glutamate neurotoxicity in the rat retina, *Life Sci.,* **44** (1989) 1963–1969.

133 Hargreaves, R. J., Hill, R. G., and Iversen, L. L., Neuroprotective NMDA antagonists: the controversy over their potential for adverse effects on cortical neuronal morphology, *Acta Neurochir. Suppl. (Wien),* **60** (1994) 15–19.

134 Gozlan, H. and Ben-Ari, Y., NMDA receptor redox sites: are they targets for selective neuronal protection? *Trends Pharmacol. Sci.,* **16** (1995) 368–374.

135 Ginsberg, M. D., Reivich, M., Frinak, S., and Harbig, K., Pyridine nucleotide redox state and blood flow of the cerebral cortex following middle cerebral artery occlusion in the cat, *Stroke,* **7** (1976) 125–131.

136 Paschen, W., Djuricic, B. M., Bosma, H.-J., and Hossmann, K.-A., Biochemical changes during graded ischemia in gerbils: regional evaluation of cerebral blood flow and brain metabolites, J. *Neurol. Sci.,* **58** (1983) 37–44.

137 Tanaka, K., Dora, E., Greenberg, J. H., and Reivich, M., Cerebral glucose metabolism during the recovery period after ischemia—its relationship to NADH-fluorescence, blood flow, ECoG and histology, *Stroke,* **17** (1986) 994–1004.

138 Slivka, A. and Cohen, G., Brain ischemia markedly elevates levels of the neurotoxic amino acid, cysteine, *Brain Res.,* **608** (1993) 33–37.

139 Andiné, P., Orwar, O., Jacobson, I., Sandberg, M., and Hagberg, H., Extracellular acidic sulfur-containing amino acids and gamma-glutamyl peptides in global ischemia: postischemic recovery of neuronal activity is paralleled by a tetrodotoxin-sensitive increase in cysteine sulfinate in the CA1 of the rat hippocampus, *J. Neurochem.,* **57** (1991) 230-236.

140 Keller, H. J., Do, K. Q., Zollinger, M., Winterhalter, K. H., and Cuenod, M., Cysteine: depolarization induced release from rat brain in vitro, *J. Neurochem.,* **52** (1989) 1801–1806.

141 Zängerle, L., Cuénod, M., Winterhalter, K. H., and Do, K. Q., Screening of thiol compounds: depolarization-induced release of glutathione and cysteine from rat brain slices, *J. Neurochem.,* **59** (1992) 181–189.

142 Aizenman, E., Jensen, F. E., Gallop, P. M., Rosenberg, P. A., and Tang, L.-H., Further evidence that pyrroloquinoline interacts with the *N*-methyl-D-aspartate receptor redox site in rat cortical neurons in vitro, *Neurosci. Lett.,* **168** (1994) 189–192.

143 Jensen, F. E., Gardner, G. J., Williams, A. P., Gallop, P. M., Aizenman, E., and Rosenberg, P. A., The putative essential nutrient pyrroloquinoline quinone is neuroprotective in a rodent model of hypoxic/ischemic brain injury, *Neuroscience,* **62** (1994) 399–406.

144 Lei, S. Z., Pan, Z. -H., Aggarwal, S. K., Chen, H.-S. V., Hartman, J., Sucher, N. J., and Lipton, S. A., Effect of nitric oxide production on the redox modulatory site of the NMDA receptor-channel complex, *Neuron,* **8** (1992) 1087–1099.

145 Fagni, L., Olivier, M., Lafon-Cazal, M., and Bockaert, J., Involvement of divalent ions in the nitric oxide-induced blockade of *N*-methyl-D-aspartate (NMDA) receptors in cerebellar granule cells, *Mol. Pharmacol.,* **47** (1995) 1239–1247.

146 Gozlan, H., Khazipov, R., Diabira, D., and Ben-Ari, Y., In CA1 hippocampal neurons, the redox state of NMDA receptors determines LTP expressed by NMDA but not by AMPA receptors, *J. Neurophysiol.,* **73** (1995) 2612–2717.

147 Chen, L. and Huang, L. Y., Protein kinase C reduces Mg^{2+} block of NMDA-receptor channels as a mechanism of modulation, *Nature,* **356** (1992) 521–523.

148 Favaron, M., Manev, H., Alho, H., Bertolino, M., Ferret, B., Guidotti, A., and Costa, E., Gangliosides prevent glutamate and kainate neurotoxicity in primary neuronal cultures of neonatal rat cerebellum and cortex, *Proc. Natl. Acad. Sci. USA,* **85** (1988) 7351–7355.

149 Carolei, A., Fieschi, C., Bruno, R., and Toffano, G., Monosialoganglioside GM1 in cerebral ischemia, *Cerebrovasc. Brain Metab. Rev.,* **3** (1991) 134–157.

150 Mayer, S. A. and Pulsinelli, W. A., Failure of GM1 ganglioside to influence outcome in experimental focal ischemia, *Stroke,* **23** (1992) 242–246.

151 Mayevsky, A., Doron, A., Manor, T., Meilin, S., Salame, K., and Ouaknine, G. E., Repetitive cortical spreading depression cycle development in the human brain: a multiparametric monitoring approach, *J. Cereb. Blood Flow Metab.,* **15(Suppl 1)** (1995) S34.

152 Lauritzen, M., Pathophysiology of the migraine aura: The spreading depression theory, *Brain,* **117** (1994) 199–210.

153 Lauritzen, M. and Hansen, A. J., The effect of glutamate receptor blockade on anoxic depolarization and cortical spreading depression, *J. Cereb. Blood Flow Metab.,* **12** (1992) 223–229.

154 Sheardown, M. J., The triggering of spreading depression in the chicken retina: a pharmacological study, *Brain Res.,* **607** (1993) 189–194.

155 Obrenovitch, T. P. and Zilkha, E., High extracellular potassium, and not extracellular glutamate, is required for the propagation of spreading depression, *J. Neurophysiol.,* **73** (1995) 2107–2114.

156 Obrenovitch, T. P., Zilkha, E., and Urenjak, J., Evidence against high extracellular glutamate directly promoting potassium-induced cortical spreading depression, *J. Cereb. Blood Flow Metab.,* (1996), in press.

157 Bliss, T. V. P. and Collingridge, G. L., A synaptic model of memory: long-term potentiation in the hippocampus, *Nature,* **361** (1993) 31–39.

158 Asztely, F., Wigstrsm, H., and Gustafsson, B., The relative contribution of NMDA receptor channels in the expression of long-term potentiation in the hippocampal CA1 region, *Eur. J. Neurosci.*, **4** (1992) 500-505.

159 Xie, X., Berger, T. W., and Barrionuevo, G., Isolated NMDA receptor-mediated synaptic responses express both LTP and LTD, *J. Neurophysiol.*, **67** (1992) 1009–1013.

160 Crépel, V., Hammond, C., Krnjevic, K., Chinestra, P., and Ben-Ari, Y., Anoxia-induced LTP of isolated NMDA receptor-mediated synaptic responses, *J. Neurophysiol.*, **69** (1993) 1774–1778.

161 Urban, L., Neill, K. H., Crain, B. J., Nadler, J. V., and Somjen, G. G., Postischemic synaptic physiology in area CA1 of the gerbil hippocampus studied in vitro, *J. Neurosci.*, **9** (1989) 3966–3975.

162 Andiné, P., Jacobson, I., and Hagberg, H., Calcium uptake evoked by electrical stimulation is enhanced postischemically and precedes delayed neuronal death in CA1 of rat hippocampus: involvement of N-methyl-D-aspartate receptors, *J. Cereb. Blood Flow Metab.*, **8** (1988) 799–807.

163 Gozlan, H., Diabira, D., Chinestra, P., and Ben-Ari, Y., Anoxic LTP is mediated by the redox modulatory site of the NMDA receptor, *J. Neurophysiol.*, **72** (1994) 3017–3022.

164 Urenjak, J., Obrenovitch, T. P., and Zilkha, E., Microdialysis application of L-*trans*-pyrrolidine-2,4-dicarboxylate (L-*trans*-PDC): effects on extracellular glutamate and local field potential, *Br. J. Pharmacol.*, **115(Suppl)** (1995) 18P.

165 Urenjak, J., Zilkha, E., and Obrenovitch, T. P., Origin of depolarizations evoked in the rat striatum by competitive inhibition of glutamate uptake with L-*trans*-pyrrolidine-2,4-dicarboxylate (L-*trans*-PDC). *Br. J. Pharmacol.*, **177(Suppl)** (1996) 67P.

166 Massieu, L., Morales-Villagrán, A., and Tapia, R., Accumulation of extracellular glutamate by inhibition of its uptake is not sufficient for inducing neuronal damage: an in vivo microdialysis study, *J. Neurochem.*, **64** (1995) 2262–2272.

167 Obrenovitch, T. P., Urenjak, J., and Zilkha, E., Evidence disputing the link between seizure activity and high extracellular glutamate, *J. Neurochem.*, **66** (1996) 2446–2454.

6

Glycine Antagonists for Treatment of Ischemic Brain Injury

Midori A. Yenari, David C. Tong, and Gregory W. Albers

1. INTRODUCTION

Pharmacologic antagonists of glutamate's *N*-methyl-D-aspartate (NMDA) receptor complex have neuroprotective properties, however, the psychomimetic and sedative properties of these agents have limited their clinical development *(1–4)*. Recent work has shown that activation of the NMDA receptor complex involves glycine. Whereas glycine is an inhibitory neurotransmitter that modulates activity in spinal cord interneurons, recent evidence has shown that there are additional glycine binding sites that can facilitate excitatory neurotransmission in the central nervous system (CNS). There is experimental evidence that glycine enhances NMDA-mediated responses in neurons by acting at a distinct site within the NMDA receptor complex. A number of compounds are known to antagonize this site, thereby inhibiting NMDA responses. These compounds offer great potential in the treatment of stroke and other causes of CNS injury. Preclinical work suggests that these agents also have neuroprotective properties and may not cause the phencyclidine (PCP)-like side-effects seen with typical competitive and noncompetitive NMDA antagonists. Although currently there is limited clinical experience with glycine site antagonists, preliminary data suggest that these agents are well tolerated in humans.

2. THE NMDA GLUTAMATE RECEPTOR AND ITS GLYCINE SITE

Glutamate's NMDA receptor complex is a ligand-gated ion channel, penetrable to calcium and regulated by both a voltage-gated and magnesium block. It consists of several different regulatory domains *5–7)* (Fig. 1). One site is the neurotransmitter recognition site that binds compounds such as glutamate, aspartate, and NMDA and is blocked by competitive NMDA antagonists such as Selfotel (CGS 19755). Another site is located within the channel itself and binds magnesium in a voltage-dependent fashion. This ion channel binds compounds only when the channel is open. Noncompetitive antagonists such as PCP, dizocilpine (MK-801), dextromethorphan and aptiganel hydrochloride (CNS 1102) act here. Other sites include zinc and polyamine sites as well as the glycine site. All of these sites are amenable to pharmacologic manipulation and are therefore potential targets for pharmacologic treatment of stroke and other types of CNS injury.

From: Clinical Pharmacology of Cerebral Ischemia *Edited by: G. J. Ter Horst and J. Korf Humana Press Inc., Totowa, NJ*

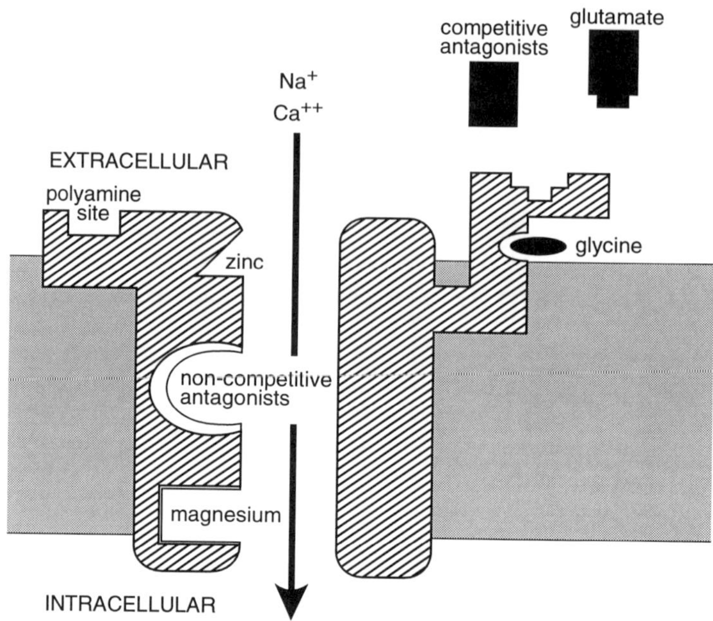

Fig. 1. Diagram of glutamate's *N*-methyl-D-aspartate receptor complex. There are separate glycine and polyamine sites.

While spinal cord glycine receptors are typically detected using radiolabeled strychnine, glycine also binds brain regions that are not bound by strychnine *(6,8–10)*. This so-called strychnine-insensitive glycine site was discovered to be part of the NMDA receptor complex. Glycine potentiates NMDA responses in both whole cell preparations and neocortical brain slices *(11–13)*, and appears necessary for receptor activation *(14–16)*. In fact, NMDA channel opening does not occur unless glycine is bound. Under physiologic conditions, this site is appears nearly saturated *(14,17,18)*. Glycine also attenuates desensitization of the NMDA receptor complex by preventing the decline in receptor response following prolonged stimulation by NMDA *(19–23)*.

There is experimental evidence that glycine plays a role in mediating excitotoxic injury during stroke. Using microdialysis techniques in various in vivo models, several groups showed that both glycine and glutamate are elevated during cerebral ischemia *(24–27)*. Studies in cultured neurons also demonstrate that glycine potentiates NMDA-induced injury *(28)*. Glycine-mediated neurotoxicity may occur in by more than one mechanism *(29)*. For example in concentrations of 300 µ*M*, glycine potentiates NMDA-induced toxicity, however in concentrations 50–100 times higher glycine may injure neurons through activation of the GABA receptor.

A number of compounds that interact with the strychnine-insensitive glycine site have been identified (Fig. 2). These compounds have helped elucidate the interactions between glutamate and glycine. Some of these agents prevent glycine from binding to the NMDA receptor and possess neuroprotective properties.

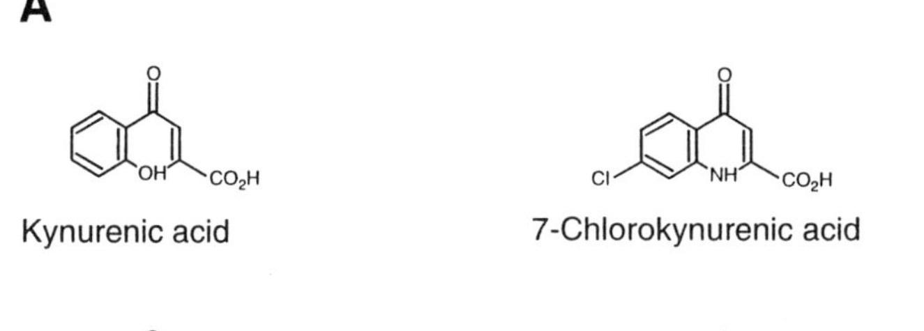

KYNURENIC ACID & DERIVATIVES

A

HA-966

B

INDOLE 2-CARBOXLIC ACID

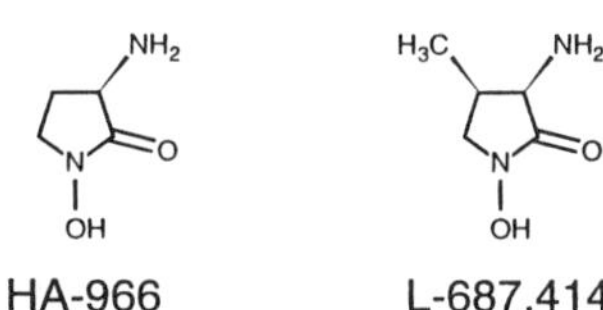

THE QUINOXALINEDIONES

CYCLIC ANALOGUES OF AMINO ACIDS

FELBAMATE

Fig. 2. Chemical structures of various glycine site antagonists by class.

3. CLASSES OF GLYCINE SITE ANTAGONISTS

3.1. Kynurenic Acid and Derivatives

Kynurenic acid (kyn) is a quinoline derivative and a metabolite of tryptophan. Found endogenously in low concentrations in the mammalian CNS *(30)*, it is also a competitive inhibitor of the glycine modulatory site *(6,31,32)*. Kynurenic acid was found to displace glycine from binding to cortical membranes *(33)*. Investigators observed that kyn NMDA could abolish NMDA responses and these responses return with the addition of excess glycine or D-serine, a glycine agonist. This reversal is observed following the application of MK-801, providing further evidence that kyn actes at a site distinct from the NMDA channel itself *(34–37)*. Derivatives of kyn were developed by substituting various carbon sites with different halogens such as fluorine, bromine, and chlorine *(38)*. Substituting a chlorine at the 7 position, 7-chloro-kynurenic acid (7-Cl-kyn), increases the affinity for binding at the glycine site. The concentrations at which the compound inhibits glycine binding by 50% (IC50) were decreased from 41 μM for kyn to 0.56 μM for 7-Cl-kyn. 7-Cl-kyn also antagonized NMDA-induced responses by 20-fold compared with kyn *(35,39)*. Additional halogen substitutions at the 5 but not 6 or 8 positions further increased the affinity for glycine site binding *(40,41)*. 5,7-dichloro-kynurenic acid (5,7-diCl-kyn) was found to be about 4 times more selective than 7-Cl-kyn *(40)*. Substitution of the hydroxyl group at the 4 position with a thiol group produced compounds that possessed even higher affinity and selectivity *(42,43)*. Kyn's IC50 as measured by glutamate-induced ileal muscle contraction in the guinea pig myenteric plexus was 160 μM and decreased to 70 μM with the thiol substituted thiokynurenate (thio-kyn). IC50 for radiolabeled glycine and N-(1-(thienyl)cyclohexyl) piperidine (TCP) binding in rat cortical membranes was 25 μM for kyn compared to 9 μM for thio-kyn. Substitution of 7-Cl-kyn with a thiol group produced 7-Cl-thio-kyn that also demonstrated greater affinity for the glycine site compared to the hydroxyl form.

3.1.1. In Vitro Studies of Neuronal Injury

In vitro studies have shown that kyn and its derivatives can reduce neuronal injury. Using a model of hypoxic degeneration produced by exposing neurons to an oxygen-free environment for 30 min, Priestley et al. *(44)* found that 7-Cl-kyn prevented cell death as measured by release of lactate dehydrogenase (LDH) and neuron specific enolase (NSE). This protective effect was dose dependent. The doses of the compound required to produce 50% protection (IC50) were 18 μM and 10 μM for LDH and NSE methods, respectively. This potency, however, was less than that of the NMDA antagonist, MK-801 in which the IC50 was 15 nM (both LDH and NSE methods). In a neuronal cell culture model of NMDA-induced toxicity, 7-Cl-kyn was found to reduce cellular injury caused by the addition of 50 μM NMDA. The IC50, as measured by LDH production, was 0.4 μM *(28)*. 7-Cl-kyn was also found to attenuate injury following a 10 min exposure to 100–300 μM of NMDA plus 100–300 μM of glycine with an IC50 of 8.6 μM (LDH method) *(29)*. In a hippocampal cell culture model of excitotoxicity, 7-Cl-kyn reversed injury induced by a 15 minute exposure to 3 mM glutamate with an IC50 of 6 μM. These neurons were still protected even if 7-Cl-kyn was added in a delayed fashion, following glutamate removal (IC50 5 μM). The addition of glycine to the medium, however, reversed the protective effect providing evidence that this compound acts through the glycine site *(45)*. The disubstituted kyn derivative 5,7-Cl-kyn was also found to protect rat cultured cortical neurons against NMDA-induced injury *(41)*. Following a 10-min exposure to 300 µ*M* NMDA, 1 µ*M* 5,7-Cl-kyn decreased neuronal injury by 55–79%, whereas 10 µ*M* decreased injury by 62–90%.

Using a model of oxygen and glucose deprivation in hippocampal slices, application of 100 µ*M* 7-Cl-kyn 10–15 min prior to a 35-min ischemic insult showed better preservation of CA1 cells compared to controls *(46)*. This protective effect was reversed if excess glycine was added to the preparation. Whereas a similar protective effect was observed in a parallel experiment using the noncompetitive NMDA receptor antagonist MK-801, the protective effect was not reversed with glycine, providing evidence that 7-Cl-kyn exerts its effect at the glycine site in this model as well.

Kyn's thiol derivatives were also found to have potent neuroprotective properties *(42)*. Thio-kyn, 7-Cl-kyn, and 7-OCH3-thio-kyn decreased glutamate-induced injury more effectively than the corresponding hydroxyl derivative. In particular, 7-Cl-thio-kyn was five times more effective than 7-Cl-kyn in reducing this injury with an IC50 of 0.19 µ*M* for 7-Cl-thio-kyn and an IC50 of 1.0 µ*M* for 7-Cl-kyn. In addition, this protective effect of 7-Cl-thio-kyn appeared to be partially attributed to inhibition of lipid peroxidation, suggesting that this compound also acted as a free radical scavenger. Non thiol substituted kyn derivatives did not appear to have this property.

3.1.2. Animals Models of Cerebral Ischemia

In vivo studies have also demonstrated neuroprotective properties of kyn and its derivatives. Germano et al. *(47)*, using a rat model of permanent MCA occlusion, showed that 300 mg/kg of kyn given intraperitoneally (ip) could significantly reduce infarct size by about 40% and improve behavioral scores if given shortly before, but not 1 h after the onset of ischemia. Monitoring of blood pressure and body temperature did not reveal differences between experimental groups; therefore, the neuroprotective efficacy did not appear to be due to hypothermia. By administration of 0.44 µg or 0.88 µg 7-Cl-kyn, Verrecchia et al. *(98)* found a significant reduction in stroke volume using a mouse model of permanent MCA occlusion. Because of poor blood-brain barrier permeability, the compound was given intracerebroventricularly (icv) 5 min prior to occlusion. A dose of 0.44 µg reduced infarct size by about 38% compared to parallel controls, and 0.88 µg reduced infarct size by 27%. Similar to the observations made in cell culture models, the protective effect was reversed if 11 µg D-cycloserine, a partial glycine agonist, was added.

Chen et al. *(49)* studied 7-Cl-thio-kyn in a rat model of permanent MCA occlusion. By giving this agent intravenously (iv) 5 min before the onset of ischemia, they found significant neuroprotective effects at doses of 20 and 30 mg/kg with reductions of infarct size by about 40%. The protective effect was seen primarily in the cortex, rather than the basal ganglia. A trend showing reduction in infarct volume was seen using a dose of 10 mg/kg, though this was not statistically significant. Since application of a potentially neuroprotective agent after the onset of injury would have more clinical utility, they administered 20 mg/kg 7-Cl-thio-kyn iv to animals either 5 min before or 15 or 60 min after occlusion. They found similar protective effects if the drug was given before or 15 min after MCA occlusion, but no protective effect when given 60 min afterward. Although difficult to translate these observations to human stroke, the "therapeutic window" for 7-Cl-thio-kyn in rats appears to be brief, somewhere between 15 to 60 min.

The kyn derivatives also appear to have efficacy in global models of cerebral ischemia. Intraventricular administration of 2 µl (500 µ*M*) 7-Cl-kyn given immediately before the onset of 20 min of bilateral carotid occlusion demonstrated 50% improvement in CA1 cell counts measured 7 d later *(50)*. Significant neuroprotection was also seen when 7-Cl-thio-kyn was given intraperitoneally to gerbils subjected to 5 min of bilateral carotid occlusion *(51)*. Hippocampal CA1 cell counts measured 7 d later showed very robust protection (approx 95% preservation) in treated animals compared to ischemic controls. Multiple injections of 100 mg/kg were given beginning at the onset of reperfusion and every 30–60 min there after for up to 6 h (a total of five doses). The investigators found that 7-Cl-thio-kyn significantly lowered body temperature and, therefore, the protective effect could have been due to the induced hypothermia rather than the direct effects of the compound itself. They therefore repeated the experiment and maintained normal body temperature for 6 h after the onset of ischemia in both treated and control groups. A significant neuroprotective effect of 54% was still observed.

3.1.3. Other Animal Models of Brain Injury and Neurodegenerative Disease

Work with glycine antagonists in models of NMDA-induced neurotoxicity also demonstrated efficacy and may have relevance to human stroke as well as clinical conditions such as Alzheimer's, Huntington's, and other neurodegenerative diseases. In one such model, 12.5 n*M* NMDA is directly injected into the striatum of neonatal rats. Neuronal injury is detected in the striatum, hippocampus, and adjacent cortex. The extent of injury is quantified by comparing hemisphere weights. By coinjecting NMDA with kyn or 7-Cl-kyn, dose dependent neuroprotection, as defined by preservation of ipsilateral hemisphere weight, was observed by postinjection d 5 *(52)*. The 7-Cl-kyn was found to have increased potency compared to kyn. One hundred nanomolars of kyn reduced injury from about 20% (controls) to 2%, whereas 40 n*M* of 7-Cl-kyn reduced injury from 17 to 3%. The IC50s of kyn derivatives were estimated at 34.1 n*M* for kyn and 17.6 n*M* for 7-Cl-kyn.

In a different model of neuronal degeneration, intrastriatal injection of 200 n*M* of the excitotoxin, quinolinate, was followed 1 h later by injection of 10–50 n*M* 7-Cl-kyn *(53,54)*. A dose dependent preservation of the enzymes choline acetyltransferase (CAT) and glutamate decarboxylase (GAD) was observed 7 d later. Complete neuroprotection was observed at a dose of 50 n*M*. This protection was reversed with the addition of the glycine agonist, D-serine. Furthermore, preservation of these enzymes was still seen if administration of 50 n*M* 7-Cl-kyn was delayed up to 2 h for GAD and up to 5 h for CAT. Hippocampal injury caused by 60 n*M* quinolinate showed complete neuroprotection from this compound as measured in the pyramidal cell layer. Striatal injury caused by other glutamate receptors subtype DL-α-amino-3-hydroxy-5-methylisoxazole-4-propionate (AMPA) or kainate were not attenuated by 7-Cl-kyn, confirming that this agent appears to be relatively selective for NMDA mediated injury.

The thiol substituted kyns that have higher affinity for strychnine-insensitive glycine sites were studied in a few models of brain injury as well. Using an in vivo model of striatal injury, direct injection of 50–80 nM 7-Cl-thio-kyn following a 250 nM injection of quinolinate significantly preserved cellular choline acetyl transferase and L-glutamic acid decarboxylase *(42)*.

Kyn also appears to have efficacy in traumatic brain injury models. In rats subjected to lateral fluid percussion, 300 mg/kg of kyn was given intravenously 15 min after injury. Using functional measures of cognitive ability, kyn-treated rats performed significantly better with less of a neurologic deficit than injured controls *(55)*. Pathologic analysis 48 h following injury, showed that kyn decreased cerebral edema formation as well as attenuating regional tissue calcium increases.

While kyn and its derivatives exhibit substantial neuroprotective properties in several different models of cerebral ischemia and brain injury, poor blood–brain barrier permeability limits their potential for clinical use. Other compounds with better systemic absorption have been developed and have recently entered clinical evaluation.

3.2. R-(+)-3-amino-1-hydroxypyrrolid-2-one (HA-966)

R-(+)-3-amino-1-hydroxypyrrolid-2-one (HA-966) is a cyclic compound with a chemical structure similar to cycloserine. It is a pyrrolidine derivative and resembles the cyclic form of gamma-aminobutyric acid (GABA) *(56,57)*. It has both partial glycine agonist and antagonist properties, but like kyn, was found to reverse NMDA responses. In a cortical wedge preparation, 200 µM HA-966 reduced NMDA-evoked responses by about 32%, but this could be reversed if 100 µM glycine were added *(6,18,58–60)*. HA-966 is not as selective for the glycine site, however, as 7-Cl-kyns is. The IC50 for glycine binding is approx 8.5 µM *(59)*. In radioligand studies, HA-966's IC50 for glycine site binding was about 50-fold greater than that of other excitatory amino acids, whereas 7-Cl-kyn's was about 270-fold greater *(35,58)*. A major clinical advantage of HA-966 compared to kyn and its derivatives is that it does penetrate the blood–brain barrier following parenteral administration *(6,58,61,62)*. Methyl derivatives of HA-966 have been synthesized and studied. R-(+)-cis-methyl-HA-966 (L687,414) was found to have about a 10 times greater affinity for the glycine site, with less partial agonist and less intrinsic efficacy *(63–65)*. An advantage of this compound is that is has reasonable bioavailability after oral administration *(65,66)*.

3.2.1. In Vitro Studies of Neuronal Injury

HA-966 blocks NMDA-induced toxicity in neuronal cell culture with an IC50 of 30 µM *(28)* and hypoxic injury with an IC50 of 150–175 µM *(44)*. Compared to the IC50s of 7-Cl-kyn, HA-966's protective effect is at least 10–100-fold less. This may be attributed in part to its partial agonist properties.

3.2.2. Animal Models of Cerebral Ischemia

Neuroprotective properties have been demonstrated with HA-966 in focal models of cerebral ischemia. This compound was given intraventricularly 5 min before the induction of ischemia in a mouse model of permanent MCA occlusion *(48)*. Significant decreases in infarct volume as measured by triphenyl tetrazolium chloride (TTC) (a mitochondrial stain) of 20 and 40% measured 24 h later were seen with doses of 20 µg and 40 µg, respectively. These investigators also conducted parallel experiments with

7-Cl-kyn and found comparable degrees of neuroprotection with both compounds (*see* Section 3.1.2.). Physiological parameters were not reported in this study and therefore it is unclear whether temperature and other variables were monitored.

Using a model of permanent MCA occlusion in rats, Gill et al *(67)* studied HA-966's methyl derivative, L687,414. They found that 14 or 30 mg/kg L687,414 given intravenously as a bolus followed by a 4-h continuous infusion of 14 or 30 mg/kg/h, respectively, immediately after occlusion, resulted in a significant reduction in infarct volume. The degree of neuroprotection did not appear dose dependent. In this study, infarct size reductions of about 34% were seen with the 14 mg/kg dose and 20% were seen with the 30 mg/kg dose. Corresponding steady-state serum levels were 25 and 60.8 µg/mL. Lower doses (17.6 mg/kg given as a single bolus and 7 mg/kg given as a bolus followed by a 4-h infusion) were not neuroprotective. Early decreases in mean arterial pressure occurred about 15–30 min into treatment in the dose ranges that were found to be neuroprotective. This decrease was on the order of 13–16 mmHg and could have potentially worsened the infarct in the treated groups. Serum glucose and arterial blood gases were monitored and not found to be different between groups, but neither body nor brain temperature was reported. Other glycine site antagonists have been shown to decrease temperature *(51,68)*; therefore, some of the neuroprotective effects of L-687,414 could have potentially been caused by hypothermia. Since infarcts can progress over several days, another caveat in this study is that infarct volume was measured only 4 h following the onset of ischemia.

3.2.3. Other Animal Models of Brain Injury

Studies conducted in in vivo models of direct neurotoxicity have also demonstrated beneficial effects of HA-966. When given intraperitoneally to perinatal rats, 15 min after direct injection of NMDA, the compound produced a reduction in brain injury in a dose-dependent fashion *(69)*. Doses of 25 and 50 mg/kg ip showed 30 and 52% reduction in NMDA-induced injury, respectively. The HA-966 was not effective in this model when quisqualate was injected, further emphasizing that its major effect is on the glycine site of the NMDA receptor.

In a rat model of neuronal degeneration, quinolinate-induced striatal injury, as measured by attenuation of CAT and GAD loss, was reduced in a dose-dependent fashion with doses of 100–500 n*M* HA-966 injected intrastriatally. When the injection of 500 n*M* HA-966 was delayed by 1, 2, 5, or 8 h, enzyme preservation was seen with delays up to 2 h for GAD and 5 h for CAT. Intrahippocampal injection of 500 n*M* HA-966, given 1 h after direct injection of 60 n*M* quinolinate, offered partial protection from pyramidal cell loss. HA-966 also reduced neurotoxicity caused by direct intrastriatal injection of NMDA, but not by non-NMDA glutamate agonists. Compared to 7-Cl-kyn, however, HA-966 was less efficacious (*see* Section 3.1.3.). Specifically, HA-966 was about 20-fold less effective than 7-CL-kyn in this model *(53,54)*.

HA-966 also showed efficacy in another model of nervous system injury induced by direct injection of 25 n*M* NMDA. HA-966 was given intraperitoneally to perinatal rats 15 min following injection of NMDA. Brains were harvested and weighed 5 d later. However, the protective benefit of HA-966 compared to noncompetitive NMDA antagonists such as Mk-801, dextromethorphan, and CGS-19755 was found to be several hundred fold less *(70)*.

3.3. Indole-Containing Compounds

Indole-2-carboxylic acid (I2CA) has also been shown to competitively inhibit the strychnine insensitive NMDA site *(71)*. Although it is a weak antagonist, halogenation at the 5 and 6 positions increases the affinity *(72)*. Optimal antagonism was found with the derivative 2-carboxy-6-chloro-3-indoleacetic acid. This compound appears to have adequate bioavailability after intraperitoneal administration and a serum half life of about 25–35 min *(73)*.

Limited work with I2CA has demonstrated that it, too, possesses neuroprotective properties. In a rat model of traumatic brain injury, I2CA in doses of 20 and 50 mg/kg given intravenously was found to decrease cognitive dysfunction and neurologic deficit when administration was begun 15 min after lateral fluid percussion *(55)*. Pathologic examination 48 h later showed decreased edema formation and reduced local tissue sodium as well as reductions in potassium, magnesium, and zinc in treated animals.

3.4. The Quinoxalinediones

Whereas kyn and its derivatives to have neuroprotective properties in animal models, their poor blood–brain barrier (BBB) permeability limits their clinical potential. The HA-966 and I2CA, have adequate BBB permeability following systemic administration, but have less affinity for the glycine site compared to kyn and derivatives. HA-966 also appears, compared to kyn and its derivatives in animal models, to be a less efficacious a neuroprotectant *(44,48,54)*. A new class of glycine antagonists, the quinoxalinediones was developed that has both high affinity for the glycine site and good BBB penetration following systemic administration *(74)*. This class includes 5-nitro-6,7-dichloro, 1,4-dihydro-2,3-quinoxalinedione (ACEA-1021) and 5-nitro-6,7-dibromo 1,4-dihydro-2,3-quinoxalinedione (ACEA 1031), that are both potent competitive antagonists of the strychnine insensitive glycine site with a Kb 6–8 n*M* in oocytes and 5–7 n*M* in neurons *(75)*.

3.4.1. In Vitro Studies in Neuronal Injury

ACEA-1021 has been studied in several in vitro and in vivo models of cerebral ischemia. In a hippocampal slice model, ACEA-1021 demonstrated dose dependent neuroprotection when given 10–15 min prior to 35 min of combined oxygen and glucose deprivation *(46)*. Brain injury 48 h later, as determined by CA1 subfield fluorescence, was markedly reduced compared to controls. The neuroprotective efficacy of ACEA-1021 was found to be comparable to that of 7-Cl-kyn in this model (*see* Section 3.1.1.). Temperature differences were carefully controlled, thereby eliminating the possibility of a hypothermic effect. ACEA-1021's protective effect was also reversed in the presence of excess glycine and this was not the case in a parallel experiment using MK-801.

3.4.2. Animal Models of Cerebral Ischemia and Brain Injury

In a rat model of transient focal cerebral ischemia, Warner et al. *(68)* demonstrated neuroprotective effects of ACEA-1021. Rats were subjected to 90 min of MCA occlusion followed by 96 h of reperfusion. Intraperitoneal treatment with ACEA-1021 was given 55 min prior to ischemia, at the end of the ischemic period, and 180 min into reperfusion. A dose-dependent reduction in infarct size and neurologic scores was seen.

Treatment with a higher doses (30 mg/kg) demonstrated about a 75% reduction in cortical infarct volume as measured by histology and improvement in neurologic scores. A lower dose 10 mg/kg showed about a 50% reduction in cortical infarct. Serum levels were measured 50 min following administration of the first dose and were about 40 μg/mL for the 10 mg/kg dose and 70 μg/mL for the 30 mg/kg dose. Other quinoxalinediones were studied as well. ACEA-1031 was also found to have neuroprotective properties using the same paradigm. Whereas ACEA-1031 in comparable doses to ACEA 1021 showed a benefit, the protective effect was less robust *(68)*. Actual brain temperatures were not measured in these studies, but rectal temperatures were slightly lower in the treated groups, raising the concern that the neuroprotective effect may have been in part due to hypothermia. Temperature differences between treated and control groups in the ACEA-1031 study appeared even more significant. However, subsequent studies have shown significant benefits of ACEA 1021 in temperature controlled parodigms. 5-chloro-7-trifluoromethyl-1,2,3,4-tetrahydroquinoxaline-2,3,-dione (ACEA-1011) was also studied, but a much lower (3 mg/kg) dose was used, and no neuroprotection was seen.

A study by Marek et al. *(76)* using another model of transient focal ischemia demonstrated a protective effect of several different quinoxalinediones when given immediately following occlusion in both cortical and subcortical regions. Doses of 10 mg/kg given as an iv bolus followed by an infusion of 7 mg/kg/h for 22 h were used. Infarction was measured by the TTC method. ACEA-1021 showed the most marked efficacy as infarct size was reduced by 89% within cortical regions and 42% within subcortical regions in treated animals. Other quinoxalinediones (5-nitro-6,7-dimethyl-1,4-dihydro-2,3 quinoxalinedione and 5-nitro-6-methyl-7-chloro-1,4-dihydro-2,3-quinoxalinedione) showed reductions of 72–75% within cortical regions and no improvement within subcortical regions.

Sauer et al. *(77)* demonstrated long-term protective effects of ACEA 1021. Using a rodent model of permanent MCA occlusion, 40 mg/kg of ACEA 1021 was given intravenously immediately following, then IP 2 h after occlusion. Significant reductions in infarct volume of 47% were seen as measured by in vivo MRI scans 2 and 28 d later as well as histopathology at 28 d. Treatment with a single dose at the onset of ischemia, however, did not result in significant neuroprotection.

Whereas the effects of administering quinoxalinediones prior to or immediately following the onset of cerebral ischemia produced very large reductions in infarct size, delayed administration of this compound also appeared promising. Using a rat model of permanent focal ischemia, doses of 3 mg/kg and 10 mg/kg of ACEA 1021 given intravenously were found to be significantly neuroprotective if given up to 1 h following the onset of ischemia *(76)*. A trend towards smaller infarcts was seen when given 3 h afterwards, and no improvement was seen when given 6 h later.

Using diffusion-weighted MRI and magnetic resonance spectroscopy to study the metabolic mechanisms underlying ACEA-1021's neuroprotective efficacy, Myseros et al. *(79)* measured cytotoxic edema and metabolic parameters in cats subjected to MCA occlusion. They found that ACEA 1021 prevented decreases in the apparent diffusion coefficient (and therefore cytotoxic edema) and lactate. Trends demonstrating preserved high energy metabolites, tissue pH and decreases in infarct size measured 5 h following occlusion were also seen.

Although there appears to be marked beneficial effects of ACEA 1021 in focal models of cerebral ischemia, in a global ischemia model, no neuroprotective effect was seen *(68)*.

ACEA 1021 has also been studied in a model of acute subdural hematoma (SDH). Following induction of SDH by subdural injection of autologous blood, brain damage, measured 4 h later, was significantly decreased by 26–39% in the treated groups *(80)*. Neuroprotection was observed when ACEA 1021 was given both 15 min before and 30 min after SDH induction. In addition, arterial blood gases, blood pressure, and temporalis muscle and body temperature were carefully monitored and not found to be different between treated and untreated groups.

(S)-9-chloro-5-(p-aminomethyl-o-(carboxymethoxy) phenylcarbamoylmethyl)-6,7-dihydro-1H,5H-pyrido(1,2,3-de)quinoxaline-2,3-dione hydrochloride trihydrate (SM-18400) is a newly developed quinoxalinedione that appears to have increased affinity for the glycine site. Preclinical work with this agent also appears promising. Yasuda et al. *(81)* and Tanaka et al. *(82)* examined this compound in models of both focal and global ischemia. When SM-18400 was given intravenously after 2 h of MCA occlusion, reduced infarct size was seen 24 h following occlusion in a dose dependent fashion. The highest dose group, 15 mg/kg followed by an infusion of 3 mg/kg/h for 25 h, showed a 72% reduction in infarct size within the cortex and 27% reduction in the striatum. In the low dose group, 7.5 mg/kg followed by 1.5 mg/kg/hr for 25 h resulted in reduction in infarct size by 42% in only the cortex. In two different models of global ischemia (rat and gerbil), SM-18400 was given at the time of reperfusion and 2 h later following 15 min of bilateral carotid occlusion. Both reduced brain edema and mortality.

3.5. 1-Aminocyclobutane-1-Carboxylic Acid (ACBC)/1-Aminocyclopropanecarboxylic Acid (ACPC)

1-aminocyclobutane-1-carboxylic acid (ACBC) and 1-aminocyclopropanecarboxylic acid (ACPC) are cyclic amino acid analogs that also interact with the glycine site. Both are antagonists with some partial agonist properties *(83–85)*. ACBC was found to antagonize the strychnine-insensitive glycine site with a Ki of 19 μM and an IC50 of 6.1 μM for TCP binding *(86)*. It also blocked glycine's potentiation of NMDA responses in *Xenopus* oocytes in a competitive manner *(87)*. ACPC was found to have an IC50 of 38 nM for glycine binding and an estimated Ki of about 32 nM *(88)*. Following parenteral administration, ACBC and ACPC were also found to cross the blood–brain barrier with half lives of 5 min or less *(89,90)*. Using high performance liquid chromatography, however, other investigators found that ACPC has a half-life of 1.5 h *(6,89)*. Furthermore, a study examining the anxiolytic and antidepressant effects of ACPC in mice showed that it was biologically available up to 6 h following parenteral and oral administration *(91)*.

In vitro studies of ACPC showed remarkable preservation of cultured cerebellar granule cells exposed to different concentrations of glutamate. The addition of 1 mM ACPC along with different concentrations of glutamate showed dose-dependent neuroprotection. Application of 5 μM glutamate reduced cell death by 90% compared to control sister cultures, whereas 37% neuroprotection was seen following the addition of 10 μM glutamate, and no neuroprotection was seen with the application of 100 μM glutamate *(92)*. Compared to the neuroprotective effects of other glycine site antagonists in cell culture, however, ACPC appears to be less effective than both 7-Cl-kyn and HA-966 in a similar model of glutamate-induced toxicity *(93)*. Since a clinical application of a neuroprotective agent might include prophylactic administration to patients at risk for brain injury, Boje et al. *(93)* studied prolonged exposure with different glycine site antagonists. Interestingly, exposing cultures to ACPC and HA-966 for 20–24 h prior to glutamate application reduced the neuroprotective efficacy of these compounds. In addition, this reduction in efficacy was not related to changes in mRNA levels of the NMDA R1 or NMDA ζ1 receptor subunits, although the authors suggested that it may be due to changes in other NMDA receptor subunit expression or an uncoupling of glutamate and glycine.

Neuroprotection seen in in vivo models was also somewhat less robust with ACPC. In a gerbil model of global ischemia, 150–600 mg/kg ACPC was given intraperitoneally 5 min after the onset of 20 min of bilateral carotid occlusion *(92)*. The optimal dose of 300 mg/kg improved 7 d survival fourfold, whereas doses of 150 mg/kg and 600 mg/kg improved survival two- and threefold, respectively. Hippocampal CA1 cell counts were significantly increased threefold with the 600 mg/kg dose. In contrast to observations made in cell culture models, chronic administration of ACPC prior to the onset of 20 min of bilateral carotid occlusion was also found to be beneficial *(94)*. Animals given daily doses of 300 mg/kg ip for 6 or 7 d prior to the onset of ischemia showed improved survival and neurologic scores, and decreased postischemic seizure frequency. Histopathology performed 7 d following the induction of ischemia showed preserved hippocampal, striatal, and cortical cell counts. Brain levels after chronic administration were 0–6.5 µg/g, whereas levels of 50 µg/g were measured 30 min after a 200 mg/kg injection.

In a model of dynorphin-A induced spinal injury, ACPC was also found to have neuroprotective efficacy. Subarachnoid injection of dynorphin-A results in ischemic injury by reducing local blood flow *(95)*. When ACPC was administered in doses of 100 or 200 mg/kg ip, the animals had improved neurologic scores by 24 h following injury. Histopathology performed at 72 h also showed decreased tissue injury. Furthermore, the addition of 800 mg/kg glycine ip reversed the beneficial effects.

3.6. 2-Phenyl-1,3-Propanediol Dicarbamate (Felbamate)

2-Phenyl-1,3-propanediol dicarbamate (felbamate), which has recently been approved for clinical use as an anticonvulsant, is felt to act as an antagonist of the strychnine-insensitive glycine site. Felbamate's binding is blocked in the presence of 5,7-dichloro-kyn, but was unaffected by noncompetitive antagonists of the NMDA receptor such as MK-801, polyamine site antagonists such as spermine or other anticonvulsants *(96)*. Felbamate also blocked NMDA currents in rat hippocampal neurons *(97)* and decreased the magnitude of glycine-enhanced intracellular calcium accumulation caused by NMDA *(98)*. In a model of audiogenic seizures, the glycine agonist, D-serine could completely reverse the anticonvulsant properties of felbamate *(98)*. More recent evidence, however, suggests that felbamate does not affect the glycine site by competitive inhibition since felbamate did not reduce specific binding of 5,7-dichloro-kyn *(99)*. Nevertheless, a few experimental studies have demonstrated that felbamate has neuroprotective properties.

In a hippocampal slice model, application of 1.3 n*M* felbamate reduced glycine induced injury to CA1 cells as measured by orthodromic and antidromic population spike recovery *(100)*. This protection was dose dependent with IC50 of 1.04–1.06 m*M* following 10 m*M* glycine exposure. Surprisingly, in this model, felbamate was not effective in reducing injury caused by glutamate. By modifying this model to evaluate hypoxia, the same investigators found that felbamate could reverse hypoxic injury and this protection could be reversed by the addition of glycine. The IC 50 for hypoxic protection was 198.7 µg/mL *(101,102)*.

In a neonatal rat model of hypoxia and ischemia, animals were subjected to permanent bilateral carotid artery occlusion followed by a 1-h exposure to 6.5% oxygen 4–5 h later. Felbamate, 300 mg/kg ip was given one hour before the onset of hypoxia. Pathological examination 72 h later showed that felbamate reduced hippocampal injury by 77% and cortical injury by 38% compared to controls. Corresponding serum levels were 115–148 µg/mL following 90–240 min of administration. Brain temperature did not differ between treated and control groups *(103)*. Posthypoxic treatment was also studied in this same model *(104)*. Doses of 100, 200, 300, 400, and 500 mg/kg were given 1 h following the onset of hypoxia. A "U" shaped dose response curve was observed. Whereas optimal neuroprotection was seen with the 300 mg/kg dose (52% reduction in cortical infarction), less dramatic protection was seen with the 200 and 400 mg/kg doses (30 and 40% reduction in cortical infarction, respectively), and no effect was seen with doses of 100 and 500 mg/kg. Maximal benefit was seen with the 300 mg/kg dose in the hippocampus where 91% of the dentate gyrus neurons were preserved. The therapeutic window of felbamate's neuroprotection was studied. By delaying the administration of 300 mg/kg felbamate, efficacy was seen if treatment commenced up to 4 but not at 6 h following the onset of hypoxia. Corresponding pharmacokinetic data showed that felbamate levels reached a plateau about 1 h following injection with plasma concentrations of about 60-120 µg/mL.

3.7. New Glycine Site Antagonists

Newer glycine site antagonists, including ZD9379 and GV150526A, have been studied in the laboratory over the past year. ZD9379 was studied in a rodent model of permanent MCA occlusion *(105)*. Treatment that was begun after 30 min of ischemia resulted in significant neuroprotection using doses of 1, 5, and 10 mg/kg with reductions in infarct size of 38, 50, and 45%, respectively. GV150526A was studied in a different rat model of permanent MCA occlusion *(78)*. In this study, this agent was compared to other potential neuroprotective agents. When given prior to the onset of ischemia, GV150526A was found to have higher levels of efficacy than MK-801, ACEA 1021, Eliprodil (a polyamine antagonist and N-type calcium channel blocker), and U92032 (a T-type calcium channel antagonist and antioxidant). When given 1 h after the onset of ischemia, all of the agents tested had similar efficacy, but only GV150526A demonstrated efficacy even when treatment was delayed 6 h. Further work in this area should prove enlightening.

4. TOXICITY OF GLYCINE SITE ANTAGONISTS

The strategy of treating stroke by limiting toxic glutamate receptor activation with competitive or noncompetitive antagonists has been an appealing one; that is theoretically relatively safe; however, significant adverse reactions have been reported in early clinical trials. Common side effects experienced by patients given these agents included dysphoria, hallucinations, dizziness and ataxia *(1–4)*. In addition, pathologic examination of animals given various types of NMDA antagonists revealed neuronal microvacuolization within regions of the retrosplenial cortex and cingulate gyrus *(106–109)*. Expression of the nonconstitutive 72-kDa heat shock protein (HSP72) was also found in these same areas *(110,111)*. Although the clinically apparent side-effects are not prohibitive and the significance of the microvacuolization remains unclear, glycine site antagonists appear to have a superior safety profile (Table 1).

4.1. Pathological Effects

Unlike NMDA antagonists, microvacuolization of neurons in the cingulate gyrus was not seen in rats given the glycine site antagonists, kyn, HA-966, and ACEA-1021. Expression of the nonconstitutive 72-kDa heat shock protein (HSP72) in animals given 10–50 mg/kg 7-Cl-thio-kyn ip was not seen, however, in animals given MK-801 (1 mg/kg), HSP72 expression in the cingulate and retrosplenial cortex was robust *(49)*.

Whereas microvacuolization was not detected in animals given glycine antagonists, recent work has disclosed new histologic findings of unknown significance. Brains of animals given 10 and 30 mg/kg ACEA 1021 examined with light microscopy revealed granulated neurons. Electron microscopy showed that these granules corresponded to mitochondrial cristae in various states of disarray with evidence of increased mitochondrial turnover *(112)*.

Studies of L-687,414 also did not appear to affect neuronal morphology or cerebral glucose metabolism in doses found to be efficacious against experimental cerebral ischemia *(113)*. MK-801, on the other hand, appeared to increase glucose metabolism within limbic structures following optimally neuroprotective doses.

4.2. Behavioral Effects: Memory, Sedation

Whereas both laboratory and clinical studies of competitive and noncompetitive NMDA antagonists have often been reported in transient behavioral abnormalities, impairment of cognition and ataxia *(1–5, 114,115)*, glycine site antagonists did not appear to cause these disturbances as dramatically. Compared to the competitive and noncompetitive NMDA antagonists, the glycine antagonists, kyn and 7-Cl-kyn did not cause impaired learning or locomotion in rats given doses in ranges that were anticonvulsant *(114,115)*.

Behavioral abnormalities in rats given 7-Cl-thio-kyn were not as severe as those seen with typical NMDA antagonists. Groups of animals were given either MK-801 (1 mg/kg ip) or 7-Cl-thio-kyn (10-50 mg/kg ip). Whereas MK-801-treated animals developed marked sedation and catatonia, 7-Cl-thio-kyn-treated animals were slightly sedated for 2 h, then had complete recovery *(49)*. In animals given high doses of 7-Cl-thio-kyn (100 mg/kg ip), no gross behavioral signs were observed compared to control animals *(51)*.

Early work showed that 1 mg/kg ip HA-966 did not produce any obvious behavioral changes in mice. With doses as high as 5–10 mg/kg, however, animals became sedated and flaccid *(116)*. Experience with even higher doses of HA-966 demonstrated more serious consequences *(69)*. Animals receiving 50 mg/kg HA-966 ip frequently became

Table 1
Significant Features of Various Glycine Antagonists

Glycine Antagonist	Advantages	Disadvantages
Kynurenic Acid (Kyn) Derivatives		
Kyn, 7-Cl Kyn, 5,7 di Cl Kyn, Thio-Kyn,7-Cl thiokyn	Very high affinity for the glycine receptor High effectiveness in cerebral ischemia neurotoxicity, and TBI models No microvacuolization at neuroprotective doses	Short time window of effectiveness (<1hr) in animal models compared to others Poor BBB penetration
Pyrrolidines		
HA-966	Can be given orally Acceptable BBB penetration No microvacuolization at neuroprotective doses	10–100-fold less potent than Kyn derivatives in comparable animal models May have reduced potency after 24 h
L687,414	Much higher affinity for glycine receptor	Causes behavioral changes and motor impairment at neuroprotective doses in animals
Indole Derivatives		
I2CA	Neuroprotective in TBI models Acceptable BBB penetration	No studies to date in cerebral ischemia models Causes ataxia at neuroprotective doses in animals
Quinoxalinediones		
ACEA-1021	Good BBB penetration Very high glycine receptor binding affinity No microvacuolization at neuroprotective doses	May require prolonged infusion for effect Not neuroprotective in global ischemia models May cause granulated neurons/mitochondrial dysfunction
SM-18400	Effective up to 2 h after occlusion in animal models	Only limited studies performed to date
Cyclic Amino Acid Analogs		
ACPC	May be effective in spinal cord injury	Less potent than 7 Cl-Kyn or HA 966 May have reduced potency at 24h
Felbamate	Effective up to 6–8 h after MCA occlusion in animals	High rate of aplastic anemia reported in seizure patients on chronic treatment
	Neuroprotective levels achievable in vivo	Bell shaped efficacy curve (i.e., high or low doses not as effective in animal models)
Others		
ZD9379	Effective in animal models	Only limited studies performed to date
GV150526A	More effective in animal models than ACEA, MK-801, etc. Effective up to 6 h after onset of ischemia	Only limited studies performed to date

sedated, often had brief periods of apnea (lasting less than 30 s), and a higher mortality rate (17%) compared to animals receiving lower doses (0%). In addition, mortality was 100% in animals receiving 100 mg/kg or more. Unlike other glycine site antagonists, the methyl derivative, L687,414, was found to cause impairments in behavior and motor function. Doses of 44 mg/kg iv and 27.3 mg/kg ip and higher in mice caused decreased performance on the rotorod test. In rats, the doses above 20 mg/kg iv caused impairment in locomotion. PCP-like behavior consisting of head waving, body rolling, and hyperlocomotion also occurred with a peak effect seen following an intravenous dose of 100 mg/kg. Compared to similar doses of MK-801, however, this effect was less dramatic *(65,66)*. Neuroprotection was achieved in rats at doses 14 and 30 mg/kg iv *(67)*.

In the study of traumatic brain injury, sham-operated animals given either 300 mg/kg kyn or 20 and 50 mg/kg I2CA intravenously, cognitive performance as measured by a water maze test was no different than in untreated shams *(55)*. Preliminary studies in unoperated rats showed that doses of I2CA in the range of 20–50 mg/kg iv produced ataxia, whereas doses of 50–100 mg/kg iv produced loss of consciousness and respiratory depression. Interestingly, the I2CA derivative, 5-fluoro-I2CA did not appear to cause these side effects at a comparable dose *(55)*.

4.3. Hemodynamic and Physiologic Effects

No effects on heart rate or respiration were reported in animal studies examining the different glycine site antagonists; however, a few groups did find transient decreases in blood pressure. A 10–50 mg/kg 7-Cl-thio-kyn dose given ip to rats transiently decreased mean arterial blood pressure from 103 ± 8 to 80 ± 2.7 mmHg for about 1–2 min *(49)*. In the study by Gill et al. *(67)*, L-687,414 given to rats with permanent MCA occlusion resulted in transient decreases in mean arterial blood pressure (MAP) in the highest dose ranges (plasma levels of about 60.8 µg/mL). The decrease in MAP was 13–16 mmHg and occurred about 15–30 min into treatment but recovered spontaneously by 60 min. In a rat model of subdural hematoma (SDH) *(80)*, animals given the highest doses of ACEA 1021 (15 mg/kg) developed profound hypotension and died when the loading dose was given too rapidly. This decrease in blood pressure was prevented by giving the loading dose over a longer period of time. Seven of 32 animals given this same dose died due to rapid onset metabolic acidosis. Since these animals were also subjected to SDH, it is not clear whether this same phenomenon would have been observed had the animals been subjected to an ischemic insult rather than the SDH. This is nevertheless an area of potential concern as some patients given high dose boluses followed by infusions of the noncompetitive NMDA antagonist, dextrorphan developed transient hypotension and depressed respirations *(1)*.

5. CLINICAL EXPERIENCE

Whereas a number of different compounds acting at the strychnine-insensitive site were found to be neuroprotective in experimental ischemia and neuronal injury models, clinical development has been hampered by problems with drug delivery. Many of these drugs penetrate the blood–brain barrier poorly, or require very high doses for efficacy. A few of these compounds do, however, have acceptable pharmacologic properties and have recently been given to humans. Clinical experience in stroke patients is limited, although a few early clinical trials are in progress.

5.1. Felbamate

There has been extensive clinical experience with felbamate, primarily in the treatment of seizures. In six clinical trials of felbamate, encompassing about 250 patients, the most commonly noted side-effects include headache and dizziness *(117–122)*. Other reported side-effects include nausea, vomiting, ataxia, fatigue, diplopia, constipation, and somnolence. These effects were usually seen in patients on multidrug anticonvulsant therapy. In one trial of patients on felbamate monotherapy, the primary adverse effects were anorexia, altered taste perception and insomnia *(121)*. A recent report has associated the drug with choreoathetosis *(123)*. Skin necrosis has also been reported, usually in association with multiple drug regimens, but also during monotherapy *(124)*. In general, the majority of these adverse effects have been minor.

Unfortunately, postmarketing surveillance has shown that felbamate is associated with an unacceptably high incidence of aplastic anemia *(125)*. As of January 1995, 33 cases of aplastic anemia were reported. The estimated incidence is 1:10,000 *(125)*. In comparison, the incidence of aplastic anemia in patients taking carbamazepine is estimated to be at most 1:200,000 *(126)*, and probably is significantly less. The felbamate related anemia may not respond to discontinuation of the drug and least eight fatalities have been reported. Other blood dyscrasias such as thrombocytopenia have also been reported *(127)*. All of the reported cases have occurred between 2.5 and 6 mo after the start of therapy.

These findings have led to the manufacturer's (Carter Wallace) voluntary withdrawal of the drug from the market. Currently, chronic felbamate treatment is indicated only in unusual circumstances such as refractory seizures in which no other alternatives are available.

Felbamate has not been used clinically for the treatment of acute stroke. Of interest, the levels of felbamate found in brain tissue obtained from temporal lobectomy patients has ranged from 13–73 µg/mL (dosages in these patients ranged from 2000–3600 mg/d) *(128)*. In rat experimental models of cerebral ischemia, brain tissue neuroprotective levels of felbamate have ranged from 35–60 µg/mL; at a dose of 250–500 mg/kg *(103,104,129)*. Therefore, potentially neuroprotective levels of felbamate are likely attainable in humans at current dosages. Because of felbamate's relatively low short term side-effect profile, these levels appear to be safe in vivo.

Based on the recent adverse experience in epilepsy patients, it seems unlikely that this drug will be tested for the treatment of stroke in the near future. However experimental data suggests that it could be potentially useful in this setting, since short term use should limit the risk of aplastic anemia. This issue will require further investigation, as well as commitment on the part of the drug's manufacturer and government regulatory authorities.

5.2. ACEA 1021

A phase I trial in normal volunteers has been completed and an early phase II trial in stroke patients is nearing completion with ACEA 1021. In the phase I trial, 30 healthy male volunteers had no serious adverse events when doses ranging from 0.013–2.0 mg/kg were given intravenously over 15 min. At the highest doses, some transient sedation, dizziness, and nausea occurred. Serum drug levels at the highest doses were above the levels that were neuroprotective in in vivo ischemia models *(130)*. Pharmacokinetic data from this study suggested that the terminal half-life was about 6 h.

5.3. ACPC

ACPC is in the early phases of clinical testing in patients undergoing coronary artery bypass, but no published data are yet available *(91,131)*.

6. CONCLUSIONS

The role of strychnine-insensitive glycine site antagonists in the treatment of stroke appears promising. Glycine receptor antagonists are characterized by a functional effect on the NMDA receptor that is comparable to direct antagonists. However, unlike the competitive and noncompetitive NMDA antagonists, their side-effect profile appears to be more favorable. Whereas NMDA antagonists have a high incidence of neuropsychiatric and behavioral side-effects, animal research and early and clinical experience suggests that these adverse effects are much less apparent with glycine antagonists (Table 1).

The initial glycine antagonists studied were hampered by poor CNS penetration and lower potency. Newer formulations with higher potency and acceptable blood–brain barrier permeability have been developed and clinical trials to evaluate these compounds are underway. The clinical effectiveness of these agents will be determined by large-scale efficacy and safety trials. Overall, however, several of these compounds apper to have the potential to become safe and effective agents for the acute treatment of brain ischemia.

REFERENCES

1 Albers, G. W., Atkinson, R. P., Kelley, R. E., and Rosenbaum, D. (on behalf of the Dextrorphan Study Group), Safety, tolerability, and pharmakokinetics of the N-methyl-D-aspartate antagonist dextrorphan in patients with acute stroke, *Stroke,* **26** (1995) 254–258.

2 Grotta, J., Clark, W., Coull, B., Pettigrew, L. C., Mackay, B., Goldstein, L. B., Meissner, I., Murphy, D., and LaRue, L., Safety and tolerability of the glutamate antagonist CGS 19755 (Selfotel) in patients with acute ischemic stroke. Results of a phase IIa randomized trial, *Stroke,* **26** (1995) 602–605.

3 Steinberg, G. K., Bell, T. E., and Yenari, M. A., Dose escalation safety and tolerance study of the NMDA antagonist, dextromethorphan in neurosurgical patients, *J. Neurosurg.,* **84** (1996) 860–866.

4 Steinberg, G. K., Perez-Pinzon, M. A., Maier, C. M., Sun, G. H., Yoon, E., Kunis, D. M., Bell, T. E., Powell, M., Kotake, A., and Giffard, R. G., CGS 19755 (selfotel): Correlation of in vitro neuroprotection, protection against experimental ischemia and CSF levels in cerebrovascular surgery patients, in J. Krieglstein and H. Oberpichler-Schwenk (eds.), *Pharmacology of Cerebral Ischemia*, Wissenschaftliche Verlagsgesellschaft mbH, Stuttgart, 1994, pp. 225–232.

5 Albers, G. W., Potential therapeutic uses of N-methyl-D-aspartate antagonists in cerebral ischemia, *Clin. Neuropharmacol.,* **13** (1990) 177–197.

6 Carter, A. J., Glycine antagonists: regulation of the NMDA receptor-channel complex by the strychnine-insensitive glycine site, *Drugs of the Future,* **17** (1992) 595–613.

7 Foster, A. C. and Fagg, G. E., Neurobiology. Taking apart NMDA receptors, *Nature,* **329** (1987) 395–396.

8 Bristow, D. R., Bowery, N. G., and Woodruff, G. N., Light microscopic autoradiographic localisation of [3H]glycine and [3H]strychnine binding sites in rat brain, *Eur. J. Pharmacol.,* **126** (1986) 303–307.

9 DeFeudis, F. V., Central glycine-receptors, *Gen. Pharmacol.,* **9** (1978) 139–144.

10 DeFeudis, F. V., Orensanz Muñoz, L. M., and Fando, J. L., High-affinity glycine binding sites in rat CNS: regional variation and strychnine sensitivity, *Gen. Pharmacol.,* **9** (1978) 171–176.

11 Forsythe, I. D., Westbrook, G. L., and Mayer, M. L., Modulation of excitatory synaptic transmission by glycine and zinc in cultures of mouse hippocampal neurons, *J. Neurosci.,* **8** (1988) 3733–3741.

12 Johnson, J. W. and Ascher, P., Glycine potentiates the NMDA response in cultured mouse brain neurons, *Nature,* **325** (1987) 529–531.

13 Thomson, A. M., Walker, V. E., and Flynn, D. M., Glycine enhances NMDA-receptor mediated synaptic potentials in neocortical slices, *Nature,* 338 (1989) 422–424.

14 Kemp, J. A. and Leeson, P. D., The glycine site of the NMDA receptor—five years on, *Trends Pharmacol. Sci.,* **14** (1993) 20–25.

15 Kleckner, N. W. and Dingledine, R., Requirement for glycine in activation of NMDA-receptors expressed in *Xenopus* oocytes, *Science,* **241** (1988) 835–837.

16 Sircar, R. and Zukin, S. R., Kinetic mechanisms of glycine requirement for N-methyl-D-aspartate channel activation, *Brain Res.,* **556** (1991) 280–284.

17 Dalkara, T., Erdemli, G., Barun, S., and Onur, R., Glycine is required for NMDA receptor activation: electrophysiological evidence from intact rat hippocampus, *Brain Res.,* **576** (1992) 197–202.

18 Fletcher, E. J., and Lodge, D., Glycine reverses antagonism of N-methyl-D-aspartate (NMDA) by 1-hydroxy-3-aminopyrrolidone-2 (HA-966) but not by D-2-amino-5-phosphonovalerate (D-AP5) on rat cortical slices, *Eur. J. Pharmacol.,* **151** (1988) 161–162.

19 Benveniste, M., Clements, J., Vyklicky, L., Jr., and Mayer, M. L., A kinetic analysis of the modulation of N-methyl-D-aspartic acid receptors by glycine in mouse cultured hippocampal neurones, *J. Physiol. (Lond.),* **428** (1990) 333–357.

20 Kemp, J. A. and Priestley, T., Effects of (+)-HA-966 and 7-chlorokynurenic acid on the kinetics of N-methyl-D-aspartate receptor agonist responses in rat cultured cortical neurons, *Mol. Pharmacol.,* **39** (1991) 666–670.

21 Lester, R. A., Tong, G., and Jahr, C. E., Interactions between the glycine and glutamate binding sites of the NMDA receptor, *J. Neurosci.,* **13** (1993) 1088–1096.

22 Mayer, M. L., Vyklicky, L., Jr., and Clements, J., Regulation of NMDA receptor desensitization in mouse hippocampal neurons by glycine, *Nature,* **338** (1989) 425–427.

23 Vyklicky, L., Jr., Benveniste, M., and Mayer, M. L., Modulation of N-methyl-D-aspartic acid receptor desensitization by glycine in mouse cultured hippocampal neurones, *J. Physiol. (Lond.),* **428** (1990) 313–331.

24 Baker, A. J., Zornow, M. H., Grafe, M. R., Scheller, M. S., Skilling, S. R., Smullin, D. H., and Larson, A. A., Hypothermia prevents ischemia-induced increases in hippocampal glycine concentrations in rabbits, *Stroke,* **22** (1991) 666–673.

25 Globus, M. Y., Busto, R., Martinez, E., Valdes, I., Dietrich, W. D., and Ginsberg, M. D., Comparative effect of transient global ischemia on extracellular levels of glutamate, glycine, and gamma-aminobutyric acid in vulnerable and nonvulnerable brain regions in the rat, *J. Neurochem.,* **57** (1991) 470–478.

26 Globus, M. Y., Ginsberg, M. D., and Busto, R., Excitotoxic index—a biochemical marker of selective vulnerability, *Neurosci. Lett.,* **127** (1991) 39–42.

27 Takagi, K., Ginsberg, M. D., Globus, M. Y., Dietrich, W. D., Martinez, E., Kraydieh, S., and Busto, R., Changes in amino acid neurotransmitters and cerebral blood flow in the ischemic penumbral region following middle cerebral artery occlusion in the rat: correlation with histopathology, *J. Cereb. Blood Flow Metab.,* **13** (1993) 575–585.

28 Patel, J., Zinkand, W. C., Thompson, C., Keith, R., and Salama, A., Role of glycine in the N-methyl-D-aspartate-mediated neuronal cytotoxicity, *J. Neurochem.,* **54** (1990) 849–854.

29 McNamara, D., and Dingledine, R., Dual effect of glycine on NMDA-induced neurotoxicity in rat cortical cultures, *J. Neurosci.,* **10** (1990) 3970–3976.

30 Moroni, F., Russi, P., Lombardi, G., Beni, M., and Carla, V., Presence of kynurenic acid in the mammalian brain, *J. Neurochem.,* **51** (1988) 177–180.

31 Henderson, G., Johnson, J. W., and Ascher, P., Competitive antagonists and partial agonists at the glycine modulatory site of the mouse N-methyl-D-aspartate receptor, *J. Physiol.,* **430** (1990) 189–212.

32 Kessler, M., Terramani, T., Lynch, G., and Baudry, M., A glycine site associated with N-methyl-D-aspartic acid receptors: characterization and identification of a new class of antagonists, *J. Neurochem.,* **52** (1989) 1319–1328.

33 Moroni, F., Pellegrini-Giampietro, D. E., Alesiani, M., Cherici, G., Mori, F., and Galli, A., Glycine and kynurenate modulate the glutamate receptors in the myenteric plexus and in cortical membranes, *Eur. J. Pharmacol.,* **163** (1989) 123–126.

34 Birch, P. J., Grossman, C. J., and Hayes, A. G., Kynurenic acid antagonises responses to NMDA via an action at the strychnine-insensitive glycine receptor, *Eur. J. Pharmacol.,* **154** (1988) 85–87.

35 Kemp, J. A., Foster, A. C., Leeson, P. D., Priestley, T., Tridgett, R., Iversen, L. L., and Woodruff, G. N., 7-Chlorokynurenic acid is a selective antagonist at the glycine modulatory site of the N-methyl-D-aspartate receptor complex, *Proc. Natl. Acad. Sci. USA,* **85** (1988) 6547–6550.

36 Ransom, R. W., and Deschenes, N. L., NMDA-induced hippocampal [3H]norepinephrine release is modulated by glycine, *Eur. J. Pharmacol.,* **156** (1988) 149–155.

37 Watson, G. B., Hood, W. F., Monahan, J. B., Lanthorm and H., T., Kynurenate antagonizes actions of N-methyl-D-aspartate through a glycine-sensitive receptor, *Neurosci. Res. Commun.,* **2** (1988) 169–174.

38 Foster, A. C., Kemp, J. A., Leeson, P. D., Grimwood, S., Donald, A. E., Marshall, G. R., Priestley, T., Smith, J. D., and Carling, R. W., Kynurenic acid analogues with improved affinity and selectivity for the glycine site on the N-methyl-D-aspartate receptor from rat brain, *Mol. Pharmacol.,* **41** (1992) 914–922.

39 Kendall, D. A. and Robinson, J. P., The glycine antagonist 7-chlorokynurenic acid blocks the effects of N-methyl-D-aspartate on agonist-stimulated phosphoinositide hydrolysis in guinea-pig brain slices, *J. Neurochem.,* **55** (1990) 1915–1919.

40 Baron, B. M., Harrison, B. L., Miller, F. P., McDonald, I. A., Salituro, F. G., Schmidt, C. J., Sorensen, S. M., White, H. S., and Palfreyman, M. G., Activity of 5,7-dichlorokynurenic acid, a potent antagonist at the N-methyl-D-aspartate receptor-associated glycine binding site, *Mol. Pharmacol.,* **38** (1990) 554–61.

41 McNamara, D., Smith, E. C., Calligaro, D. O., PJ, O. M., McQuaid, L. A., and Dingledine, R., 5,7-Dichlorokynurenic acid, a potent and selective competitive antagonist of the glycine site on NMDA receptors, *Neurosci. Lett.,* **120** (1990) 17–20.

42 Moroni, F., Alesiani, M., Facci, L., Fadda, E., Skaper, S. D., Galli, A., Lombardi, G., Mori, F., Ciuffi, M., Natalini, B., and Pellicciari, R., Thiokynurenates prevent excitotoxic neuronal death in vitro and in vivo by acting as glycine antagonists and as inhibitors of lipid peroxidation, *Eur. J. Pharmacol.,* **218** (1992) 145–151.

43 Moroni, F., Alesiani, M., Galli, A., Mori, F., Pecorari, R., Carla, V., Cherici, G., and Pellicciari, R., Thiokynurenates: a new group of antagonists of the glycine modulatory site of the NMDA receptor, *Eur. J. Pharmacol.,* **199** (1991) 227–232.

44 Priestley, T., Horne, A. L., McKernan, R. M., and Kemp, J. A., The effect of NMDA receptor glycine site antagonists on hypoxia-induced neurodegeneration of rat cortical cell cultures, *Brain Res.,* **531** (1990) 183–188.

45 Shalaby, I., Chenard, B., and Prochniak, M., Glycine reverses 7-Cl kynurenate blockade of glutamate neurotoxicity in cell culture, *Eur. J. Pharmacol.,* **160** (1989) 309–311.

46 Newell, D. W., Barth, A., and Malouf, A. T., Glycine site NMDA receptor antagonists provide protection against ischemia-induced neuronal damage in hippocampal slice cultures, *Brain Res.,* **675** (1995) 38–44.

47 Germano, I. M., Pitts, L. H., Meldrum, B. S., Bartkowski, H. M., and Simon, R. P., Kynurenate inhibition of cell excitation decreases stroke size and deficits, *Ann. Neurol.,* **22** (1987) 730–734.

48 Verrecchia, C., Lakhmeche, N., Boulu, R. G., and Plotkine, M., Neuroprotective effect induced by glycine receptor antagonists in a model of focal cerebral ischaemia in mice, *J. Cereb. Blood Flow Metab.,* **15** (1995) S368.

49 Chen, J., Graham, S., Moroni, F., and Simon, R., A study of the dose dependency of a glycine receptor antagonist in focal ischemia, *J. Pharmacol. Exp. Ther.,* **267** (1993) 937–941.

50 Wood, E. R., Bussey, T. J., and Phillips, A. G., A glycine antagonist reduces ischemia-induced CA1 cell loss in vivo, *Neurosci. Lett.,* **145** (1992) 10–14.

51 Pellegrini-Giampietro, D. E., Cozzi, A., and Moroni, F., The glycine antagonist and free radical scavenger 7-Cl-thio-kynurenate reduces CA1 ischemic damage in the gerbil, Neuroscience, **63** (1994) 701–709.

52 Uckele, J. E., McDonald, J. W., Johnston, M. V., and Silverstein, F. S., Effect of glycine and glycine receptor antagonists on NMDA-induced brain injury, *Neurosci. Lett.,* **107** (1989) 279–283.

53 Foster, A. C., Donald, A. E., Willis, C. L., Tridgett, R., Kemp, J. A., and Priestley, T., The glycine site on the NMDA receptor: pharmacology and involvement in NMDA receptor-mediated neurodegeneration, *Adv. Exp. Med. Biol.,* **268** (1990) 93–100.

54 Foster, A. C., Willis, C. L., and Tridgett, R., Protection against N-methyl-D-aspartate receptor-mediated neuronal degeneration in rat brain by 7-chlorokynurenate and 3-amino-1-hydroxypyrrolid-2-one, antagonists at the allosteric site for glycine, *Eur. J. Neurosci.,* **2** (1989) 270–277.

55 Smith, D. H., Okiyama, K., Thomas, M. J., and McIntosh, T. K., Effects of the excitatory amino acid receptor antagonists kynurenate and indole-2-carboxylic acid on behavioral and neurochemical outcome following experimental brain injury, *J. Neurosci.,* **13** (1993) 5383–5392.

56 Bonta, I. L., De Vos, C. J., and Grijsen, H., 1-hydroxy-3-amino-pyrrolidone-2 (HA-966). II. Extrapyramidal aspects, *Arch. Int. Pharmacodyn. Ther.,* **182** (1969) 394,395.

57 Bonta, I. L., De Vos, C. J., and Hillen, F. C., 1-hydroxy-3-amino-pyrrolidone-2 (HA-966). I. Behaviour and motor effects, *Arch. Int. Pharmacodyn. Ther.,* **182** (1969) 391–393.

58 Foster, A. C. and Kemp, J. A., HA-966 antagonizes N-methyl-D-aspartate receptors through a selective interaction with the glycine modulatory site, *J. Neurosci.,* **9** (1989) 2191–2196.

59 Keith, R. A., Mangano, T. J., Meiners, B. A., Stumpo, R. J., Klika, A. B., Patel, J., and Salama, A. I., HA-966 acts at a modulatory glycine site to inhibit N-methyl-D-aspartate-evoked neurotransmitter release, *Eur. J. Pharmacol.,* **166** (1989) 393–400.

60 Reynolds, I. J., Harris, K. M., and Miller, R. J., NMDA receptor antagonists that bind to the strychnine-insensitive glycine site and inhibit NMDA-induced Ca2+ fluxes and [3H]GABA release, *Eur. J. Pharmacol.,* **172** (1989) 9–17.

61 Bonta, I. L., De Vos, C. J., Grijsen, H., Hillen, F. C., Noach, E. L., and Sim, A. W., 1-Hydroxy-3-amino-pyrrolidone-2(HA-966): a new GABA-like compound, with potential use in extrapyramidal diseases, *Br. J. Pharmacol.,* **43** (1971) 514–535.

62 Wood, P. L., Emmett, M. R., Rao, T. S., Mick, S., Cler, J., Oei, E., and Iyengar, S., In vivo antagonism of agonist actions at N-methyl-D-aspartate and N-methyl-D-aspartate-associated glycine receptors in mouse cerebellum: studies of 1-hydroxy-3-aminopyrrolidone-2, *Neuropharmacology,* **29** (1990) 675–679.

63 Foster, A. C., Donald, A. E., Grimwood, S., Leeson, P. D., and Williams, B. J., Activities of 4-methyl derivatives of HA-966 at the glycine site of the N-methyl-D-aspartate receptor from rat brain, *Br. J. Pharmacol.*, **102** (1991) 64P.

64 Kemp, J. A., Priestley, T., Marshall, G. R., Leeson, P. D., and Williams, B. J., Functional assessment of the actions of 4-methyl derivatives of HA-966 at the glycine site of the N-methyl-D-aspartate receptor, *Br. J. Pharmacol.*, **102** (1991) 65P.

65 Tricklebank, M. D., Bristow, L. J., Hutson, P. H., Leeson, P. D., Rowley, M., Saywell, K., Singh, L., Tattersall, F. D., Thorn, L., and Williams, B. J., The anticonvulsant and behavioural profile of L-687,414, a partial agonist acting at the glycine modulatory site on the N-methyl-D-aspartate (NMDA) receptor complex, *Br. J. Pharmacol.*, **113** (1994) 729–736.

66 Saywell, K., Singh, L., Oles, R. J., Vass, P. D., Leeson, P. D., Williams, B. J., and Tricklebank, M. D., The anticonvulsant properties in the mouse of the glycine/NMDA receptor antagonist, L-687,414, *Br. J. Pharmacol.*, **102** (1991) 66P.

67 Gill, R., Hargreaves, R. J., and Kemp, J. A., The neuroprotective effect of the glycine site antagonist 3R-(+)-cis-4-methyl-HA966 (L-687,414) in a rat model of focal ischaemia, *J. Cereb. Blood Flow Metab.*, **15** (1995) 197–204.

68 Warner, D. S., Martin, H., Ludwig, P., McAllister, A., Keana, J. F., and Weber, E., In vivo models of cerebral ischemia: effects of parenterally administered NMDA receptor glycine site antagonists, *J. Cereb. Blood Flow Metab.*, **15** (1995) 188–196.

69 McDonald, J. W., Uckele, J., Silverstein, F. S., and Johnston, M. V., HA-966 (1-hydroxy-3-aminopyrrolidone-2) selectively reduces N-methyl-D-aspartate (NMDA)-mediated brain damage, *Neurosci. Lett.*, **104** (1989) 167–170.

70 McDonald, J. W. and Johnston, M. V., Pharmacology of N-methyl-D-aspartate-induced brain injury in an in vivo perinatal rat model, *Synapse*, **6** (1990) 179–188.

71 Huettner, J. E., Indole-2-carboxylic acid: a competitive antagonist of potentiation by glycine at the NMDA receptor, *Science*, **243** (1989) 1611–1613.

72 Gray, N. M., Dappen, M. S., Cheng, B. K., Cordi, A. A., Biesterfeldt, J. P., Hood, W. F., and Monahan, J. B., Novel indole-2-carboxylates as ligands for the strychnine-insensitive N-methyl-D-aspartate-linked glycine receptor, *J. Med. Chem.*, **34** (1991) 1283–1292.

73 Rao, T. S., Gray, N. M., Dappen, M. S., Cler, J. A., Mick, S. J., Emmett, M. R., Iyengar, S., Monahan, J. B., Cordi, A. A., and Wood, P. L., Indole-2-carboxylates, novel antagonists of the N-methyl-D-aspartate (NMDA)-associated glycine recognition sites: in vivo characterization, *Neuropharmacology*, **32** (1993) 139–147.

74 Cai, S. X., Dinsmore, C., Gee, K. R., Glenn, A. G., huang, I. C., Johnson, B. L., Kher, S. M., Lu, Y., Oldfield, P. L., Marck, P., Zheng, H., Weber, B., and Keana, J. F. W., Synthesis and activity of substituted quinoxaline 2,3 diones as antagonists for the glycine/NMDA receptor, *Soc. Neurosci. Abs.*, **19** (1993) 718.

75 Woodward, R. M., Huettner, J. E., Guastella, J., Keana, J. F., and Weber, E., In vitro pharmacology of ACEA-1021 and ACEA-1031: systemically active quinoxalinediones with high affinity and selectivity for N-methyl-D-aspartate receptor glycine sites, *Mol. Pharmacol.*, **47** (1995) 568–581.

76 Marek, P., Eilon, G., Keana, J., and Weber, E., The neuroprotective effect of three novel NMDA receptor glycine site antagonists in a rat model of focal cerebral ischemia, *Soc. Neurosci. Abs.*, **20** (1994) 184.

77 Sauer, D., D'Amato, F., and Allegrini, P. R., The glycine site NMDA antagonist ACEA 1021 attenuates brain damage determined 4 weeks after MCA occlusion in rats, *J. Cereb. Blood Flow Metab.*, **14** (1995) S145.

78 Pietra, C., Ziviani, L., Valerio, E., Tarter, G., and Reggiani, A., Neuroprotection by GV150526A: comparison with reference compounds, *Stroke*, **27** (1996) 189.

79 Myseros, J., Oetting, G. M., Bullock, M., Corwin, F., and Fatouros, P., Focal ischemia after therapy with a glutamate antagonist-preserved tissue energy metabolism and reductin in infarct size, *J. Cereb. Blood Flow Metab.,* **15** (1995) S367.

80 Tsuchida, E., and Bullock, R., The effect of the glycine site-specific N-methyl-D-aspartate antagonist ACEA1021 on ischemic brain damage caused by acute subdural hematoma in the rat, *J. Neurotrauma,* **12** (1995) 279–88.

81 Yasuda, H., Tanaka, H., Ohtani, K., Kato, T., Maruoka, Y., Kawabe, A., Kawashima, C., Sakamoto, H., Tojo, S., and Nakamura, M., Neuroprotective effects of a novel NMDA glycine site antagonist, SM-18400, in the rat model of cerebral ischemia, *Soc. Neurosci. Abs.,* **21** (1995).

82 Tanaka, H., Yasuda, H., Kato, T., Maruoka, Y., Kawabe, A., Kawasima, C., Ohtani, K., and Nakamura, M., Neuroprotective and anticonvulsant actions of SM-18400, a novel strychnine-insensitive glycine site antagonist, *J. Cereb. Blood Flow Metab.,* **15** (1995) S431.

83 Compton, R. P., Hood, W. F., and Monahan, J. B., Evidence for a functional coupling of the NMDA and glycine recognition sites in synaptic plasma membranes, *Eur. J. Pharmacol.,* **188** (1990) 63–70.

84 Nadler, V., Kloog, Y., and Sokolovsky, M., 1-Aminocyclopropane-1-carboxylic acid (ACC) mimics the effects of glycine on the NMDA receptor ion channel, *Eur. J. Pharmacol.,* **157** (1988) 115,116.

85 Watson, G. B. and Lanthorn, T. H., Pharmacological characteristics of cyclic homologues of glycine at the N-methyl-D-aspartate receptor-associated glycine site, *Neuropharmacology,* **29** (1990) 727–730.

86 Hood, W. F., Sun, E. T., Compton, R. P., and Monahan, J. B., 1-Aminocyclobutane-1-carboxylate (ACBC): a specific antagonist of the N-methyl-D-aspartate receptor coupled glycine receptor, *Eur. J. Pharmacol.,* **161** (1989) 281–282.

87 Watson, G. B., Bolanowski, M. A., Baganoff, M. P., Deppeler, C. L., and Lanthorn, T. H., Glycine antagonist action of 1-aminocyclobutane-1-carboxylate (ACBC) in Xenopus oocytes injected with rat brain mRNA, *Eur. J. Pharmacol.,* **167** (1989) 291–294.

88 Marvizon, J. C., Lewin, A. H., and Skolnick, P., 1-Aminocyclopropane carboxylic acid: a potent and selective ligand for the glycine modulatory site of the N-methyl-D-aspartate receptor complex, *J. Neurochem.,* **52** (1989) 992–994.

89 Miller, R., La Grone, J., Skolnick, P., and Boje, K. M., High-performance liquid chromatographic assay for 1-aminocyclopropanecarboxylic acid from plasma and brain, *J. Chromatogr.,* **578** (1992) 103–108.

90 Rao, T. S., Cler, J. A., Compton, R. P., Emmett, M. R., Mick, S., Sun, E. T., Iyengar, S., and Wood, P. L., Neuropharmacological characterization of 1-aminocyclopropane-1-carboxylate and 1-aminocyclobutane-1-carboxylate, ligands of the N-methyl-D-aspartate-associated glycine receptor, *Neuropharmacology,* **29** (1990) 305–309.

91 Trullas, R., Folio, T., Young, A., Miller, R., Boje, K., and Skolnick, P., 1-aminocyclopropanecarboxylates exhibit antidepressant and anxiolytic actions in animal models, *Eur. J. Pharmacol.,* **203** (1991) 379–385.

92 Fossom, L. H., Von Lubitz, D. K., Lin, R. C., and Skolnick, P., Neuroprotective actions of 1-aminocyclopropanecarboxylic acid (ACPC): a partial agonist at strychnine-insensitive glycine sites, *Neurol. Res.,* **17** (1995) 265–269.

93 Boje, K. M., Wong, G., and Skolnick, P., Desensitization of the NMDA receptor complex by glycinergic ligands in cerebellar granule cell cultures, *Brain Res.,* **603** (1993) 207–214.

94 Von Lubitz, D. K., Lin, R. C., McKenzie, R. J., Devlin, T. M., McCabe, R. T., and Skolnick, P., A novel treatment of global cerebral ischaemia with a glycine partial agonist, *Eur. J. Pharmacol.,* **219** (1992) 153–158.

95 Long, J. B. and Skolnick, P., 1-Aminocyclopropanecarboxylic acid protects against dynorphin A-induced spinal injury, *Eur. J. Pharmacol.,* 261 (1994) 295–301.

96 McCabe, R. T., Wasterlain, C. G., Kucharczyk, N., Sofia, R. D., and Vogel, J. R., Evidence for anticonvulsant and neuroprotectant action of felbamate mediated by strychnine-insensitive glycine receptors, *J. Pharmacol. Exp. Ther.,* **264** (1993) 1248–1252.

97 Rho, J. M., Donevan, S. D., and Rogawski, M. A., Mechanism of action of the anticonvulsant felbamate: opposing effects on N-methyl-D-aspartate and gamma-aminobutyric acidA receptors, *Ann. Neurol.,* **35** (1994) 229–234.

98 White, H. S., Harmsworth, W. L., Sofia, R. D., and Wolf, H. H., Felbamate modulates the strychnine-insensitive glycine receptor, *Epilepsy Res.,* **20** (1995) 41–48.

99 Subramaniam, S., Rho, J. M., Penix, L., Donevan, S. D., Fielding, R. P., and Rogawski, M. A., Felbamate block of the N-methyl-D-aspartate receptor, *J. Pharmacol. Exp. Ther.,* **273** (1995) 878–886.

100 Wallis, R. A., Panizzon, K. L., and Nolan, J. P., Glycine-induced CA1 excitotoxicity in the rat hippocampal slice, *Brain Res.,* **664** (1994) 115–125.

101 Wallis, R. A. and Panizzon, K. L., Glycine reversal of felbamate hypoxic protection, *Neuroreport,* **4** (1993) 951–954.

102 Wallis, R. A., Panizzon, K. L., Fairchild, M. D., and Wasterlain, C. G., Protective effects of felbamate against hypoxia in the rat hippocampal slice, *Stroke,* **23** (1992) 547–551.

103 Wasterlain, C. G., Adams, L. M., Hattori, H., and Schwartz, P. H., Felbamate reduces hypoxic-ischemic brain damage in vivo, *Eur. J. Pharmacol.,* **212** (1992) 275–278.

104 Wasterlain, C. G., Adams, L. M., Schwartz, P. H., Hattori, H., Sofia, R. D., and Wichmann, J. K., Posthypoxic treatment with felbamate is neuroprotective in a rat model of hypoxia-ischemia, *Neurology,* **43** (1993) 2303–1230.

105 Takano, K., Formato, J. E., Tatlisumak, T., Carano, R. A. D., Sotak, C. H., Bare, T. M., Patel, J. B., and Fisher, M., Effects of the glycine antagonist, ZD9379, on experimental focal brain ischemia in rats, *Stroke,* **27** (1996) 188.

106 Fix, A. S., Horn, J. W., Wightman, K. A., Johnson, C. A., Long, G. G., Storts, R. W., Farber, N., Wozniak, D. F., and Olney, J. W., Neuronal vacuolization and necrosis induced by the noncompetitive N-methyl-D-aspartate (NMDA) antagonist MK(+)801 (dizocilpine maleate): a light and electron microscopic evaluation of the rat retrosplenial cortex, *Exp. Neurol.,* **123** (1993) 204–215.

107 Nakki, R., Koistinaho, J., Sharp, F. R., and Sagar, S. M., Cerebellar toxicity of phencyclidine, *J. Neurosci.,* **15** (1995) 2097–2108.

108 Olney, J. W., Labruyere, J., and Price, M. T., Pathological changes induced in cerebrocortical neurons by phencyclidine and related drugs [see comments], *Science,* **244** (1989) 1360–1362.

109 Olney, J. W., Labruyere, J., Wang, G., Wozniak, D. F., Price, M. T., and Sesma, M. A., NMDA antagonist neurotoxicity: mechanism and prevention, *Science,* **254** (1991) 1515–1518.

110 Sharp, F. R., Butman, M., Aardalen, K., Nickolenko, J., Nakki, R., Massa, S. M., Swanson, R. A., and Sagar, S. M., Neuronal injury produced by NMDA antagonists can be detected using heat shock proteins and can be blocked with antipsychotics, *Psychopharmacol. Bull.,* **30** (1994) 555–560.

111 Sharp, F. R., Jasper, P., Hall, J., Noble, L., and Sagar, S. M., MK-801 and ketamine induce heat shock protein HSP72 in injured neurons in posterior cingulate and retrosplenial cortex, *Ann. Neurol.,* **30** (1991) 801–809.

112 Auer, R. N., Keana, F. W., and Weber, E., New structural findings in brain treated with a glycine antagonist, ACEA-1021, *J. Cereb. Blood Flow Metab.,* **15** (1995) S430.

113 Hargreaves, R. J., Rigby, M., Smith, D., and Hill, R. G., Lack of effect of L-687,414 ((+)-cis-4-methyl-HA-966), an NMDA receptor antagonist acting at the glycine site, on cerebral glucose metabolism and cortical neuronal morphology, *Br. J. Pharmacol.,* **110** (1993) 36–42.

114 Chiamulera, C., Costa, S., and Reggiani, A., Effect of NMDA- and strychnine-insensitive glycine site antagonists on NMDA-mediated convulsions and learning, *Psychopharmacology (Berl.),* **102** (1990) 551–552.

115 Koek, W. and Colpaert, F. C., Selective blockade of N-methyl-D-aspartate (NMDA)-induced convulsions by NMDA antagonists and putative glycine antagonists: relationship with phencyclidine-like behavioral effects, *J. Pharmacol. Exp. Ther.,* **252** (1990) 349–357.

116 Menon, M. K., Demonstration of the antimyoclonic effect of 1-hydroxy-3-amino-pyrrolidone-2 (HA-966) using a new animal model, *Life Sci.,* **28** (1981) 2865–2868.

117 Bourgeois, B. F., Felbamate in the treatment of partial-onset seizures, *Epilepsia,* **35(Suppl. 5)** (1994) S58–61.

118 Faught, E., Sachdeo, R. C., Remler, M. P., Chayasirisobhon, S., Iragui-Madoz, V. J., Ramsay, R. E., Sutula, T. P., Kanner, A., Harner, R. N., Kuzniecky, R., Kramer, L. D., Kamin, M., and Rosenberg, A., Felbamate monotherapy for partial-onset seizures: an active-control trial, *Neurology,* **43** (1993) 688–692.

119 Felbamate, S. G., Efficacy of felbamate in childhood epileptic encephalopathy (Lennox-Gastaut syndrome). The Felbamate Study Group in Lennox-Gastaut Syndrome, *N. Engl. J. Med.,* **328** (1993) 29–33.

120 Leppik, I. E., Dreifuss, F. E., Pledger, G. W., Graves, N. M., Santilli, N., Drury, I., Tsay, J. Y., Jacobs, M. P., Bertram, E., Cereghino, J. J., Cooper, G., Sahlroot, J. T., Sheridan, P., Ashworth, M., Lee, S. I., and Sierzant, T. L., Felbamate for partial seizures: results of a controlled clinical trial, *Neurology,* **41** (1991) 1785–1789.

121 Sachdeo, R., Kramer, L. D., Rosenberg, A., and Sachdeo, S., Felbamate monotherapy: controlled trial in patients with partial onset seizures [see comments], *Ann. Neurol.,* **32** (1992) 386–392.

122 Theodore, W. H., Raubertas, R. F., Porter, R. J., Nice, F., Devinsky, O., Reeves, P., Bromfield, E., Ito, B., and Balish, M., Felbamate: a clinical trial for complex partial seizures, *Epilepsia,* **32** (1991) 392–397.

123 Kerrick, J. M., Kelley, B. J., Maister, B. H., Graves, N. M., and Leppik, I. E., Involuntary movement disorders associated with felbamate, *Neurology,* **45** (1995) 185–187.

124 Travaglini, M. T., Morrison, R. C., Ackerman, B. H., Haith, L. R., Jr., and Patton, M. L., Toxic epidermal necrolysis after initiation of felbamate therapy, *Pharmacotherapy,* **15** (1995) 260–264.

125 Pennell, P. B., Ogaily, M. S., and Macdonald, R. L., Aplastic anemia in a patient receiving felbamate for complex partial seizures, *Neurology,* **45** (1995) 456–460.

126 Young, N., and Alter, B., *Aplastic Anemia: Acquired and Inherited.,* Saunders, Philadelphia, 1994.

127 Ney, G. C., Schaul, N., Loughlin, J., Rai, K., and Chandra, V., Thrombocytopenia in association with adjunctive felbamate use, *Neurology,* **44** (1994) 980,981.

128 Adusumalli, V. E., Wichmann, J. K., Kucharczyk, N., Kamin, M., Sofia, R. D., French, J., Sperling, M., Bourgeois, B., Devinsky, O., Dreifuss, F. E., Kuzniecky, R. I., Faught, E., and Rosenfeld, W., Drug concentrations in human brain tissue samples from epileptic patients treated with felbamate, *Drug Metab. Dispos.,* **22** (1994) 168–70.

129 Adusumalli, V. E., Wichmann, J. K., Kucharczyk, N., and Sofia, R. D., Distribution of the anticonvulsant felbamate to cerebrospinal fluid and brain tissue of adult and neonatal rats, *Drug Metab. Dispos.,* **21** (1993) 1079–1085.

130 Whitehouse, M. J., Navalta, L., Densel, M., Ballas, M., Davies, P. J. W., Enterling, D., Kurth, H.-J., and Weber, E., Initial clinical experience with ACEA 1021: safety, tolerance, and pharmacokinetcs in healthy voluntees, Grossman, C. J., and Hayes, A. G., Kynurenic acid antagonises responses to NMDA via an action at the strychnine-insensitive glycine receptor, *Eur. J. Pharmacol.,* **154** (1988) 85–87.

131 Maccecchini, M., ACPC, a partial agonist of the NMDA receptor, as a neuroprotectant in CABG, *Stroke,* **26** (1995) 163.

7

Free Radical-Mediated Cerebral Damage After Hypoxia/Ischemia and Stroke

Tomohiro Matsuyama

1. INTRODUCTION

Free radicals are not biological curiosities or anomalies but are byproducts of cellular physiology generated by specific enzymes, by autoxidation, and by energy transfer reactions in all life forms using oxygen. Under physiological conditions, sophisticated antioxidant defense systems prevent the reactive O_2 metabolites from potential injurious interactions with cell components critical for the viability of cells. However, under the conditions of hypoxia/ischemia or reoxygenation/reperfusion, the equilibrium between the O_2 radical and its antioxidant defense system is lost. This imbalance leads to radical-induced cell damage *(1)*. This chapter contains the mechanism of toxicity posed by oxygen stresses and the responses of brain cells to these common etiologies that may happen in stroke.

2. MECHANISM OF ISCHEMIA/HYPOXIA INJURY

2.1. Reactive Oxygen Species (Table 1)

The sequential four-electron reduction of oxygen, which occurs during normal oxidative mechanisms, results in the production of highly reactive, potentially tissue damaging, oxygen species including superoxide anion ($O2^-$), the hydroxyradical ($\cdot OH$), hydrogen peroxide (H_2O_2)*(2)*, and nitric oxide (NO) *(3)*. The brain contains rich amounts of unsaturated fatty acids and catecholamines, which are thought to be target molecules for peroxidation induced by reactive oxygen species (free radicals). Therefore, in the case that the huge amount of free radicals are produced by the neural cells, a free radical-induced cell damage happens easily. The mechanism for free radical production, however, still remains unclear because there are lots of difficulties in measuring radical species in vivo. Here, the characteristics of reactive oxygen species and their role in the cell pathology are described.

2.2. Superoxide

Superoxide anion (O_2^-) is a product of the radiolysis of oxygenated water, and at present, it is understood to be a common product of both spontaneous and enzyme-catalyzed oxidations. It is generated by the mitochondrial respiratory chain reaction to

From: Clinical Pharmacology of Cerebral Ischemia *Edited by:* G. J. Ter Horst and J. Korf Humana Press Inc., Totowa, NJ

Table 1
Reactive Oxygen Species

$O_2{\cdot}^-$	Superoxide radical
H_2O_2	Hydrogen peroxide
$\cdot OH$	Hydroxyl radical
$^1O_2$	Singlet oxygen
ROOH	Hydroperoxide[a]
ROO·	Peroxyl radical[a]
RO·	Alkoxyl radical[a]
NO	Nitric oxide
$ONOO^-$	Peroxynitrite

[a]R = lipid.

oxygen and hydrogen peroxide *(4,5)* or through xanthine-xanthine oxidase pathway *(6)*. Normally, superoxide dismutases (SODs) act to dismute the superoxide. Therefore, the concentration of intracellular superoxide is very low under homeostatic conditions in general. However, when the mitchondria are injured or the abnormal increase in uptake of oxygen is induced by vigorous hyperoxic or hyperbaric therapy, superoxide acts as cytotoxic byproduct of a variety of normal processes in aerobic metabolism. It also can react with hydrogen peroxide to generate singlet oxygen and hydroxyradicals, which are even more reactive and cytotoxic than superoxide or hydrogen peroxide *(6)*. Superoxide is known to inactivate both creatine kinase in cardiac muscle cells and glutamine synthetase which converts glutamate to glutamine in brain *(7–9)*. In addition, both superoxide and hydroxyradicals can increase the release of glutamate, which is known as an excitotoxic amino acid (EAA) and inducer of neuronal injury *(10)*.

2.3. Nitric Oxide

Nitric oxide (NO) is now known to be produced in a variety of mammalian cells such as neurons, platelets, macrophages, renal mesangial cells, and vascular smooth muscle cells. It was at first identified as an endothelium-derived relaxing factor (EDRF) *(3,11,12)*. In blood vessels, it acts as a strong vasodilator via the activation of intracellular cyclic GMP, and in platelets it inhibits platelet-aggregation. NO also has been proposed as a neurotransmitter both in the central *(13,14)* and peripheral *(15)* nervous systems. In the hippocampus, it is suggested that NO liberated from postsynaptic neurons travel back to presynaptic terminals to cause long-term potentiation (LTP), which is a type of synaptic plasticity that is thought to contribute to certain forms of learning in mammals *(16–18)*.

NO is synthesized from L-arginine as a base by nitric oxide synthase (NOS) *(19)*. It reacts with hemoglobin, and its stable decomposition products are nitrite (NO_2^-) and nitrate (NO_3^-). NO also reacts with superoxide to produce peroxynitrite anion ($ONOO^-$) *(20)*, which is even more toxic than superoxide (Fig. 1). Therefore, in the area in which the superoxide is generated abundantly, for example in the ischemic/hypoxic brain upon reperfusion/reoxygenation, the formation of peroxynitrite as well as other radicals accelerate the brain damage.

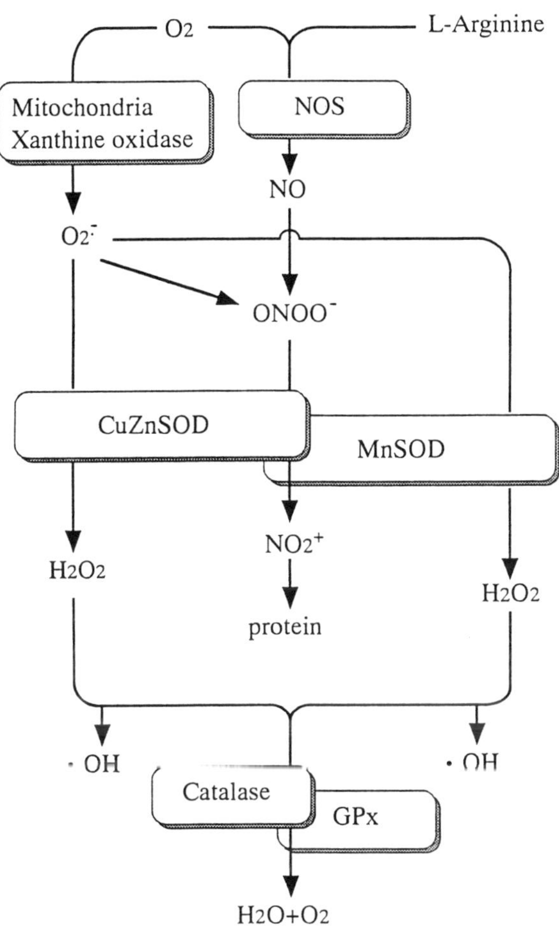

Fig. 1. Schematic representation of interaction between $O_2{}^{·-}$ and NO, and their metabolism. Most of mammalian cells possess two types of SOD (CuZnSOD and MnSOD) and NOS. The amount and function of radical species generated in the cells are considered to depend on the amount of these scavenging enzymes.

2.4. Excitatory Amino Acids and Free Radicals

Several pathophysiological mechanisms for cerebral ischemia/hypoxia injury have been proposed, such as increased excitotoxicity, intracellular calcium overload, inhibition of protein synthesis, and alteration of gene expression. An ischemic–hypoxic episode causes the release of excitatory amino acids (EAA) such as glutamate and aspartate from the presynaptic terminals, which synaptically stimulate the neuron to increase of Ca^{2+} influx. Such EAA-calcium theory has long received considerable attention. Recent studies have proposed that free radicals have a very important role in regulating this mechanism. In vitro studies have shown that superoxide stimulates the release of EAA, which in turn increases the production of free radicals *(10)*. NO also mediates the action of the excitatory neurotransmitter glutamate *(21,22)* via the NMDA receptor-mediated neurotransmitter release *(23)*. Thus, these two metabolic events (free radical formation

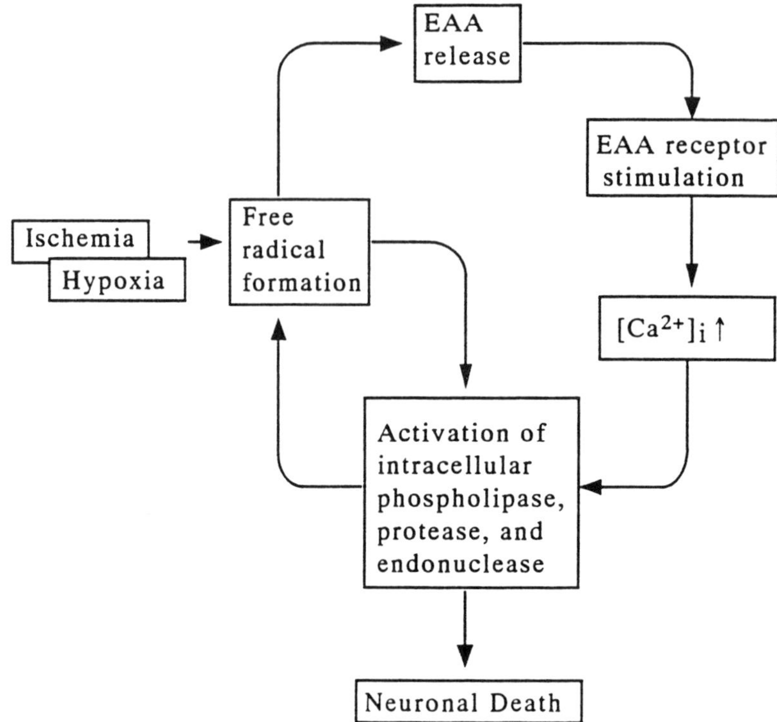

Fig. 2. Hypothetical vicious cycle between free radical formation and excitotoxicity. Since SOD mimetic prevents the excitotoxic cell death but not the Ca^{2+} influx induced by excitotoxic amino acids, the site of its action is considered to center on postreceptor events. Therefore, free radicals generated in the cell are suggested to activate directly the postreceptor signal transduction system *(31)*.

and excitatory amino acid release) are correlated. Moreover, they cooperated and form a vicious cycle that leads to necrotic or apoptotic neuronal death (Fig. 2).

Since Demopouros et al. *(24)* has reported a possible role of free radicals in neuronal death after brain ischemia, a number of in vivo studies have suggested that the formation of free radicals in the brain is an important step in the sequence of events that link cerebral blood flow reduction to excitotoxicity, during the acute ischemic attack and when blood and oxygen eventually return to the brain upon reperfusion *(25–30)*. These studies have not provided direct evidence for the effects of free radicals on the excitotoxic brain damage. For this, one need to measure formation of such reactive oxygen species directly in the brain. A recent work has shown that inactivation of the TCA cycle enzyme, aconitase, can be used as a marker of intracellular O_2^- levels *(31)*. The same study revealed that the cell-permeable SOD mimetic decreases both aconitase inactivation and cell death produced by excitotoxic amino acids, implying direct evidences for O_2^- generation in the pathway of excitotoxic injury (Fig. 2).

2.5. Cellular Sources of Oxygen-Free Radicals in Brain

Because of their very short life-span, it is very difficult to measure the free radical reaction in vivo. Several studies have attempted to detect the reaction by directly measuring the formation of free radicals or metabolites. These include the detection of radical formation following global ischemia, by electron spin resonance spectrometry (ESR) *(32)*, and the measuring of lipoperoxide *(33)* and conjugated diens formation *(34)*. NO formation by nitric oxide synthase (NOS) can be examined by enzymatic histochemical method using NADPH-diaphorase *(35)*.

By using the mentioned techniques, however, one could not determine the precise site of free radical generation in the brain following ischemia/hypoxia episodes. Brain parenchyma has very low amounts of xanthine oxidase *(36)* which is an important enzymatic origin for generation of superoxide. On the other hand, endothelial cells contain rich amounts of the enzyme xanthine oxidase. Therefore, one of the candidates generating free radicals may be the endothelium which has an adhesive interaction with the activated leukocytes *(37)*. The leukocyte itself is also an important candidate for the generation of free radicals, and the leukocyte-endothelial cell interaction has been attributed an important role in the ischemia/reperfusion injury of the brain *(38)*.

Neurons and glial cells also can generate oxygen-free radicals. Neurons are rich in mitochondria, and the mitochondrial respiratory chain including the ubiquinone and NADH dehydrogenase complex can be intracellular sources of superoxide. Other intracellular sources include microsomal cytochrome P450 system, plasma-membrane NAD(P)H oxidases, nuclear membrane and cell membrane-bound enzymes such as cyclooxygenase, and lipoxygenase, which causes lipid peroxidation *(39,40)*.

Thus, such reactive oxygen species are normally generated in all the cells by specific enzymes, by autooxidation, and by energy transfer reactions. Because the free radical toxicity might be the result from "imbalanced" conditions between the generation of free radicals and antioxidant defense system *(2)*, measuring characteristics of the cellular antioxidant defense system may help to establish concepts about free radical-dependent neural injury after cerebrovascular accidents on ischemia/hypoxia.

3. PATHOLOGY OF FREE RADICAL TOXICITY

3.1. Neuronal Cell Injury

Neuronal damage following hypoxic-ischemic insults has been studied morphologically. In the transient global cerebral ischemia/hypoxia model, there are two types of cell injury, that is acute ischemic cell change and delayed neuronal death *(41,42)*. Morphologically, acute ischemic cell change is defined as a change in the cytoplasm micro-organelles. Ultrastructurally, swollen mitochondria are found that appears as microvacuoles at a light microscopic level, also dilations of tubules, vesicles, and cisternae of the endoplasmic reticulum, and an increase in ribosomes are observed. The damaged neurons are surrounded by swollen astrocytic processes.

In contrast, delayed neuronal cell death typically affects the hippocampal CA1 pyramidal neurons, and starts 2–4 d after the ischemic episode produced by carotid artery occlusion in gerbils and rats. The ultrastructural examination shows that the affected neuron contains a picnotic nucleus and cytoplasmic condensation, and also mitochondrial disarrangement is seen. The damaged neurons are surrounded by the processes of microglia, and sometimes the microglia invade the damaged area to remove the debris of the degenerating neurons. This morphology is clearly distinct from that occurring in acute ischemic cell change *(43)*, suggesting that different mechanisms are responsible for these two types of cell death following ischemic insults *(44)*.

3.1.1. Necrosis and Apoptosis

Apoptosis is a morphologically distinct form of cell death that is involved in many physiological and pathological processes *(45–49)*. Because apoptosis is an integral part of the developmental program, and is frequently the end-result of a temporal course of cellular events, it is sometimes referred to as programmed cell death (PCD). The conventional features of apoptosis include:

1. Intracellular compaction of nuclear chromatin and cytoplasm into sharply circumscribed, uniformly dense masses surrounded by lysosomal membranes (apoptotic bodies).
2. Endonuclease digestion, double-stranded cleavage of DNA to produce fragments that are multiples of approx 185 bp (DNA laddering).
3. Cell death, which requires activation of a "cell death" gene program and therefore is inhibited by RNA and protein synthesis inhibitors (for details, *see* Chapter 1).

In contrast, necrosis is a cell death that is caused by energy failure owing to lack of oxygen and glucose supply. It has been long believed that necrosis is the main cause of neuronal degeneration after cerebrovascular accidents, and both neurons and glial cells died in the infarct brain. However, the selective vulnerability of the hippocampal neurons (CA1 pyramidal cells) to ischemia could not be explained solely by the energy failure. Neurons that are prone to die following the ischemic insult, maintain their energy level as normal during the early reperfusion period, suggesting that these neurons survived the insult but undergo other metabolic changes that trigger a cell death program. Recent studies have proposed that apoptotic cell death is involved in a neuronal cell death caused by the ischemic insult *(50–52)*. Thus, delayed neuronal death in hippocampal CA1 neurons has been considered to be a candidate of neuronal apoptosis. Indeed, the inhibition of protein synthesis by cyclophosphamide has been reported to ameliorate the delayed neuronal death in gerbil *(52)*. Moreover, DNA laddering has been observed in the hippocampus following transient ischemia *(53,54)*. However, the neurons exhibiting delayed neuronal death have shown molphogically distinct features from the cells showing typical apoptosis in peripheral tissues *(54)*. In addition, although the DNA fragmentation has been observed within the cells of the postischemic brain by using *in situ* nick-end labeling of biotinylated dUTP mediated by terminal deoxytransferase (TUNEL reaction) *(55)*, no evidence has been presented that the apoptotic cells are of neuronal origin. Nevertheless, one can imagine that neurons have their own feature of apoptosis, and some oncogenes associated with induction of apoptosis such as c-*fos* and c-*jun*, have been found in the vulnerable hippocampal neurons following ischemia *(56–58)*.

Many of the chemical and physical treatments capable of inducing apoptosis are known to evoke oxidative stress. For example, both ionizing and ultraviolet radiation can provoke apoptosis, and at the same time generate free radicals such as H_2O_2 and ·OH *(59)*. Duvall and Willy *(60)* have suggested that the H_2O_2 may induce either apoptosis or necrosis, depending on its doses applied to the cells. Some agents other than free radicals, such as antineoplastic agents and cell toxic cytokines, may induce apoptosis by eliciting free radical formation. It is well established that both TNF/TNF receptor and Fas ligand/Fas antigen can activate pathways that lead to apoptosis in various cell types *(61)*. Fas antigen mRNA is induced in some neural cells after ischemic insult *(62,63)* (Fig. 3). Stimulation of the TNF receptor results in a rapid rise

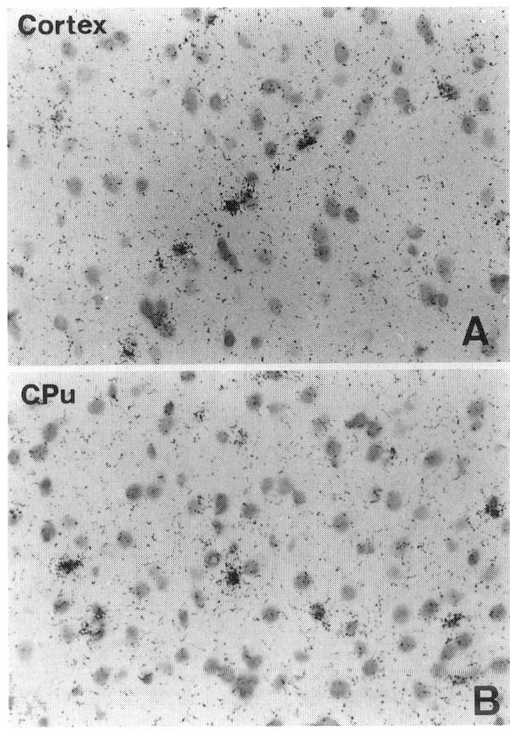

Fig. 3. Photomicrographs showing Fas antigen hybridization in layer IV of the cerebral cortex **(A)** and the caudate putamen (CPu) **(B)** from a mouse 6 h after transient global forebrain ischemia for 30 min.

in the levels of intracellular free radicals *(64)*. In addition, the apoptosis induced through these pathways can be blocked by increasing the ability of cell to scavenge or detoxify the free radicals, for example, by treatment with SOD, catalase, and glutathione peroxidase *(65)*. This implies that oxidative stress may be a cause of decreased ability of a cell to scavenge or detoxify free radicals, and that such oxidative stress is closely associated with mechanisms that trigger cell death programs leading to apoptosis. A recent study has shown that NO also acts as an inducer of apoptosis in macrophages and monocytes *(66)*. How can oxidative stress induce apoptosis? It is possible that intracellularly generated free radicals may result in the activation of genes responsible for apoptosis, conceivably through an oxidative stress-responsive nuclear transcription factor such as NF-κB *(67)*.

3.1.2. Ischemic Preconditioning (Ischemic Tolerance)

Ischemic tolerance has been first reported by Kitagawa and Kirino *(68,69)* using a gerbil model of global cerebral ischemia. Ischemic preconditioning is a phenomenon observed after a brief episode of nonlethal ischemic insult in which vulnerable neurons acquire an endogenous tolerance to the ischemic insult that is otherwise lethal to them. It has been proposed that two major factors are prerequisites for the development of ischemic tolerance: ischemic stress strong enough to perturb the energy metabolism of the neurons, and a long enough interval between nonlethal and lethal ischemia for induction of gene expression and protein synthesis. In addition to brief ischemia, also prolonged but mild hypoperfusion *(70)*, hyperthermic stress *(71,72)* and metabolic stress that generates oxygen-derived free radicals *(73)* have induced tolerance to subsequent lethal ischemia in hippocampal neurons.

The mechanism underlying this ischemic tolerance has remained unclear. Because the tolerance is shown to be well correlated with both the biosynthesis of and the amount of stress proteins such as the 72 kD-heat shock protein (HSP72) *(74,75)*, one can speculate that this phenomenon is induced by some genetic alteration that may change the cellular phenotype from a vulnerable to a tolerant one. HSP72 production has been shown in various kinds of stress as well as heat stress. Oxidative stress with copper chelating agents such as diethyl dithiocarbamate (DDC), increase the intracellular content of thiobarbituric acid-reactive substance, a reliable index of lipid peroxidation, and also it induces the HSP72 synthesis in the brain *(73)*.

Although the precise mechanism by which heat shock proteins might attenuate cell injury remains unclear, it seems essential for survival of neuronal cells to induce synthesis of HSP72, prior to and/or after lethal ischemia *(68–73)*. HSP72 binds ATP and may conserve energy for cellular metabolism under pathophysiological conditions. Heat shock proteins have been associated with stabilization of the cytoskeleton *(76)*, augmented activity of intracellular antioxidants *(77)*, and preservation of cytosolic enzymatic activity *(78)*. Thus, experimental evidence linking cytoprotection with induction of HSP72 is promising *(79,80)*.

On the other hand, ischemic tolerance has been detected in various brain regions such as the cerebral cortex, as well as in the hippocampus *(70)*. Brief occlusion of the middle cerebral artery induces tolerance in cortical neurons to a subsequent episode of potentially lethal ischemia *(81)*. Since this tolerance corresponds to the phenomenon of cortical spreading depression rather than to the expression of HSP72 in the cortex, one can suggest that the expression of HSP72 is more likely one of trigger processes that cause ischemic tolerance. Besides heat shock proteins, immediate early genes such as *fos* and *jun* and their products induced after transient global ischemia *(82)* have also been proposed to be responsible for ischemic tolerance. These proteins act as transcription factors at specific DNA target sites such as AP-1 (activator protein-1) and CRE (cyclic AMP-responsive element) *(83,84)*. Among them, c-*jun* protein is likely to participate in the control of rescue or survival genes following severe neuronal injury irrespective of the pathogenetic mechanism *(57,85)*.

3.2. Glial Cell Response

Recent findings suggest that glial cells may be more actively involved in brain function than has been previously thought. Glial cells are classified as astrocytes, oligodendrocytes, and microglia. Astrocytes are strategically located to exert neurotrophic functions, and oligodendrocytes to form myelin surrounding the axons. Microglia are considered to be the macrophages of the central nervous system. Because glial cells are intimately associated with most neurons, one can consider the significance of active neuron-glia interaction. Recent studies have shown that astrocytes produce a number of growth factors that can promote neurite outgrowth and cell survival. For example, nerve growth factor (NGF) and basic fibroblast growth factor (bFGF) promote cell survival and neurotransmitter synthesis *(86,87)*, and can protect neurons against

Table 2
Cytokines Produced by Neural Cells

Astrocytes	TNF-α, TGF-β, IL-1, IL-5, IL-6, IL-8, G-CSF, M-CSF, GM-CSF, IFN-α, INF-β
Microglia	TNF-α, TGF-β, IL-1, IL-5, IL-6

Cytokine Receptors Expressed on Neural Cells

Astrocytes	IL-6R, IL-7R, GM-CSFR, M-CSFR
Microglia	IL-2R, IL-3R, IL-4R, IL-6R, GM-CSFR, M-CSFR
Oligodendrocytes	IL-2R, IL-3R, IL-4R, M-CSFR
Neuron	IL-6R, GM-CSFR

ischemic damage *(88,89)*. Glial cells also produce a variety of inflammatory cytokines and express their receptors *(90)*. Some cytokines such as tumor necrosis factor-α (TNF-α) and interleukin-1 (IL-1) can exert both neurotrophic and neurotoxic effects *(91,92)*. Thus, glial cells are implicated to have a variety of functions to maintain the extracellular environment of the central nervous system, but it is also possible that when the glial cells were exposed to ischemic/hypoxic insult, that they may turn to accelerate the brain injury rather than to exert the beneficial effects on it.

3.2.1. Proliferation and Activation of Glial Cells After Ischemia

Both astrocytes and microglia showed a proliferation in the injured area after an ischemic insult. In the model of global cerebral ischemia, which showed delayed neuronal death in the CA1 sector of the hippocampus, the number of GFAP-positive astrocytes increased throughout the hippocampus. However, this glial response is noticeable at late periods of recirculation around 24–48 h after ischemia. In contrast, microglia activation becomes visible within 30 min of reperfusion, when the microglia response was detected by the lectin reaction using griffonia simplicifolia B4-isolectin *(93)*. These findings suggest that microglial activation precedes the astrocyte response. The significance of the time-lag of proliferation among them remains unclear, but early studies have suggested the presence of microglial regulation of astrocytic hyperplasia *(94)*.

3.2.2. Cytokine Network

Evidences have been accumulated that glial cells contribute to form a cytokine network in the central nervous system *(90–92)*. Cytokines are reported to have a multiple function such as proliferative and/or mitogenic effects on either glial cells or neurons. They stimulate or suppress the synthesis of other cytokines, and can induce major histocompatibility class (MHC) antigens both in astrocytes and microglia, implicating that they act as antigen-presenting cells (APC) and exert immune responses within the central nervous system *(90)*. Cytokines synthesized in neural cells and their expressing receptors are shown in Table 2.

Astrocytes and microglia are expressing similar cytokines and receptors, and often express both the ligand and receptor of one cytokine within single cells, suggesting that these cytokines act either in a paracrine or autocrine manner. However, these cells do not always have common stimulating factors for the production of similar cytokines.

For example, both astrocytes and microglia produce IL-6 *(95,96)*. Granulocyte-macrophage colony-stimulating factor (GM-CSF) stimulates the synthesis of IL-6 in microglia but not that in astrocytes, whereas TNF-α induces IL-6 only in astrocytes *(97)*. In addition, they have different intervals of cytokine production after the same stimulus such as lipopolysaccharide (LPS). For example, the production of TNF-α by microglia has been reported to precede the TNF-α synthesis by astrocytes following LPS stimulation in culture *(98)*. A detailed and precise network of cytokines has not been fully analyzed in the central nervous system. However, in vitro studies have revealed a part of the mechanism, that is, IL-1 and TNF-α produced by microglia induce the synthesis of IL-6 in astrocytes, which in turn produce GM-CSF to stimulate the synthesis of IL-6 in microglia *(97,98)*. However, it is still equivocal whether this network functions in vivo. In the rat closed head injury model, it has been shown that the temporal profile of IL-6 synthesis lagged 2–4 h behind that of TNF-α in the injured brain *(99)*.

The effects of ischemia/hypoxia on the cytokine network in vivo have been reported by some authors. In general, cytokines show an increase in their levels after ischemia/hypoxia injury, in good agreement with the in vitro experiments *(91,100)*. However, the cell population that is generating cytokines is not clearly defined in the brain. It is very difficult to detect cytokines histochemically, probably because of their very small amount of synthesis by the cells. Nevertheless, some cytokines have been demonstrated in the glial cells and even neuronal cells in the brain *(100,101)*.

In a mouse model of transient global cerebral ischemia, which can produce the hippocampal neuronal death similar to the delayed neuronal death in rat and gerbil model, both astrocytes and microglia have shown to be immunopositive for TNF-α (Fig. 4) *(101)*. In that model, the TNF-α-positive glial cells are visualized as early as 1.5 h after reperfusion, suggesting very rapid synthesis of TNF-α following ischemia/reperfusion insult. These cells might be microglia, because TNF-α immunopositive cells could be distinguished from the GFAP immunoreactive astrocytes. GFAP-immunoreactive astrocytes also synthesize TNF-α for about 1 d after ischemic insult (Fig. 5). The presence of time-lag seen between the microglia and astrocytes, which is in good agreement with results obtained from in vitro experiment *(97,98)*, can explain the target cell-dependent and time-dependent function of TNF-α in the injured brain by ischemia/reperfusion.

TNF is a cytokine with multiple biological activities *(102)*, including microglial activation *(103)*, blood–brain barrier (BBB) interruption *(104)*, cell adhesion *(105)*, cytotoxicity *(106)*, and release of other cytokines *(97)*. The rapid induction of TNF-α gene activation by ischemic insults *(100,101)* which may follow a typical cytokine response, suggests the involvement of some immediate early genes such as c-*fos* and NF-κB that might be important in initiating the cytokine metabolic cascade.

3.2.3. Oxygen-Reactive Proteins

The effects of oxygen deprivation followed by reoxygenation have been extensively studied on cultured astrocytes. The period of hypoxia is associated with upregulation of glucose transporter *(107)*, activation of IL-6 and TNF-α *(108)*, and activation of AP-1 *(109)*. In contrast, during reoxygenation, oxygen-free radicals appear to induce synthesis and elaboration of synthesis of IL-6, RNA splicing factors such as RA301 *(110)*,

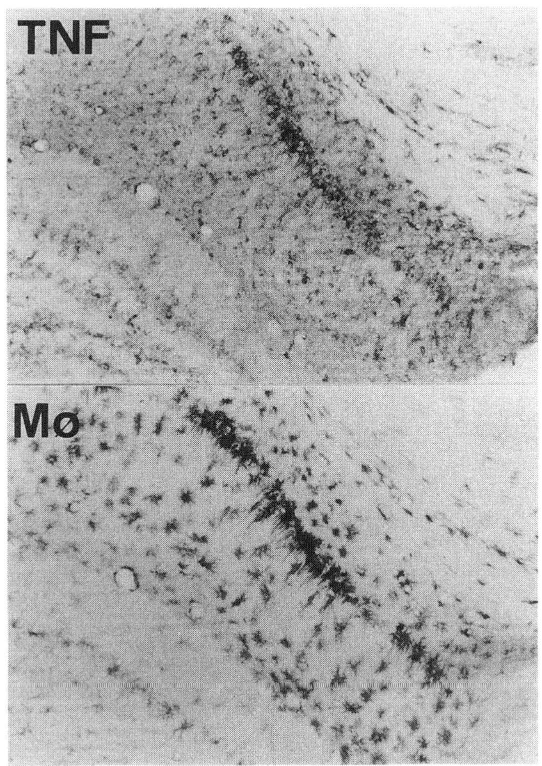

Fig. 4. Photomicrographs showing TNF-α immunoreactivite cells (TNF) and microglia that are stained with antimacrophage antibody (Mø) in the hippocampus from a mouse 3 d after transient forebrain ischemia for 30 min. Microglia that migrated in the CA1 pyramidal layer express TNF-α immunoreactivity.

and stress proteins such as oxygen regulated proteins (ORP) in astrocytes *(111)*. One mechanism through which such reactive oxygen species impact on the biosynthetic apparatus is through activation of the transcription factor NF-κB *(112)*. The ORP is localized to the endoplasmic reticulum, and is induced in gerbil brain following ischemic insult, suggesting an important role in astrocyte adaptation to oxygen deprivation *(111)*. Inhibition of RA301 expression with an antisense probe reduced the release of IL-6 by astrocytes subjected to hypoxia/reoxygenation, suggesting that this rapidly induced factor regulates the elaboration of cytokine synthesis *(110)*. The presence of ORPs and RA301 in injured astrocytes places it at the critical locus for controlling *de novo* protein synthesis and processing, especially of rapid elaborating proteins such as cytokines.

3.3. Endothelial Injury

During cerebral ischemia and reperfusion, alteration of endothelial cell interactions, coagulation system activation, and leukocyte-endothelial cell interactions are a few events affecting microvascular integrity. The subject of oxygen free radical generation during vascular occlusion/reperfusion is quite relevant to the discussion of microvascular phenomena and has been reviewed *(113–115)*. Oxygen radicals such as $O_2{}^-$ and

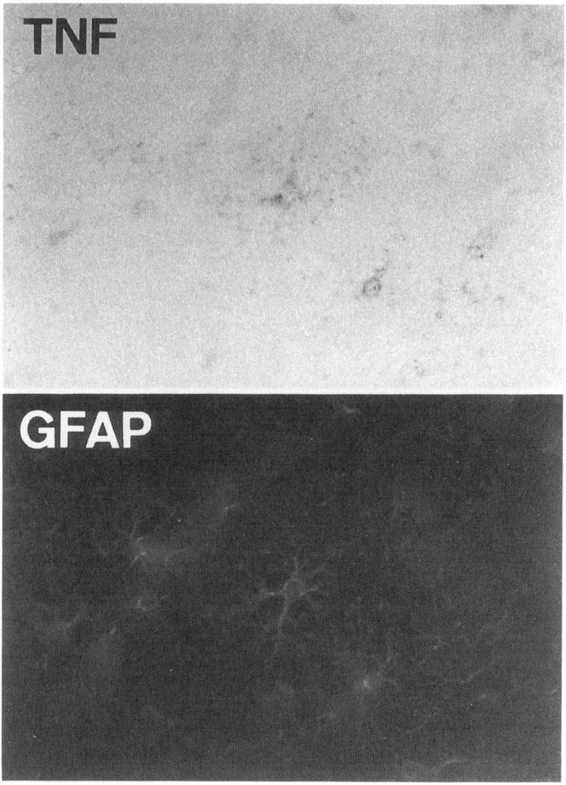

Fig. 5. Astrocyte that is double-positive for TNF-α and GFAP, located in the hippocampus from a mouse 1 d after transient forebrain ischemia for 30 min.

·OH, may be generated from capillary endothelial cells *(116)* and activated polymor-phonuclear (PMN) leukocytes *(117,118)* during and/or after ischemia, and in tissues containing activated leukocytes and macrophages during inflammation *(119)*.

In the vascular endothelium, Ca^{2+}-dependent conversion of xanthine dehydrogenase to xanthine oxidase occurs during ischemia *(116,120,121)*. Hypoxanthine transport and xanthine oxidase activity have been identified in cerebral microvessels *(122)*. Membrane lipid peroxidation results from xanthine oxidase activity, which together with free radicals generated from other sources such as the mitochondrial respiratory chain, leads to significant alteration in endothelial cell function. Vascular consequences of free radical generation include arteriolar dilatation *(123)*, disturbances of permeability *(124)*, and direct injury of vascular endothelium and myointimal cells *(125)*.

The interaction of activated PMN leukocytes with postcapillary endothelium generates oxygen free radicals at the contact region *(126)*. This may in turn facilitate their interaction and lead to transmigration of leukocytes. One report examining transmigration of PMN leukocytes across cerebral microvascular endothelial cells in vitro has appeared *(127)*. Cytokines such as TNF-α are also generated by endothelial cells subjected to ischemia and reperfusion. Cultured endothelial cells exposed to cytokines provide some indication of endothelial responses following ischemic injury. TNF-α, IL-1β and interferon (IFN)-β up-regulate surface adhesion molecule expression such as intercellular adhesion molecule (ICAM)-1 on human cerebral endothelial cells *(128)*, release the chemotactic factors, and promote transendothelial leukocytes migration *(102)*. The appearance of TNF-α as intermediaries in the development of ischemia has been seen in the very early stages of ischemia and reperfusion as well as in the later inflammatory stage *(101)*. Thus, ischemia/reperfusion stress facilitates the ischemic injury both by leukocyte migration and activation of leukocytes and endothelial cells, subsequently forming the inflammatory stage of cerebral ischemic injury. Oxygen radical generation may contribute to cytokine release and expression of adhesion molecules both on leukocytes and endothelium.

3.4. Damage of Blood–Brain Barrier

The blood–brain barrier mainly consists of cerebral endothelial cells and perivascular structures such as astrocytes. PMN leukocyte activation and endothelial conversion of hypoxanthin to xanthine by xanthine oxidase may generate oxygen free radicals and cause lipid peroxidation of endothelial cell membranes. Xanthine oxidase has been identified richly in brain microvessels *(122)*. The findings of superoxide dismutase, glutathione peroxidase, and catalase in rat isolated cerebral microvessels suggest the existence of some protective mechanism against membrane lipid oxidative injury *(129)*. Exposure of cultured human cerebral endothelial cells to exogenous H_2O_2 has been shown to alter their permeability *(130)*. These findings imply that alterations in cerebral endothelial permeability may result from free radical-mediated injury.

4. THE EFFECTS OF ISCHEMIA/HYPOXIA ON ENDOGENOUS ANTIOXIDANT ENZYMES

Free radicals damage cellular membranes as well as DNA *(131)*. The cells has many strategies to prevent the propagation of radical reactions. It is possible to scavenge reactive oxygen directly with endogenous antioxidant enzymes such as superoxide dismutase (SOD), catalase (CAT), or glutathione peroxidase (GPx). SOD catalyzes the conversion of O_2^{-}:

$$2O_2^{-} + 2H^+ \rightarrow H_2O_2 + O_2$$

The heme-containing enzyme catalase transforms H_2O_2 into water and oxygen. GPx also reduces peroxides. The selenium-dependent GPx reduces H_2O_2 as well as organic hydroperoxides. An important reaction is the transformation of H_2O_2 into a hydroxy radical ($\cdot$OH), which is extremely reactive, by traces of transition ions such as iron and copper. This Fenton reaction proceeds as follows:

$$Fe^{2+} + H_2O_2 \rightarrow Fe^{3+} + \cdot OH + OH^-$$

Although it is still uncertain whether $\cdot$OH or an iron-oxygen complex is the ultimate reactive initiating species, it is evident that safe storage of iron ions is crucial *(132)*. In many cases, these scavenger enzymes function as the effective protection against oxygen free radicals providing the cells with the use of radical biochemistry without the risk of uncontrolled reactions in the physiologic state. However, it is uncertain whether these enzymes can act beneficially or not on the injured neuronal cells suffering ischemic/hypoxic damage. This chapter describes the alteration in the synthesis of these endogenous enzymes on the ischemia/reperfusion injury, which may or may not cause neuronal cell death.

4.1. Superoxide Dismutases

4.1.1. Functional Significance of SOD on Ischemia/Reperfusion

The influence of ischemia and of ischemia followed by reperfusion on the endogenous antioxidant defense system against oxygen-dependent generation of free radicals is of growing clinical interest. However, the contribution of endogenous SOD to ischemic brain damage remains equivocal. Chan et al. *(133)* reported a significant decrease in SOD activity in rat brain following global cerebral ischemia, suggesting that the depletion of SOD indicates a vulnerability to reperfusion, with an influx of superoxide radicals following global ischemia. Other investigators *(134,135)* have shown that SOD activity did not change during the acute period of focal cerebral ischemia. There are still limited controlled data that define the cellular and molecular events on the intracellular antioxidant defense system following ischemia/hypoxia insults.

Mammalian cells possess two intracellular forms of SODs and a third extracellular form of SOD with distinct distributions characterized by their catalytic metal ion requirements *(4,136)*. Copper–zinc SOD (CuZnSOD) is found predominantly in the cytosol, and is called as cytosol SOD or SOD1. In contrast, manganese SOD (MnSOD) is located in the mitochondria and identified as mitochondrial SOD or SOD2 *(4,137–139)*. Extracellular SOD (SOD3) contains copper and zinc, and exists in the interstitial space of tissues *(136)*. MnSOD is inducible, whereas CuZnSOD is constitutively produced, but biosynthesis of both SODs is promoted by oxidative stress *(4,138)*. It is still unclear whether these three kinds of SODs share the role intracellularly as the first enzymes involved in the antioxidant defense system. There have been some reports demonstrating the altered expression of each SOD at the cellular level in vivo following cerebral ischemia/hypoxia.

CuZnSOD is contained in most of the mammalian cell population, including the neurons and glia. Alternatively, MnSOD shows a more or less heterogenous distribution. The cellular localization of these SODs in the brain has been determined both by *in situ* hybridization and by immunocytochemistry *(30,140–143)*. In the hippocampus, CuZnSOD mRNA and protein are exclusively located in the pyramidal neurons in CA sectors (CA1-3) and the granule cells in dentate gyrus (Fig. 6). Glial cells contain both the mRNA and protein, but the intensity of labeling is weaker than that in neuronal cells. In contrast, MnSOD is less observed in CA1 pyramidal neurons than in other hippocampal neurons *(141,142)* (Figs. 6 and 7). The CA1 pyramidal cells are known to show the delayed neuronal death following transient global ischemia in rats and gerbils. Therefore, this heterogeneity may explain the vulnerability of the CA1 neurons to the ischemic insult as a cause of oxygen-derived free radical generation. Almost no level of MnSOD mRNA or protein has been observed in glial cells of the normal brain (Fig. 8A).

The time course of the alteration of the SOD level has been examined in detail in the global cerebral ischemia model *(140,141)*. The model of transient occlusion of bilateral common carotid artery in gerbil can produce almost complete ischemia in forebrain and makes it possible to analyze the effect of an ischemic insult on the cellular antioxidant defense system in vivo. An ischemic insult of 5 min (lethal ischemic insult) causes the delayed neuronal death in CA1 pyramidal neurons, while an insult of 2 min

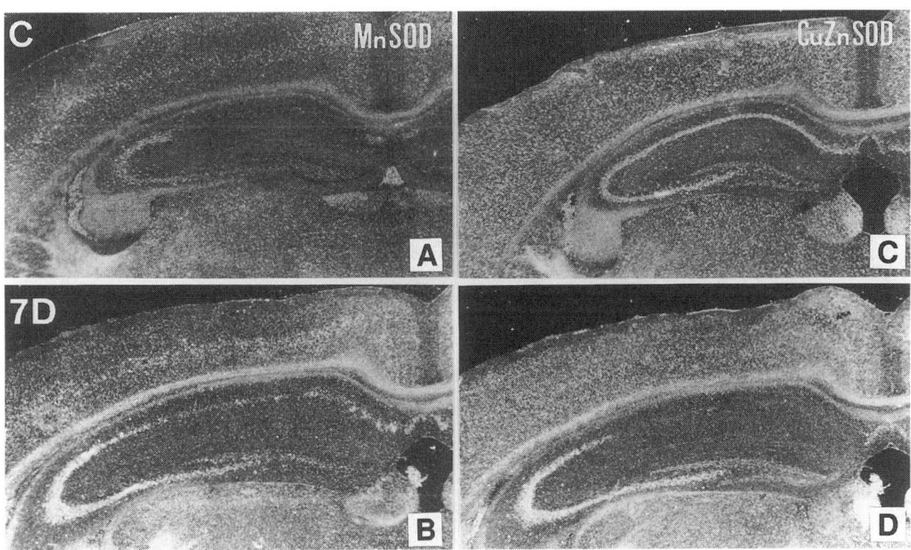

Fig. 6. Photomicrographs showing MnSOD hybridization **(A,B)** and CuZnSOD hybridization **(C,D)** in gerbil hippocampus from control brain (c) and from brain 7 days after 5 min of ischemia (B,D). Note the difference in distribution between the MnSOD- and CuZnSOD-labeled cells in control hippocampus. Strong hybridization for MnSOD was induced in the glial cells (microglia) in the CA1 pyramidal layer at 7 d after ischemia (B), whereas the adjacent section showed no hybridization for CuZnSOD with loss of pyramidal neurons (D).

(sublethal ischemic insult) results in the induction of tolerance to the subsequent lethal ischemia by the CA1 pyramidal neurons *(68,69)*. The examination of SOD-synthesis differences in the affected neurons in relation to their viability after ischemia/hypoxia, may allow the analysis of the crucial role of the endogenous SOD in the cell defense systems in vivo.

A lethal ischemic insult results in the transient increase in mRNA levels of both CuZnSOD and MnSOD without the subsequent increase in their protein levels in the reperfusion period. This suggests the decrease of SOD activity in the affected neurons, when they may need SODs to keep their antioxidant defense system for scavenging of free radicals generated upon reperfusion. The discrepancy between the expression of SOD mRNAs and the proteins also may be explained by a persistent and selective deficit in translation activity in CA1 neurons following ischemia *(144,145)*. This translation block of intracellular protein synthesis is rather commonly observed in other endogenous proteins such as HSP72 *(146,147)*. In contrast to the change in the neuronal SODs, glial cell-expressing SODs, especially MnSOD, show the increase in their protein levels already in the acute phase of reperfusion, and the expression is observable even before the neuronal cell death could be seen (Fig. 8). The hippocampal concentrations of SOD have been reported to increase following lethal ischemia *(148)*. The increase in hippocampal SOD may reflect the change of the glial expression of SOD.

Contrary to the lethal insult, sublethal ischemic insults produce a different pattern of expressions of CuZnSOD and MnSOD *(149)*. Although the insult does not affect the levels of CuZnSOD mRNA and protein, it increases the expression of MnSOD pro-

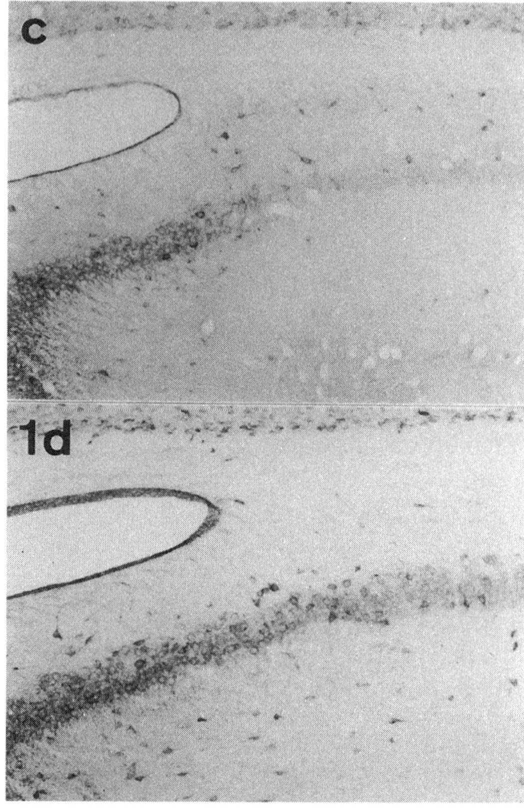

Fig. 7. Photomicrographs showing MnSOD immunoreactivity in gerbil hippocampus from control brain (c) and from 1 d after 2 min of ischemia (1 d). The induced MnSOD protein in the vulnerable CA1 neurons after the sublethal ischemia may contribute to the acquisition of tolerance to subsequent lethal ischemic insult.

teins as well as the mRNAs in the vulnerable CA1 neurons, indicating that the mild nonlethal ischemic insult may trigger the MnSOD synthesis in the vulnerable neurons (Fig. 7). The overexpression of MnSOD is lasting for a longer period than the HSP72 induction seen in the same cell population. Because the nonlethal ischemic insult induces tolerance in CA1 neurons, MnSOD but not CuZnSOD may have an important role in the acquisition of tolerance by the vulnerable neurons to oxygen-derived free radicals, which might be generated by the subsequent stronger ischemic insult. In fact, after acquisition of tolerance, subsequent lethal ischemia does not affect the expression of the SODs very much, that is, both CuZnSOD and MnSOD can be detected at a significant level in the CA1 pyramidal neurons during the whole reperfusion period. These findings suggest that neuronal injury may be eliminated when the endogenous SOD is abundant enough to scavenge the free radicals generated upon ischemic insult, and the balance between the amount of free radical and SOD activity may decide the fate of affected neurons. For instance, an increased expression of CuZnSOD in transgenic mice offered protection against free radicals generated upon focal ischemic injury *(150)*.

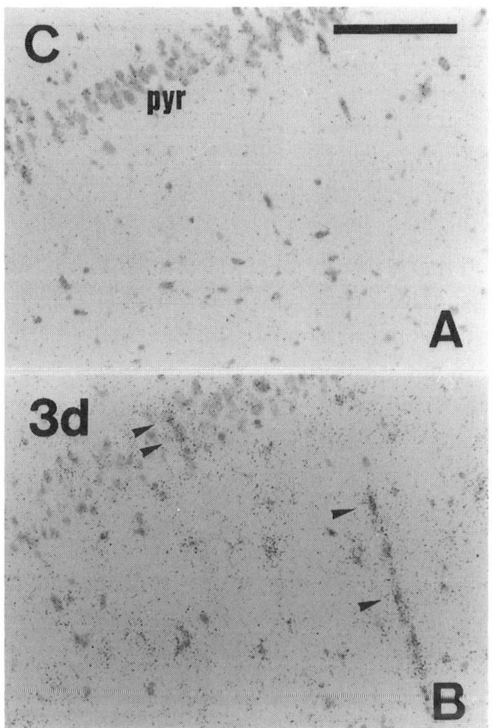

Fig. 8. Photomicrographs showing MnSOD hybridization in CA1 region of the hippo-campus from control brain (**A**) and from brain at 3 d after 5 min of ischemia (**B**). Small labeled cells are distributed in both the pyramidal layer (pyr) and stratum radiatum after 3 d (**B**). Arrows indicate labeled cells, which exhibited a definite polarity in the direction of the apical dendrites of pyramidal cells. The location of these cells is similar to that of microglial cells.

4.1.2. Mechanism of SOD Induction

The severity of ischemia/reperfusion neuronal cell injury is known to depend on the duration of the ischemic period, and probably on the production of free radicals. However, the events occurring during the ischemia have not thoroughly been analyzed. During the ischemia, a number of events occur intracellularly, including genetic alterations. The superinduction of numerous rapidly inducible mRNAs has been observed following the pharmacologic suppression of protein synthesis *(151,152)*, and such immediate early genes (IEG) including c-*fos* and c-*jun* are known to be induced dramatically following focal and global cerebral ischemia *(82,153)*. It has been reported that these genes affect various steps of proteoneogenesis, implying that oxidative stress triggers protein synthesis. Oxidative stress has been proposed as a mediator of apoptosis by activating so called "death genes" *(65)*. On the other hand, oxygen-derived free radicals have been reported to make proteolysis of various endogenous enzymes including CuZnSOD *(154)*. The finding that the free radicals may either activate or inactivate the enzymes, in turn, interferes with the analysis of the precise mechanism of the SOD induction.

The free radical production in phagocytes is associated with the induction of heat shock response *(155,156)*. Thus, the oxidative stress is involved in a variety of events such as ischemic/hypoxic and/or reperfusion/reoxygenation stress, heat shock stress, chemically induced stress produced by diehyldithiocarbamate (DDC) *(73)* and pharmacologically excitotoxin-induced stress *(10,23)*. Although both CuZnSOD and MnSOD are reported to be induced by oxidative stress, the mechanism on the factors directly related to the induction is still under investigation.

It is generally believed that CuZnSOD is a constitutive enzyme, whereas mitochondrial MnSOD can be induced by a variety of stimuli *(138)*. Several studies have reported differences in the regulation of CuZnSOD and MnSOD in some diseased condition. A discrepancy between CuZnSOD and MnSOD has been observed both in motor neurons after nerve transection *(157)* and in Parkinsonian substantia nigra *(158)*. In gerbils, heat shock stress and chemically induced stress as well as mild sublethal ischemia induces MnSOD but not CuZnSOD in the affected neurons *(73,149)*. Reduction of oxidative stress by iv antioxidant enzyme supplementation also results in the similar discrepancy between MnSOD and CuZnSOD mRNA following lethal ischemic insult *(141)*. The candidates responsible for triggering such differential expression may involve cytokines, because some cytokines such as TNF and IL-1 are known to induce the mRNA for MnSOD, but not the mRNAs for other antioxidant or mitochondrial enzymes including CuZnSOD, glutathione peroxidase, and cytochrome c oxidase in human lung cell line *(159)* and rat pulmonary epithelial cells *(160)*. Such cytokines are expressed predominantly in glial cells in the brain, and are induced very quickly following ischemia/reperfusion *(101)*. Many investigators have argued that the elevation of oxygen free radical levels directly increases the concentration of cellular antioxidant enzymes. Others, however, have demonstrated that transcriptional induction of SOD cannot be a direct consequence of elevated free radical concentrations. More likely, according to their observation is that SOD transcription is stimulated by a component of the acute inflammatory response such as cytokines *(160)*. However, it is also possible that cytokines stimulate MnSOD synthesis by the increase in production of $O_2{}^-$ at mitochondria *(159)*. Although the mechanism of SOD gene production is unclear, recent reports demonstrates that cellular calcium-activated, phospholipid-dependent protein kinase C can play an important role in the induction of MnSOD mRNA transcription *(161)*.

4.2. Glutathione Peroxidase and Catalase

SOD protects cells from the toxicity of $O_2{}^-$, whereas catalase (CAT) and glutathione peroxidase (GPx) scavenge H_2O_2. If the activity or amount of SOD is increased without concomitant increase of GPx and/or CAT, then H_2O_2 accumulates and reacts with transition metals in the Fenton's reaction to produce hydroxyl radical ($\cdot$OH), which is one of the most noxious radical species to the cell membrane. Thus, the glutathione system seems to be essential to efficient intracellular management of partially reduced oxygen and lipid peroxidation products *(162)*. In the rat model of transient ischemia/hypoxia *(163)*, up-regulation of GPx has been shown in affected brain areas such as hippocampus and cortex. In the same model, however, the areas with increased GPx mRNA levels are not protected from delayed neuronal death, suggesting that the GPx protein and its enzyme activity are not capable of scavenging all radicals produced in the ischemia/hypoxia areas. A similar phenomenon is observed in the case of SOD mRNA following lethal/irreversible ischemic insult in gerbil *(140,141)*. In the lung, both glutathione peroxidase and reductase activities have been reported to be resistant to the effects of both hypoxia-hypoperfusion and reoxygenation, although MnSOD decreased significantly *(164)*. Differential regulations of CuZnSOD and GPx mRNA expression have also been found in murine brains *(165)*. Because the CAT activity in the rat forebrain is very low *(166)*, GPx and SOD are most likely to be the key enzymes of intracellular antioxidant defense systems in the forebrain. Thus, the cooperation of SOD with GPx may be protective against oxidative stress only when the cerebral injury is reversible.

5. THE EFFECTS OF ISCHEMIA/HYPOXIA ON NITRIC OXIDE SYNTHASES (NOS)

5.1. NOS Expression in Brain

NO is synthesized from L-arginine by the enzyme NO synthase. NO is involved in the regulation of cerebral blood flow *(167)*, in mediating glutamate neurotoxcicity *(21–23)*, in long-term potentiation *(16–18)* and in cortical spreading depression (CSD) *(168)*. NOS is not only located in the endothelium, but also in neurons, perivascular nerves, and astrocytes *(169,170)*. Clearly there are several distinct NOS enzymes. Recent studies have revealed at least three isoforms of NOS: neural NOS, endothelial NOS, and macrophage NOS. These isoforms have been cloned along with their amino acid sequences *(14,19,171,172)*. The former two isoforms are constitutively expressed in each cell (constitutive NOS; cNOS) and stimulated with calcium/calmodulin, whereas macrophage NOS is induced at the transcriptional level by various cytokines (inducible NOS; iNOS) and is not subjected to regulation by calcium ions *(169,173)*. During inflammation, NO is thought to mediate the cytotoxicity of activated macrophages and leukocytes *(174)*. In these cells, NO is synthesized by the inducible isoform of NOS (iNOS), an enzyme that produces high, potentially toxic, levels of NO *(175)*. iNOS expression has also been observed in astrocytes, microglia, and vascular cells *(176,177)*.

Histologically, NOS is localized by either using the reactivity with NADPH-diaphorase (NADPH-d) which has been identical to neural and endothelial NOS, or using immunocytochemistry for nNOS *(14,179,180)*. Electrochemical measurement has revealed that NO increases in brain following focal cerebral ischemia in rat *(181)*. It is suggested that NOS activity also elevates during focal ischemia and reperfusion in the rat *(182)*. The increase in NO concentration in ischemic brain may be due to the elevation of enzymatic activity of the constitutive neural NOS *(22,183)*, but it is also conceivable that iNOS induced in the migrated neutrophils contributes to the alteration of NOS activity in the postischemic brain *(176)*.

5.2. Effects of NOS Inhibition on Brain Injury

A number of studies have been accumulated describing the effects of NOS inhibitors on cerebral circulation and injury: For example, effects on cerebral blood flow (CBF) increase accompanying hypercapnia *(184,185)* or cortical depression *(186)*, and glutamate-mediated neurotoxicity *(187)*. In animal models of focal cerebral ischemia, inhibition of NO synthesis has been reported to reduce infarction volume and ischemic injury *(183,188)*. However, these studies may have somewhat controversial results, which probably depend on the selectivity for the NOS isoform and/or the duration of the action of inhibitors tested.

There are several compounds that have been reported as NOS inhibitors. N^G-nitro-L-arginine methyl ester (L-NAME) and N-nitro-L-arginine (L-NA) are nonselective NOS inhibitors, and 7-nitroindazole (7-NI) is a selective nNOS inhibitor. Aminoguanidine is an iNOS inhibitor, which has also been reported to reduce the extent of tissue damage produced by MCA occlusion in rats *(189)*. The beneficial effects of these compounds are likely owing to the reduction of NO, high levels of which are known to be cytotoxic. NO depresses cellular energy production by inhibiting the glycolytic and the mitochondrial enzyme *(190,191)*. NO leads to DNA damage, activates a nuclear enzyme and may contribute to neurotoxicity *(192)*. Furthermore, NO combines with O_2^- to produce peroxynitrite, a stable oxidant that generates toxic free radicals *(20,193)* (Fig. 1). It is also possible that the beneficial effects of iNOS inhibitor may be due to other mechanisms such as interference of the adhesion between activated neutrophils to endothelial cells and of the glial cell response, because these cells express a high level of iNOS in the postischemic brain in the rat *(176)*.

NOS is distributed throughout the brain and expressed in most of the neural cell populations as well as endothelium, similarly to SOD. The combination of NOS inhibitors and SOD is of interest as a therapeutic tool of a cerebral ischemic/hypoxic injury from the aspect of the reduction of peroxynitrite generation. CuZnSOD transgenic mice have been shown to have smaller infarct areas than the control animals following MCA occlusion *(194)*, suggesting neuroprotecting effects of SOD. In the same transgenic animals, however, additional donation of a selective nNOS inhibitor has failed to reduce the infarct size. This may suggest that in the presence of sufficient SOD, SOD alone is enough to protect against cerebral ischemia and even ameriolates NO neurotoxicity by reducing the generation of peroxynitrite.

6. PROTECTION OF FREE RADICAL-MEDIATED BRAIN INJURY

The therapeutic potential of antioxidant enzymes such as SODs has been proposed for numerous clinical situations. Since these enzymes have a very short circulating half-life of only 6–9 min *(195)*, their scavenging action of free radicals may not be expected so much when they are administered intravenously. In addition, SOD can be inhibited by H_2O_2 *(196)*, catalase by O_2^- *(197)*, and GPx by O_2^- *(198)*. Consequently, in vivo studies must take into account the pharmacodynamics and tissue distribution of circulating enzyme. Nevertheless, a huge number of studies have shown the beneficial effects of antioxidant drugs on oxidative injury in various organs. Some of them have attempted to modify the native enzymes to keep their long circulation half-lives and/or to allow them to be taken up by neurons *(199)*. In even such cases, however, one must consider the possibility of forming antigenicity of the injected enzymes.

6.1. Superoxide Dismutases

Previous studies have shown that exogenously applied CuZnSOD may ameliorate brain injury in cerebral ischemia models. Conjugated enzymes include SOD-conjugated with divinyl ether-maleic acid copolymer (pyran-SOD) *(200)*, polyethylene glycol (PEG)-conjugated SOD and catalase *(201)*, and liposome-entrapped SOD and catalase *(195)*. These enzymes dramatically increase circulating half-life up to 2–40 h without minimal change in enzymatic activity *(199,200)*. It has been reported that pyran-SOD has protective effects against delayed neuronal death in gerbil, and PEG-SOD and PEG-catalase reduce infarct volume. Liposome-entrapped SOD has also been reported to reduce cerebral infarction in focal cerebral ischemia in rats *(202)*.

Recombinant-human SOD (r-hSOD) *(203)* should have a shorter half-life than the modified SODs, but possess potential applicability for clinical therapeutics from the aspect of the antigenicity. In the brain, r-hSOD can reduce the vasogenic edema *(204)* and the hippocampal neuronal injury of delayed neuronal death following global cerebral ischemia in gerbil *(30,141,205)*. The protecting effects have been observed when it is injected before, but not after, the ischemic insult, supporting the role of free radicals generated during the ischemia in the progression of delayed neuronal death following ischemia/reperfusion. However, the in vivo results obtained by SOD treatment are more or less compromising, because few direct evidences are noted for the effect of exogenously applied SOD on reducing cellular oxidative stress in vivo. In addition, the native SOD has a limited ability to pass through the blood–brain barrier *(30,202)* and to be taken up by neurons *(206)*. Thus, determining therapeutic uses of r-hSOD in cerebral ischemia and reperfusion needs to be approached with extreme caution.

6.2. Other Antioxidant Components

A number of synthesized compounds have been taken into consideration as antioxidant pharmacotherapeutic agents. Both vitamin C and vitamin E analogs are included in good antioxidants *(207,208)*. U74006F has been developed as a potent inhibitor of iron-dependent lipid peroxidation and appears to be effective in preventing radical damage to the central nervous system *(209,210)*. Cu(II)$_2$(3,5-DIPS)$_4$ is a copper complex whose enzymatic activity mimics SOD *(211)*. LY231617 has been developed as an antioxidant that can pass through the blood–brain barrier. It inhibits both iron-dependent lipid peroxidation and arachidonic cascade, and reduces global ischemic neuronal injury in rats *(212)*. L-deprenyl is an irreversible inhibitor of monoamine oxidase-B (MAO-b) and has been reported to reduce cerebral damage in a transient ischemia/hypoxia model of rat *(213)*.

Although such antioxidant compounds including SOD act as blockers of radical processes, it should be noted that these compounds have pro-oxidant action as well *(211)*. The possible effects of these compounds functioning as pro-oxidants has not fully been elucidated in vivo. Furthermore, it remains to be answered whether such antioxidants rather might disintegrate a delicately integrated physiologic system or not. Ideally, the antioxidant compound should locate and function only at target sites that generate excess amounts of free radicals. From this aspect, treatment with transgenic technique to overexpress the endogenous antioxidant enzyme should also be targeted to locations with undesirable excessive radical formation, although intrinsically increased expression of these enzymes have beneficial effects on cerebral ischemic injury *(150)*.

7. CONCLUSIONS

It is now convincing that free radical generation is involved in ischemic stroke, and may trigger most of the pathological reactions in the stroke brain. Basic research mentioned above may clarify the toxic effects of free radicals on the cerebral injury, but few researchers have mentioned the physiological contribution of free radicals to maintenance of the normal function of the central nervous system. In addition, it cannot be excluded that the radical reaction may also be part of mechanisms that protect the neural cells outside the severely affected zone. Application of antioxidant drugs for the treatment of cerebrovascular disease is limited at present, because the precise localization of added antioxidant drugs is difficult to control and is also largely unknown for current delivery systems. Although the timely administration of antioxidant, for instance, on reconstitution of blood flow in acute ischemic stroke patients, may exert the beneficial effects on ischemia/reperfusion injury, one must consider the following possible alterations of intra- and/or intercellular events such as genetic and/or immunoreactions. The ultimate effect of ischemia/reperfusion injury on neuronal function and whether such events serve a protective function to limit the ischemic injury are under active study.

REFERENCES

1 Priestley, J., Experiments and observations on different kinds of air. In *The Discovery of Oxygen,* Simpkin, Marshall, Hamilton, & Kent, London, Vol. 2, Pt. 1., 1984, pp. 29–103.

2 Flank, L., Oxidant injury to pulmonary endothelium. In S. I. Said (ed.) *The Pulmonary Circulation and Acute Lung Injury,* 2nd Ed., Futura, Mount Kisco, NY, 1991, pp. 403–432.

3 Palmer, R. M. J., Ashton, D. S., and Moncada, S., Vascular endothelial cells synthesize nitric oxide from L-arginine, *Nature,* **333** (1988) 664–666.

4 Banister, J. V., Banister, W. H., Rotilio, G., Aspect of the structure, function, and applications of superoxide dismutase, *CRC Crit. Rev. Biochem.,* **22** (1987) 111–180.

5 Weisiger, R. A. and Fridovich, I., Superoxide dismutase. Organelle specificity, *J. Biol. Chem.,* **248** (1973) 3582–3592.

6 Freeman, B. A. and Crapo, J. D., Biology of disease: free radicals and tissue injury, *Lab. Invest.,* **47** (1982) 412–426.

7 McCord, J. M., Free radicals and myocardial ischemia: overview and outlook, *Free Radical Biol. Med.,* **4** (1988) 9–14.

8 Murphy, T. H., Miyamoto, M., Sartre, A. , Schnaar, R. L., and Coyle, J. T., Glutamate toxicity in a neural cell line involves inhibition of cystine transport leading to oxidative stress, *Neuron,* **2** (1989) 1547–1558.

9 Schor, N. F., Inactivation of mammalian glutamate synthetase by oxygen radicals, *Brain Res.,* **456** (1988) 17–21.

10 Pellegrini-Giampietro, D. E., Cherici, G., A.lesiani, M., Carla, V., and Moroni, F., Excitatory amino acid release and free radial formation may cooperate in the genesis of ischemia-induced neuronal damage, *J. Neurosci.,* **10** (1990) 1035–1041.

11 Furchgott, R. F. and Zawadzki, J. V., The obligatory role of endothelial cells in the relaxation of arterial smooth muscle by acetylcholine, *Nature,* **288** (1980) 373–376.

12 Palmer, R. M. J., Ferrige, A. G. and Moncada, S., Nitric oxide release account for the biological activity of endothelium-derived relaxing factor, *Nature,* **327** (1987) 524–526.

13 Knowles, R. G., Palacios, M., Palmer, R. M. J., and Moncada, S., Formation of nitric oxide from L-arginine in the central nervous system: a transduction mechanism for stimulation of the soluble granylate cyclase, *Proc. Natl. Acad. Sci. USA* **89** (1989) 5159–5162.

14 Bredt, D. S. and Snyder, S. C., Nitric oxide, a novel neuronal messenger, *Neuron,* **8** (1992) 3–11.

15 Rajfer, J., A.ronson, W. J., Bush, P. A., Dorey, F. J., and Ignarro, L. J., Nitric oxide as a mediator of relaxation of the corpus cavernosum in response to nonadrenergic, non cholinergic neurotransmission, *N. Engl. J. Med.,* **326** (1992) 90–94.

16 Schuman, E. M. and Madison, D. V., A requirement for the intercellular messenger nitric oxide in long-term potentiation, *Science,* **254** (1991) 1503–1506.

17 Zhuo, M., Small, S. A., Kandel, E. R., and Hawkins, R. D., Nitric oxide and carbon monoxide produce activity-dependent long-term synaptic enhancement in hippocampus, *Science,* **260** (1993) 1946–1950.

18 Hawkins, R. D., Kandel, E. R., and Siegelbaum, S. A., Learning to modulate transmitter release: themes and variations in synaptic plasticity, *Annu. Rev. Neurosci.,* **16** (1993) 625–665.

19 Bredt, D. S., Hwang, P. H., Glatt, C., Jowenstein, C., Reed, R. R., and Snyder, S. H., Cloned and expressed nitric oxide synthase structure resembles cytochrome P-450 reductase, *Nature,* **352** (1991) 714–718.

20 Ischiropoulos, H., Zhu, L., and Beckman, J. S., Peroxynitrite formation from macrophage-derived nitric oxide, *Arch. Biochem. Biophys.,* **298** (1992) 446–451.

21 Bredt, D. S. and Snyder, S. H., Nitric oxide mediates glutamate-linked enhancement of cGMP levels in the cerebellum, *Proc. Natl. Acad. Sci. USA.* **86** (1989) 9030–9033.

22 Garthwaite, J., Glutamate, nitric oxide and cell-cell signaling in the nervous system, *Trends Neurosci.,* **14** (1991) 60–67.

23 Montague, P. R., Gancayco, C. D., Winn, M. J., Marchase, R. B., and Friedlander, M. J., Role of NO production in NMDA receptor-mediated neurotransmitter release in cerebral cortex, *Science,* **263** (1994) 973 977.

24 Demopoulos, H. B., Flamm, E. S., Pietronigro, D. D., and Seligman, M. L., The free radical pathology and the microcirculation in the major central nervous system disorders, *Acta Physiol. Scand.,* **492(Suppl)** (1980) 91–119.

25 Flamm, E. S., Demopoulos, H. B., Seligman, M. L., Poser, G. R., and Ransohoff, J., Free radicals in cerebral ischemia, *Stroke,* **9** (1978) 445–451.

26 McCord, J. M., Oxygen-derived free radicals in postischemic tissue injury, *N. Engl. J. Med.,* **312** (1985) 159–163.

27 Siesjo, B. K., Cell damage in the brain: a speculative synthesis, *J. Cereb. Blood Flow Metab.,* **1** (1981) 155–185.

28 Raichle, M. E., The pathophysiology of brain ischemia, *Ann. Neurol.,* **13** (1983) 2–10.

29 Hall, E. D. and Braughler, J. M., Central nervous system trauma and stroke. II. Physiological and pharmacological evidence for involvement of oxygen radicals and lipid peroxidation, *Free Radical Biol. Med.,* **6** (1989) 303–313.

30 Uyama, O., Matsuyama, T., Michishita, H., Nakamura, H., and Sugita, M., Protective effects of human recombinant superoxide dismutase on transient ischemic injury of CA1 neurons in gerbils, *Stroke,* **23** (1992) 75–81.

31 Patel, M., Day, B. J., Crapo, J. D., Fridovich, I., and McNamara, J. O., Requirement for superoxide in excitotoxic cell death, *Neuron,* **16** (1996) 345–355.

32 Kirsch, J. R., Phelan, A. M., Lange, D. G., and Traystman, R. J., Reperfusion induced free radical formation following global ischemia, *Ped. Res.,* **21** (1987) 202A.

33 Bromont, C., Merie, C., and Bralet, J., Increased lipid peroxidation in vulnerable brain regions after transient ischemia in rats, *Stroke,* **20** (1989) 918–924.

34 Watson, B. D., Busto, R., Goldberg, W. J., Santiso, M., Yoshida, S., and Ginsberg, M. D., Lipid peroxidation in vivo induced by reversible global ischemia in rat brain, *J. Neurochem.,* **42** (1984) 268–274.

35 Yoshida, T., Limmroth, V., Irikura, K., and Moskowitz, M. A., The NOS inhibitor, 7-nitroindazole, decreases focal infarct volume but not the response to topical acethylcholine in pial vessels, *J. Cereb. Blood Flow Metab.,* **14** (1994) 924–929.

36 Jarasch, E. D., Grund, C., Bruder, G., Heid, H. W., Keenan, T. W., and Franke, W. W., Localization of xanthine oxidase in mammary-gland epithelium and capillary endothelium, *Cell,* **25** (1981) 67–82.

37 Springer, T. A., Adhesion receptors of the immune system, *Nature,* **346** (1990) 425–433.

38 del Zoppo, G. J., Microvascular changes during cerebral ischemia and reperfusion, *Cerebrovasc. Brain Metab. Rev.,* **6** (1994) 47–96.

39 Halliwell, B. and Gutterridge, J. M. C., *Free Radicals in Biology and Medicine,* Oxford University Press, London, 1985.

40 Digniseppi, J. and Fridovich, I., The toxicology of molecular oxygen, *CRC Crit. Rev. Toxicol.,* **12** (1984) 315–341.

41 Brown, A. W. and Brierley, J. B., Anoxic-ischaemic cell change in rat brain. Light microscopic and fine-structural observations, *J. Neurol. Sci.,* **16** (1972) 59–84.

42 Kirino, T., Delayed neuronal death in the gerbil hippocampus following ischemia, *Brain Res.,* **239** (1982) 57–69.

43 Kirino, T. and Sano, K., Fine structure of delayed neuronal death following ischemia in the gerbil hippocampus, *Acta Neuropathol. (Berl),* **62** (1984) 209–218.

44 Matsuyama, T., Tsuchiyama, M., Nakamura, H., Matsumoto, M., and Sugita, M., Hilar somatostatin neurons are more vulnerable to an ischemic insult than CA1 pyramidal neurons, *J. Cereb. Blood Flow Metab.,* **13** (1993) 229–234.

45 Cohen, J. J., Apoptosis, *Immunol. Today,* **14** (1993) 126–130.

46 Fesus, L., Davies, P. J. A., and Piacentini, M., Apoptosis: molecular mechanisms in programmed cell death, *Eur. J. Cell Biol.,* **56** (1991) 170–177.

47 Lockshin, R. A. and Zakeri, Z., Programmed cell death and apoptosis. In T. D. Tomei and F. O. Cope (eds) *Apoptosis: The Molecular Basis of Cell Death,* Cold Spring Harbor Laboratory, Cold Spring Harbor, NY, 1991, pp. 47–60.

48 Schwartz, L. M. and Osborne, B. A., Programmed cell death, apoptosis and killer genes, *Immunol. Today,* **14** (1993) 582–590.

49 Wyllie, A. H., Glucocorticoid-induced thymocyte apoptosis is associated with endogenous endonuclease activation, *Nature,* **284** (1980) 555–556.

50 Li, Y., Chopp, M., Jiang, N., and Zaloga, C., In situ detection of DNA fragmentation after focal cerebral ischemia in mice, *Mol. Brain Res.,* **28** (1995) 164–168.

51 Linnik, M. D., Zobrist, R. H., and Hatfield, M. D., Evidence supporting a role for programmed cell death in focal cerebral ischemia in rats, *Stroke,* **24** (1993) 2002–2009.

52 Shigeno, T., Yamasaki, Y., Kato, G., Kusaka, K., Mima, T., Takakura, K., Graham, D. I., and Furukawa, S., Reduction of delayed neuronal death by inhibition of protein synthesis, *Neurosci. Lett.,* **120** (1990) 117–119.

53 Okamoto, M., Matsumoto, M., Ohtsuki, T., Taguchi, A, Mikoshiba, K., Yanagihara, T., and Kamada, T., Internucleosomal DNA cleavage involved in ischemia-induced neuronal death, *Biochem. Biophys. Res. Commun.,* **196** (1993) 1356–1362.

54 Nitatori, T., Sato, N., Waguri, S., Karasawa, Y., Araki, H., Shibanai, K., Kominami, E., and Uchiyama, Y., Delayed neuronal death in the CA1 pyramidal cell layer of the gerbil hippocampus following transient ischemia is apoptosis, *J. Neurosci.,* **15** (1995) 1001–1011.

55 Li, Y., Chopp, M., Jiang, N., Yao, F., and Zaloga, C., Temporal profile of in situ DNA fragmentation after transient middle cerebral artery occlusion in the rat, *J. Cereb. Blood Flow Metab.,* **15** (1995) 389–397.

56 Colotta, F., Polentarutti, N., Sironi, M., and Mantovani, A., Expression and involvement of c-fos and c-jun protooncogenes in programmed cell death induced by growth factor deprivation in lymphoid cell lines, *J. Biol. Chem.,* **267** (1992) 18,278–18,283.

57 Dragunow, M., Young, D., Hughes, P., MacGibbon, G., Lawlor, P., Singleton, K., Sirimanne, E., Beiharz, E., and Gluckman, P., Is c-jun involved in nerve cell death following status epilepticus and hypoxic-ischaemic brain injury? *Mol. Brain Res.,* **18** (1993) 347–352.

58 Smyne, R. J., Vendrell, M., Hayward, M., Baker, S. J., Miao, G. G., Schilling, K., Robertson, L. M., Curran, T., and Morgan, J. I., Continuous c-fos expression precedes programmed cell death in vivo, *Nature,* **363** (1993) 166–169.

59 Halliwell, B. and Gutteridge, J. M. C., Role of free radicals and catalytic metal ions in human disease: an overview, *Meth. Enzymol.,* **186** (1990) 1–85 .

60 Duvall, E. and Wyllie, A. H., Death and the cell, *Immunol. Today,* **7** (1986) 115–119.

61 Nagata, S. and Golstein, P., The Fas death factor, *Science,* **267** (1995) 1449–1456.

62 Matsuyama, T., Hata, R., Tagaya, M., Yamamoto, Y., Nakajima, T., Furuyama, J., Wanaka, A., and Sugita, M., Fas antigen mRNA induction in postischemic murine brain, *Brain Res.,* **657** (1994) 342–346.

63 Matsuyama, T., Hata, R., Yamamoto, Y., Tagaya, M., A.kita, H., Uno, H., Wanaka, A., Furuyama, J., and Sugita, M., Localization of Fas antigen mRNA induced in postischemic murine forebrain by in situ hybridization, *Mol. Brain Res.,* **34** (1995) 166–172.

64 Larrick, J. W. and Wright, S. C., Cytotoxic mechanism of tumor necrosis factor-α, *FASEB J.,* **4** (1990) 3215–3223.

65 Buttke, T. M. and Sandstrom, P. A., Oxidative stress as a mediator of apoptosis, *Immunol. Today,* **15** (1994) 7–10.

66 Albina, J. E., Cui, S., Mateo, R. B., and Reichner, J. S., Nitric oxide-mediated apoptosis in murine peritoneal macrophages, *J. Immunol.,* **150** (1993) 5080–5085.

67 Schreck, R., Albermann, K. A. J., and Baeuerle, P. A., Nuclear factor kappa B: an oxidative stress-responsive transcription factor of eukaryotic cells (a review), *Free Radical Res. Comm.,* **17** (1992) 221–237.

68 Kitagawa, K., Matsumoto, M., Tagaya, M., Hata, R., Ueda, H., Niinobe, M., Handa, N., Fukunaga, R., Kimura, K., Mikoshiba, K., and Kamada, T., 'Ischemic tolerance' phenomenon found in the brain, *Brain Res.,* **528** (1990) 21–24.

69 Kirino, T., Tsujita, Y., and Tamura, A., Induced tolerance to ischemia in gerbil hippocampal neurons, *J. Cereb. Blood Flow Metab.,* **11** (1991) 299–307.

70 Kitagawa, K., Matsumoto, M., Kuwabara, K., Tagaya, M., Ohtsuki, T., Hata, R., Ueda, H., Handa, N., Kimura, K., and Kamada, T., 'Ischemic tolerance' phenomenon detected in various brain regions, *Brain Res.,* **561** (1991) 203–211.

71 Kitagawa, K., Matsumoto, M., Tagaya, M., Kuwabara, K., Hata, R., Handa, N., Fukunaga, R., Kimura, K., and Kamada, T., Hyperthermia-induced neuronal protection against ischemia injury in gerbils, *J. Cereb. Blood Flow Metab.,* **11** (1991) 449–452.

72 Chopp, M., Chen, H., Ho, K.-L., Dereski, M. O., Brown, E., Hetzel, F. W., and Welch, K. M. A., Transient hyperthermia protects against subsequent forebrain ischemic cell damage in the rat, *Neurology,* **39** (1989) 1396–1398.

73 Ohtsuki, T., Matsumoto, M., Kuwabara, K., Kitagawa, K., Suzuki, K., Taniguchi, N., and Kamada, T., Influence of oxidative stress on induced tolerance to ischemia in gerbil hippocampal neurons, *Brain Res.,* **599** (1992) 246–252.

74 Li, G. C. and Werb, Z., Correlation between synthesis of heat shock proteins and development of thermotolerance in Chinese hamster fibroblasts, *Proc. Natl. Acad. Sci. USA* **79** (1982) 3218–3222.

75 Johnston, R. N. and Kucey, B. L., Competitive inhibition of hsp70 gene expression causes thermosensitivity, *Science,* **242** (1988) 1551–1554.

76 Lindquist, S., The heat-shock response, *Annu. Rev. Biochem.,* **55** (1986) 1151–1191.

77 Currie, R. W., Karmazyn, M., Kloc, M., and Mailer, K., Heat-shock response is associated with enhanced postischemic ventricular recovery, *Circ. Res.,* **63** (1988) 543–549.

78 Nguyen, V. T., Morange, M., and Bensaude, O., Protein denaturation during heat shock and related stress, *J. Biol. Chem.,* **264** (1989) 10,487–10,492.

79 Pelham, H. R. B., Speculations on the functions of the major heat shock and glucose-regulated proteins, *Cell,* **46** (1986) 959–961.

80 Schlesinger, M. J., Heat shock proteins, *J. Biol. Chem.*, **265** (1990) 12,111–12,114.

81 Glazier, S. S., O'Rourke, D. M., Graham, D. I., and Welsh, F. A., Induction of ischemic tolerance following brief focal ischemia in rat brain, *J. Cereb. Blood Flow Metab.*, **14** (1994) 545–553.

82 Kiessling, M., Stumm, G., Xie, Y., Herdegen, T., Aguzzi, A., Bravo, R., and Gass, P., Differential transcription and translation of immediate early genes in the gerbil hippocampus after transient global ischemia, *J. Cereb. Blood Flow Metab.*, **13** (1993) 914–924.

83 Halazonetis, T. D., Georgopoulos, K., Greenberg, M. E., and Leder, P., c-jun dimerizes with itself and with c-fos, forming complexes of different DNA-binding affinities, *Cell*, **55** (1988) 917–924.

84 Rauscher, F. J., III, Voulalas, P. J., Franza, B. R., Jr., and Curran, T., Fos and jun bind cooperatively to the AP-1 site: reconstitution in vitro, *Genes Dev.*, **2** (1988) 1687–1699.

85 Sommer, C., Gass, P., and Kiessling, M., Selective c-jun expression in CA1 neurons of the gerbil hippocampus during and after acquisition of an ischemia-tolerant state, *J. Cereb. Blood Flow Metab.*, **15(Suppl 1)** (1995) S66.

86 Hefti, F., Dravid, A., and Hartikka, J., Chronic intraventricular injections of nerve growth factor elevate hippocampal choline acetyltransferase activity in adult rats with partial septhippocampal lesions, *Brain Res.*, **293** (1984) 305–311.

87 Cheng, B. and Mattson, M. P., Glucose deprivation elicits neurofibrillary tangle-like antigenic changes in hippocampal neurons: prevention by NGF and bFGF, *Exp. Neurol.*, **117** (1992) 114–123.

88 Sigeno, T., Mima, T., Takakura, K., Graham, D. I., Kato, G., Hashimoto, Y., and Furukawa, S., Amelioration of delayed neuronal death in the hippocampus by nerve growth factor, *J. Neurosci.*, **11** (1991) 2914–2919.

89 Nozaki, K., Finklestein, S. P., and Beal, M. F., Basic fibroblast growth factor protects against hypoxia-ischemia and NMDA neurotoxicity in neonatal rats, *J. Cereb. Blood Flow Metab.*, **13** (1993) 221–228.

90 Benveniste, E. N., Inflammatory cytokines within the central nervous system: sources, function, and mechanism of action, *Am. J. Physiol.*, **263** (1992) C1–C16.

91 Feuerstein, G. Z., Liu, T., and Barone, F. C., Cytokines, inflammation, and brain injury: role of tumor necrosis factor-α, *Cerebrovasc. Brain Metab. Rev.*, **6** (1994) 341–360.

92 Rothwell, N. J. and Hopkins, S. J., Cytokines and the nervous system II: actions and mechanisms of action, *Trends Neurosci.*, **18** (1995) 130–136.

93 Morioka, T., Kalehua, A. N., and Streit, W. J., The microglial reaction in the rat dorsal hippocampus following transient forebrain ischemia, *J. Cereb. Blood Flow Metab.*, **11** (1991) 966–973.

94 Giulian, D. and Baker, T. J., Peptides released by ameboid microglia regulate astroglial proliferation, *J. Cell Biol.*, **101** (1985) 2411–2415.

95 Chao, C. C., Hu, S., Close, K., Choi, C. S., Molitor, T. W., Noviac, W. J., and Peterson, P. K., Cytokine release from microglia: differential inhibition by pentoxifylline and dexamethasone, *J. Infect. Dis.*, **166** (1992) 847–53.

96 Alosi, F., Care, A., Borsellino, G., Gallo, P., Rosa, S., Bassani, A., Cabibbo, A., Testa, U., Levi, G., and Paschle, C., Production of hemolymphopoietic cytokines (IL-6, IL-8, corony-stimulating factors) by normal human astrocytes in response to IL-1β and tumor necrosis factor-α, *J. Immunol.*, **149** (1992) 2358–2366.

97 Sawada, M., Suzumura, A., and Marunouchi, T., TNF alpha induces IL-6 production by astrocytes but not microglia, *Brain Res.*, **583** (1992) 296–299.

98 Sawada, M., Kondo, N., Suzumura, A., and Marunouchi, T., Production of tumor necrosis factor-alpha by microglia and astrocytes in culture, *Brain Res.*, **491** (1989) 394–397.

99 Shohami, E., Novikov, M., Bass, R., Yamin, A., and Gallily, R., Closed head injury triggers early production of TNFα and IL-6 by brain tissue, *J. Cereb. Blood Flow Metab.,* **14** (1994) 615–619.

100 Liu, T., Clark, R., McDonnell, P. C., Young, P. R., White, R. F., Barone, F. C., and Feuerstein, G. Z., Tumor necrosis factor-α expression in ischemic neurons, *Stroke,* **25** (1994) 1481–1488.

101 Uno, H., Matsuyama, T., Tagaya, M., Hata, R., Akita, H., Nakamura, H., and Sugita, M., Cerebral ischemia triggers early production of tumor necrosis factor-α by microglia, *J. Cereb. Blood Flow Metab.,* **15(Suppl 1)** (1995) S118.

102 Semenzato, G., Tumor necrosis factor: a cytokine with multiple biological activities, *Br. J. Cancer,* **61** (1990) 354–361.

103 Aldersson, P. B., Perry, V. H., and Gordon, S., Intracerebral injection of pro-inflammatory cytokines or leukocyte chemotaxins induces minimal myelononocytic cell recruitment to the parenchyma of the central nervous system, *J. Exp. Med.,* **176** (1992) 255–259.

104 Sharief, M. K. and Thompson, E. J., In vivo relationship of tumor necrosis factor-α to blood-brain barrier damage in patients with active multiple sclerosis, *J. Neuroimmunol.,* **38** (1992) 27–34.

105 Male, D., Pryce, F., and Rahman, J., Comparison of the immunological properties of rat cerebral and aortic endothelium, *J. Neuroimmunol.,* **30** (1990) 161–168.

106 Piani, D., Spranger, M., Frei, K., Schaffer, A., and Fontana, A., Macrophage-induced cytotoxicity of N-methyl-D-aspartate receptor positive neurons involves excitatory amino acids rather than reactive oxygen intermediates and cytokines, *Eur. J. Immunol.,* **22** (1992) 2429–2436.

107 Loike, J. D., Cao, L., Brett, J., Ogawa, S., Silverstein, S. C., and Stern, D., Hypoxia induces glucose transporter expression in endothelial cells, *Am. J. Physiol.,* **263** (1992) C326 333.

108 Yan, S.-F., Tritto, I., Pinsky, D., Liao, H., Huang, J., Fuller, G., Brett, J., May, L., and Stern, D., Induction of interleukin 6 (IL-6) by hypoxia in vascular cells, *J. Biol. Chem.,* **270** (1995) 11,463–11,471.

109 Mayer, M., Schreck, R., and Baeuerle, P., H_2O_2 and antioxidants have opposite effects on activation of NF-κB and AP-1 in intact cells: AP-1 as secondary antioxidant-responsive factor, *EMBO J.,* **12** (1993) 2005–2015.

110 Matsuo, N., Ogawa, S., Imai, Y., Takagi, T., Tohyama, M., Stern, D., and Wanaka, A., Cloning of a novel RNA binding polypeptide (RA301) induced by hypoxia/reoxygenation, *J. Biol. Chem.,* **270** (1995) 28,216–28,222.

111 Maeda, Y., Matsumoto, M., Hori, O., Kuwabara, K., Ogawa, S., Yan, S. D., Ohtsuki, T., Kinoshita, T., Kamada, T., and Stern, D. M., Hypoxia/Reoxygenation-mediated induction of astrocyte interleukin 6: paracrine mechanism potentially enhancing neuron survival, *J. Exp. Med.,* **180** (1994) 2297–2308.

112 Shreck, R., Rieber, P., and Baeuerle, P. A., Reactive oxygen intermediates as apparently widely used messengers in the activation of the NF-κB transcription factor and HIV-1, *EMBO J.,* **10** (1991) 2247–2258.

113 Braughler, J. M. and Hall, E. D., Central nervous system and stroke: I. Biochemical considerations for oxygen radical formation and lipid peroxydation, *Free Radical Biol. Med.,* **6** (1989) 289–301.

114 Traystman, R. J., Kirsch, J. R., and Koehler, R. C., Oxygen radical mechanisms of brain injury following ischemia and reperfusion, *J. Appl. Physiol.,* **71** (1991) 1185–1195.

115 del Zoppo, G. J., Microvascular changes during cerebral ischemia and reperfusion, *Cerebrovasc. Brain Metab. Rev.,* **6** (1994) 47–96.

116 Jarasch, E.-D., Bruder, G., and Heid, H. M., Significance of xanthine oxidase in capillary endothelial cells, *Acta Physiol. Scand.,* **548(Suppl)** (1986) 39–46.

117 Babior, B. M., Curnutte, J. T., and McMurrich, B. J., The particulate superoxide-forming system from human neutrophils: properties of the system and further evidence supporting its participation in the respiratory burst, *J. Clin. Invest.,* **58** (1976) 989–996.

118 Britigan, B. E., Rosen, G. M., Chai, Y., and Cohen, M. S., Do human neutrophils make hydroxy radical? Determination of free radicals generated by human neutrophils activated with a soluble or particulate stimulus using electron paramagnetic response spectrometry, *J. Biol. Chem.,* **261** (1986) 4426–4431.

119 Petrone, W. F., English, D. K., Wong, K., and McCord, J. M., Free radicals and inflammation: superoxide dependent activation of a neutrophil chemotactic factor in plasma, *Proc. Natl. Acad. Sci. USA* **77** (1980) 1159–1163.

120 Battelli, M. G., Della Corte, E., and Stirpe, F., Xanthine oxidase type D (dehydrogenase) in the intestine and other organs of the rat, *Biochem. J.,* **126** (1972) 747–749.

121 Parks, D. A., and Granger, D. N., Xanthine oxidase: biochemistry, distribution and physiology, *Acta Physiol. Scand.,* **548(Suppl)** (1986) 87–99.

122 Betz, A. L., Identification of hypoxanthine transport and xanthine oxidase activity in brain capillaries, *J. Neurochem.,* **44** (1985) 574–579.

123 Kontos, H. A., Wei, E. P., Povlishock, J. T., and Christman, C. W., Oxygen radicals mediate the cerebral arteriolar dilation from arachidonate and bradykinin in cats, *Circ. Res.,* **55** (1984) 295–303.

124 Olesen, S.-P., Free oxygen radicals decrease electrical resistance on microvascular endothelium in brain, *Acta Physiol. Scand.,* **129** (1987) 181–187.

125 Wei, E. P., Christman, C. W., Kontos, H. A., and Povlishock, J. T., Effects of oxygen radicals on cerebral arterioles, *Am. J. Physiol.,* **248** (1985) H157–H162.

126 Suematsu, M., Kurose, I., Asako, H., Miura, S., and Tsuchiya, M., In vivo visualization of oxyradical-dependent photoemission during endothelium-granulocyte interaction in microvascular beds treated with platelet-activating factor, *J. Biochem.,* **106** (1989) 355–360.

127 Dorovini-Zis, K., Bowman, P. D., and Prameya, R., Adhesion and migration of human polymorphonuclear leukocytes across cultured bovine brain microvessel endothelial cells. *J. Neuropathol. Exp. Neurol.,* **51** (1992) 194–205.

128 Wong, D. and Dorovini-Zis, K., Up-regulation of intercellular adhesion molecule-1 (ICAM-1) expression in primary cultures of human brain microvessel endothelial cells by cytokines and lipopolysaccharide, *J. Neuroimmunol.,* **39** (1992) 11–22.

129 Tayarani, I., Chaudiere, J., Lefauconnier, J.-M., and Bourre, J.-M., Enzymatic protection against peroxidative damage in isolated brain capillaries, *J. Neurochem.,* **48** (1987) 1399–1402.

130 MacCarron, R. M., Uematsu, S., Merkel, N., Long, D., Bembry, J., and Spatz, M., The role of arachidonic acid and oxygen radicals on cerebromicrovascular endothelial permeability, *Acta Neurochir.,* **51** (1990) 61–64.

131 Halliwell, B. and Aruoma, O. I., DNA damage by oxygen-derived species. Its mechanism and measurement in mammalian systems, *FEBS Lett.,* **281** (1991) 9–19.

132 Halliwell, B., Oxygen radicals and metal ions: potential antioxidant intervention strategies, *Ann. Intern. Med.,* **107** (1987) 526–545.

133 Chan, P. H., Chu, L., and Fishman, R. A., Reduction of activities of superoxide dismutase but not of glutathione peroxidase in rat brain regions following decapitation ischemia, *Brain Res.,* **439** (1988) 388–390.

134 Paschen, W., Linn, F., and Csiba, L., Superoxide dismutase activity in experimental focal cerebral ischemia, *Exp. Neurol.,* **90** (1985) 611–618.

135 Michowiz, S. D., Melamed, E., Pikarsky, E., and Rappaport, Z. H., Effect of ischemia induced by middle cerebral artery occlusion on superoxide dismutase activity in rat brain, *Stroke,* **21** (1990) 1613–1617.

136 Karlsson, K. and Marklund, S. L., Extracellular superoxide dismutase in the vascular system of mammals, *Biochem. J.,* **255** (1988) 223–228.

137 Fridovich, I., Superoxide dismutase, *Ann. Rev. Biochem.,* **44** (1975) 147–159.

138 Hassan, H. M., Biosynthesis and regulation of superoxide dismutases, *Free Radical Biol. Med.,* **5** (1988) 377–385.

139 Kawaguchi, T., Noji, S., Uda, T., Nakashima, Y., Takeyasu, A., Kawai, Y., Takagi, H., Tohyama, M., and Taniguchi, N., A monoclonal antibody against COOH-terminal peptide of human liver manganese superoxide dismutase, *J. Biol. Chem.,* **264** (1989) 5762–5767.

140 Matsuyama, T., Michishita, H., Nakamura, H., Tsuchiyama, M., Shimizu, S., Watanabe, K., and Sugita, M., Induction of copper-zinc superoxide dismutase in gerbil hippocampus after ischemia, *J. Cereb. Blood Flow Metab.,* **13** (1993) 135–144.

141 Matsuyama, T., Shimizu, S., Nakamura, H., Michishita, H., Tagaya, M., and Sugita, M., Effects of recombinant superoxide dismutase on manganese superoxide dismutase gene expression in gerbil hippocampus after ischemia, *Stroke,* **25** (1994) 1417–1424.

142 Akai, F., Maeda, M., Suzuki, K., Inagaki, S., Takagi, H., and Taniguchi, N., Immunocytochemical localization of manganese superoxide dismutase (MnSOD) in the hippocampus of the rat, *Neurosci. Lett.,* **115** (1990) 19–23.

143 Ceballos, I., Javoy-Agid, F., Hirsch, E. C., Dumas, S., Kamoun, P. P., Sinet, P. M., and Agid, Y., Localization of copper-zinc superoxide dismutase mRNA in human hippocampus by in situ hybridization, *Neurosci. Lett.,* **105** (1989) 41–46.

144 Kiessling, M., Dienel, G. A., Jacewicz, M., and Pulsinelli, W. A., Protein synthesis in postischemic rat brain: a two-dimensional electrophoretic analysis, *J. Cereb. Blood Flow Metab.,* **6** (1986) 642–649.

145 Thilmann, R., Xie, Y., Kleihues, P., and Kiessling, M., Persistent inhibition of protein synthesis precedes delayed neuronal death in postischemic gerbil hippocampus, *Acta Neuropathol. (Berl),* **71** (1986) 88–93.

146 Vass, K., Welch, W. J., and Nowak, T. S., Jr., localization of 70 kDa stress protein induction in gerbil brain after ischemia, *Acta Neuropathol. (Berl),* **77** (1988) 128–135.

147 Nowak, T. S., Jr., Localization of 70kDa stress protein mRNA induction in gerbil brain after ischemia, *J. Cereb. Blood Flow Metab.,* **11** (1991) 432–439.

148 Ohtsuki, T., Matsumoto, M., Suzuki, K., Taniguchi, N., and Kamada, T., Effect of transient forebrain ischemia on superoxide dismutases in gerbil hippocampus, *Brain Res.,* **620** (1993) 305–309.

149 Matsuyama, T., Shimizu, S., Nakamura, H., Hata, R., Terayama, Y., Yajima, Y., and Sugita, M., Induction of superoxide dismutase in gerbil hippocampus before and after acquisition of ischemic tolerance, *Soc. Neurosci. Abstr.,* **21** (1995) 513.

150 Kinouchi, H., Epstein, C. J., Mizui, T., Carlson, E., Chen, S. F., and Chan, P. H. , Attenuation of focal cerebral ischemic injury in transgenic mice overexpressing CuZn superoxide dismutase, *Proc. Natl. Acad. Sci. USA,* **88** (1991) 11,158–11,162.

151 Greenberg, M. E., Hermanowski, A. L., and Ziff, E. B., Effect of protein synthesis inhibitors on growth factor activation of c-fos, c-myc, and actin gene transcription, *Mol. Cell Biol.,* **6** (1986) 1050–1057.

152 Ginty, D. D., Marlowe, M., Pekala, P. H., Seidel, E. R., Multiple pathways for the regulation of ornithine decarboxylase in intestinal epithelial cells, *Am. J. Physiol.,* **258** (1990) G454–G460.

153 An, G., Lin, T.-N., Liu, J.-S., Xue, J.-J., He, Y.-Y., and Hsu, C. Y., Expression of c-fos and c-jun family genes after focal cerebral ischemia, *Ann. Neurol.,* **33** (1993) 457–464.

154 Salo, D. C., Pacifici, R. E., Lin, S. W., Giulive, C., and Davies, K. J. A., Superoxide dismutase undergoes proteolysis and fragmentation following oxidative modification and inactivation, *J. Biol. Chem.,* **265** (1990) 11,919–11,927.

155 Maridonneau-Parini, I., Clerc, J., and Polla, B. S., Heat shock inhibits NADPH oxidase in human neutrophils, *Biochem. Biophis. Res. Commun.* **154** (1988) 179–186.

156 Clerget, M. and Polla, B. S., Erythrophagocytosis induces heat shock protein synthesis by human monocytes-macrophages, *Proc. Natl. Acad. Sci. USA* **87** (1990) 1081–1085.

157 Yoneda, T., Inagaki, S., Hayashi, Y., Nomura, T., Takagi, H., Differential regulation of manganese and copper/zinc superoxide dismutases by the facial nerve transection, *Brain Res.*, **582** (1992) 342–345.

158 Saggu, H., Cooksey, J., Dexter, D., Wells, F. R., Lees, A., Jenner, P., and Marsden, C. D., A selective increase in particulate superoxide dismutase activity in parkinsonian substantia nigra, *J. Neurochem.*, **53** (1989) 692–697.

159 Wong, H. W. and Goeddel, D. V., Induction of manganese superoxide dismutase by tumor necrosis factor: possible protective mechanism, *Science*, **242** (1988) 941–944.

160 Visner, G. A., Dougall, W. C., Wilson, J. M., Burr, I. A., and Nick, H. S., Regulation of manganese superoxide dismutase by lipopolysaccharide, interleukin-1, and tumor necrosis factor, *J. Biol. Chem.*, **265** (1990) 2856–2864.

161 Fujii, J. and Taniguchi, N., Phorbol ester induced manganese-superoxide dismutases in tumor necrosis-resistant cell, *J. Biol. Chem.*, **266** (1991) 23,142–23,146.

162 Meister, A. and Anderson, M., Glutathione. *Annu. Rev. Biochem.*, **52** (1983) 711–760.

163 Ter Horst, G. J., Knollema, S., Stuiver, B., Hom, H., Yoshimura, S., Ruiters, M. H. J., and Korf, J., Differential glutathione peroxidase mRNA up-regulation in rat forebrain areas after transient hypoxia-ischemia, *Ann. NY Acad. Sci.*, **738** (1995) 329–333.

164 Jackson, R. M. and Veal, C. F. Effect of hypoxia and reoxygenation on lung glutathione system, *Am. J. Physiol.*, **259** (1990) H518–H524.

165 De Haan, J., Newman, J. D., and Kola, I., Cu/Zn superoxide dismutase mRNA and enzyme activity, and susceptibility to lipid peroxidation, increases with aging in murine brains, *Mol. Brain Res.*, **13** (1992) 179–187.

166 Moreno, S. and Muganini, E., Immunocytochemical localization of catalase in rat brain, *Soc. Neurosci. Abstr.*, **18** (1992) 673.1.

167 Estrada, C., Mengual, E., and Gonzalez, C., Local NADPH-diaphorase neurons innervate pial arteries and lie close or project to intracerebral blood vessels: a possible role for nitric oxide in the regulation of cerebral blood flow, *J. Cereb. Blood Flow Metab.*, **13** (1993) 978–984.

168 Nelson, C. W., Wei, E. P., Povlishock, J. T., Kontos, H. A., and Moskowittz, M. A., Oxygen radicals in cerebral ischemia, *Am. J. Physiol.*, **263** (1992) H1356–H1362.

169 Bredt, D. S., Hwang, P. M., and Snyder, S. H., Localization of nitric oxide synthase indicating a neural role for nitric oxide, *Nature*, **347** (1990) 768–770.

170 Murphy, S., Simmons, M. L., Agullo L., Garcia, A., Feinstein, D. L., Galea, E., Reis, D. J., Minc-Golomb, D., Schwartz, J. P., Synthesis of nitric oxide in CNS glial cells, *Trends Neurosci.*, **16** (1993) 323–328.

171 Lamas, S., Marsden, P. A., Li, G. K., Tempst, P., and Michel, T., Endothelial nitric oxide synthase: molecular cloning and characterization of a distinct constitutive enzyme isoform, *Proc. Natl. Acad. Sci. USA* **89** (1992) 6348–6352.

172 Lowenstein, C. J., Glatt, C. S., Bredt, D. S., and Snyder, S. H., Cloned and expressed macrophage nitric oxide synthase contrasts with the brain enzyme, *Proc. Natl. Acad. Sci. USA* **89** (1992) 6711–6715.

173 Nathan, C., Nitric oxide as a secretory product of mammalian cells, *FASEB J.*, **6** (1992) 3051–3064.

174 Stuehr, D. J. and Nathan, C. F., Nitric oxide. A macrophage product responsible for cytostasis and respiratory inhibition in tumor target cells, *J. Exp. Med.*, **169** (1989) 1543–1555.

175 Morris, S. J. and Billiar, T. R., New insights into the regulation of inducible nitric oxide synthesis, *Am. J. Physiol.*, **266** (1994) E829–E839.

176 Iadecola, C., Zhang, F., Xu, S., Casey, R., and Ross, M. E., Inducible nitric oxide synthase gene expression in brain following cerebral ischemia, *J. Cereb. Blood Flow Metab.*, **15** (1995) 378–384.

177 Murphy, S., Simmons, M. L., Agullo, L., Garcia, A., Feinstein, D. L., Galea, E., Reis, D. J., Minc, G. D., and Schwartz, J. P., Synthesis of nitric oxide in CNS glial cells, *Trends Neurosci.*, **16** (1993) 323–328.

178 Tomimoto, H., Nishimura, M., Suenaga, T., Nakamura, S., Akiguchi, I., Wakita, H., Kimura, J., and Mayer, B., Distribution of nitric oxide synthase in the human cerebral blood vessels and brain tissues, *J. Cereb. Blood Flow Metab.*, **14** (1994) 930–938.

179 Iadecola, C., Beitz, A. M., Renno, W., Xu, X., Mayer, B., and Zhang, F., Nitric oxide synthase-containing neural process on large cerebral arteries and microvessels, *Brain Res.*, **606** (1993) 148–155.

180 Nozaki, K., Mosckowitz, M. A., Maynard, K. I., Koketsu, N., Dawson, T. M., Bredt, D. S., and Snyder, S. H., Possible origins and distribution of immunoreactive nitric oxide synthase-containing nerve fibers in cerebral arteries, *J. Cereb. Blood Flow Metab.*, **13** (1993) 70–79.

181 Malinski, T., Bailey, F., Zhang, Z. G., and Chopp, M., Nitric oxide measured by a porphyrinic microsensor in rat brain after transient middle cerebral artery occlusion, *J. Cereb. Blood Flow Metab.*, **13** (1993) 355–358.

182 Kumura, E., Kosaka, H., Shiga, T., Yoshimina, T., and Hayakawa, T., Elevation of plasma nitric oxide end products during focal cerebral ischemia and reperfusion in the rat, *J. Cereb. Blood Flow Metab.*, **14** (1994) 487–491.

183 Nowicki, J. P., Duval, D., Poignet, H., and Scatton, B., Nitric oxide mediates neuronal death after focal ischemia in the mouse, *Eur. J. Pharmacol.*, **204** (1991) 339,340.

184 Iadecola, C., Does nitric oxide mediate the increases in cerebral blood flow elicited by hypercapnia? *Proc. Natl. Acad. Sci. USA* **89** (1992) 3913–3916.

185 Wang, Q., Paulson, O. B., and Lassen, N. A., Effect of nitric oxide blockade by NG nitro L-arginine on cerebral blood flow response to changes in carbon dioxide tension, *J. Cereb. Blood Flow Metab.*, **12** (1992) 947–953.

186 Zhang, Z. G., Chopp, M., Maynard, K. I., and Moskowitz, M. A., Cerebral blood flow changes during cortical spreading depression are not altered by inhibition of nitric oxide synthesis, *J. Cereb. Blood Flow Metab.*, **14** (1994) 939–943.

187 Moncada, S. and Higgs, E. A., The L-arginine-nitric oxide pathway, *N. Engl. J. Med.*, **329** (1993) 2002–2012.

188 Yoshida, T., Limmroth, V., Irikura, K., and Moskowitz, M. A., The NOS inhibitor, 7-nitro-indazole, decreases focal infarct volume but not the response to topical acetylcholine in pial vessels, *J. Cereb. Blood Flow Metab.*, **14** (1994) 924–929.

189 Iadecola, C., Zhang, F., and Xu, X., Inhibition nitric oxide synthase ameliorates cerebral ischemic damage, *Am. J. Physiol.*, **268** (1995) R286–R292.

190 Morris, S. J. and Billiar, T. R., New insights into the regulation of inducible nitric oxide synthesis, *Am. J. Physiol.*, **266** (1994) E829–E839.

191 Stamler, J. S., Redox signaling: nitrosylation and related target interactions of nitric oxide, *Cell,* **78** (1994) 931–936.

192 Bredt, D. and Snyder, S. A., Nitric oxide: a physiologic messenger molecule, *Ann. Rev. Biochem.*, **63** (1994) 175–195.

193 Beckman, J. S., Beckman, T. W., Chen, J., Marshall, P. A., Freeman, B A., Apparent hydroxyl radical production by peroxynitrite: implications for endothelial injury from nitric oxide and superoxide, *Proc. Natl. Acad. Sci. USA* **87** (1990) 1620–1624.

194 Kamii, H., Mikawa, S., Kinouchi, H., Yoshimoto, T., Reola, L., Carlson, E., Epstein, C. J., and Chan, P. H., Effects of nitric oxide synthase inhibition on brain infarction in SOD-1 transgenic mice following transient focal cerebral ischemia, *J. Cereb. Blood Flow Metab.*, **15(Suppl 1)** (1995) S89.

195 Turrens, J. F., Crapo, J. D., and Freeman, B. A., Protection against oxygen toxicity by intravenous injection of liposome-entrapped catalase and superoxide dismutase, *J. Clin. Invest.*, **73** (1984) 87–95.

196 Hodgson, E. K. and Fridovich, I., The interaction of bovine erythrocyte superoxide dismutase with hydrogen peroxide: inactivation of the enzyme, *Biochemistry*, **14** (1975) 5294–5299.

197 Kono, Y. and Fridovich, I., Superoxide radical inhibits catalase, *J. Biol. Chem.*, **257** (1982) 5751–5754.

198 Blum, J. and Fridovich, I., Inactivation of glutathione peroxidase by superoxide radical, *Arch. Biochem. Biophys.*, **240** (1985) 500–508.

199 Beckman, J. S. and Freeman, B. A., Antioxidant enzymes as mechanistic probes of oxygen-dependent toxicity. In *Physiology of Oxygen Radicals*, 1986, pp. 39–53.

200 Kitagawa, K., Matsumoto, M., Oda, T., Niinobe, M., Hata, R., Handa, N., Fukunaga, R., Isaka, Y., Kimura, K., Maeda, H., Mikoshiba, K., and Kamada, T., Free radical generation during brief period of cerebral ischemia may trigger delayed neuronal death, *Neuroscience*, **35** (1990) 551–558.

201 Liu, T. H., Beckman, J. S., Freeman, B. A., Hogan, E. L., and Hsu, C. Y., Polyethylene glycol-conjugated superoxide dismutase and catalase reduce ischemic injury, *Am. J. Physiol.*, **256** (1989) H589–H593.

202 Imaizumi, S., Woolworth, V., Fishman, R. A., and Chan, P. H., Liposome-entrapped superoxide dismutase reduces cerebral infarction in cerebral ischemia in rats, *Stroke*, **21** (1990) 1312–1317.

203 Kajihara, J., Enomoto, M., Nishijima, K., Yabuuchi, M., and Katoh, K., Comparison of properties between human recombinant and placental copper-zinc SOD, *J. Biochem.*, **104** (1988) 851–854.

204 Uyama, O., Shiratsuki, N., Matsuyama, T., Nakanishi, T., Matsumoto, Y., Yamada, T., Narita, M., and Sugita, M., Protective effects of superoxide dismutase on acute reperfusion injury of gerbil brain, *Free Radical Biol. Med.*, **8** (1990) 205–208.

205 Tagaya, M., Matsumoto, M., Kitagawa, K., Niinobe, M., Ohtsuki, T., Hata, R., Ogawa, S., Handa, N., Mikoshiba, K., and Kadada, T., Recombinant human superoxide dismutase can attenuate ischemic neuronal damage in gerbils, *Life Sci.*, **51** (1992) 253–259.

206 Saiz, C. J., Kessler, J. A., Bennett, M. V. L., and Spry, D., Superoxide dismutase protects cultured neurons against death by starvation, *Proc. Natl. Acad. Sci. USA* **84** (1987) 3056–3059.

207 Burton, G. W. and Ingold, K. U., Vitamin, E., Application of the principles of physical organic chemistry to the exploration of its structure and function, *Acc. Chem. Res.*, **19** (1986) 194–201.

208 Kato, K., Terao, S., Shimamoto, N., and Hirata, M., Studies on scavengers of active oxygen species. 1. Synthesis and biological activity of 2-O-alkylascorbic acids, *Med. Chem.*, **31** (1988) 793–798.

209 Braughler, J. M., Pregenzer, J. F., Chase, R. L., Duncan, L. A., Hacobsen, E. J., and McCall, J. M., Novel 21-amino steroids as potent inhibitors of iron-dependent lipid peroxidation, *J. Biol. Chem.*, **262** (1987) 10,438–10,440.

210 Young, W., Wojak, J. C., and De Crescito, V., 21-aminosteriod reduces ion shifts and edema in the rat middle cerebral artery occlusion model of regional ischemia, *Stroke*, **19** (1988) 1013–1019.

211 Bast, A., Hasnen, G. R. M. M., and Doelman, C. J. A., Oxidants and antioxidants: state of the art, *Am. J. Med.*, **91(Suppl 3C)** (1991) 2–13.

212 Clements, J. A., Saunders, R. D., Ho, P. P., Phebus, L. A., and Panetta, J. A., The antioxidant LY 231617 reduces global ischemic neuronal injury in rats, *Stroke*, **24** (1993) 716–723.

213 Knollema, S., Aukema, W., Hom, H., Korf, J., and Ter Horst, G. J., L-deprenyl reduces brain damage in rats exposed to transient hypoxia/ischemia, *Stroke*, **26** (1995) 1883–1887.

8

Genomic Responses Following Cerebral Ischemia

Christoph Wiessner and Konstantin-Alexander Hossmann

1. INTRODUCTION

About 20 yr ago, it was shown that within a few hours after complete cerebral ischemia the enzymes ornithin decarboxylase and S-adenosylmethionin decarboxylase were induced in the cortex, despite an inhibition of the overall rate of protein synthesis *(1)* that persists for prolonged periods after the ischemic insult *(2)*. About 10 yr later, the postischemic synthesis of the heat shock protein HSP70, and glial fibrillary acidic protein (GFAP) was reported *(3–7)*. Meanwhile, the number of studies concentrating on ischemia-induced synthesis of specific proteins has been exploding. Nevertheless, it can be anticipated that at present the exploration of potentially important genes in vivo is far from complete.

The genes already investigated belong to various groups that could have potentially important functions in an ischemia-challenged brain, such as protection against environmental stress (heat shock proteins such as HSP70, trophic factors, immediate-early genes) *(8,9)*, involvement in death programs which may be activated by ischemia *(10)*, long term alteration of the cellular phenotype (immediate early genes such as c-*fos* and c-*jun*, neurotrophic factors) *(8,11)* regulation of glia/microglia activation (cytokines, cell surface proteins, structural proteins such as GFAP) *(12,13)*, and inflammatory events including cell-cell interaction, cell adhesion and cell attraction (cytokines, chemokines, adhesion molecules) *(14,15)*.

The ultimate goal in investigating ischemia-induced genomic alterations in brain cells is to contribute to the understanding of the degenerative processes and to define targets for highly specific modulation of these cellular responses. For example, a close association between gene inductions and programmed cell death (PCD) has been described *(16–20)*. PCD is well known to occur in neurons during development of the central nervous system *(21)*. Therefore, theoretically, this naturally occurring death program is intrinsic to each neuron and could be activated under pathological conditions, for example by disturbances of calcium homeostasis, free radical formation, and impairment of protein synthesis. These events are well known to occur after cerebral ischemia *(2,22,23)* and have also been implicated in PCD *(24)*. If inductions of genes with crucial functions along this pathway could be identified in the ischemia-challenged brain, it may be possible to interfere with cell degeneration at a very late stage.

From: Clinical Pharmacology of Cerebral Ischemia *Edited by:* G. J. Ter Horst and J. Korf *Humana Press Inc., Totowa, NJ*

Alterations in gene expression occur also in astrocytes and microglia in the course of activation after the ischemic insult *(25)*. As for neurons, the underlying idea in describing genomic alterations in glial cells is to contribute to the understanding of the activation process and to define targets for specific modulation. Astrocytic activation, for example, may interfere with reorganization of synaptic connections and microglia have a considerable cytotoxic potential if heavily activated. Therefore, it may be possible to limit tissue damage and to improve the functional recovery after ischemia by transiently suppressing astrocytic and microglial activation.

2. GENOMIC RESPONSES TO TRANSIENT FOREBRAIN ISCHEMIA

Transient global ischemia results in a form of delayed neuronal death that selectively destroys pyramidal neurons of the hippocampal CA1 sector in rodents *(26)* as well as in humans *(27)*. Selective neuronal vulnerability, however, is not specific to ischemic insults, but is also found in neurodegenerative diseases and certain types of epilepsy *(28)*. The highly ordered laminar structure of the hippocampus, the highly synchronized response of CA1 neurons to the insult, and the high reproducibility of the phenomenon makes delayed neuronal death after ischemia an extremely valuable model to investigate mechanisms responsible for this type of neuronal cell death. In addition, it offers a unique possibility to investigate regenerative attempts of the tissue to a precisely localized stereotyped brain lesion.

2.1. General Features of Ischemia-Induced Genomic Responses in Neurons

In general, expression of any gene in the first hours after global ischemia is not predictive or indicative for survival or degeneration of any cell population *(25)*. These responses include a very early induction of immediate early genes, such as c-*fos*, c-*jun*, and *mkp-1*, *Krox-24*, and *fosB*, which is already obvious at 15 or 30 min after ischemia *(29–34)*. In addition, the mRNA coding for the inducible member of the heat shock protein family, *hsp70*, is induced with a similar temporal and spatial pattern *(35)*. This general stress response also includes several other typical stress proteins such as ornithin decarboxylase *(36)*, *grp78 (25)*, *hsp90 (37)*, and heme oxygenase *(38)*. The early responses also include the expression of brain-derived neurotrophic factor (BDNF), nerve growth factor (NGF), and basic fibroblast growth factor *(39)*. Induction of BDNF and NGF, however, is strictly confined to the dentate gyrus and not found in any other region outside the hippocampus *(40)*.

The functional relevance of these early gene inductions to the pathophysiology of transient forebrain ischemia is unclear at present. They may represent adaptational attempts of ischemia-challenged cells (which are unsuccessful in the case of vulnerable cell populations), but may also be epiphenomena of disturbances of intracellular calcium homeostasis and other signaling cascades immediately following the ischemic insult.

In support for a role of gene activation in adaptational responses are phenomena of induced tolerance in global cerebral ischemia. Brief ischemic insults, which do not cause injury, can induce tolerance to subsequent, more severe insults *(41,42)*. Initially, the heat shock protein HSP70 has been thought to be responsible for this protective effect *(43,44)*. Recent studies, however, suggest that HSP70 is not required for ischemic tolerance, but that immediate early genes such as c-*jun*, *junB*, and *junD* or other yet undefined proteins are more likely to be involved *(45,46)*. In line with these studies are experiments showing that thermal conditioning of primary neuronal cultures protected from subsequent more severe heat shock, but did not induce HSP70 in neurons (Vogel et al., in press).

A number of studies have reported specifically elevated mRNA levels for several genes in dying CA1 neurons 1–2 d after the ischemic insult, i.e., at times rather shortly preceding cell death. In rat models quite consistently a late CA1-specific expression for c-*fos* and c-*jun* has been described *(25,31,47)*. Elevated c-*fos* and c-*jun* mRNA levels in the CA1 sector at times shortly preceding cell degeneration have also been described following hypoxic/ischemic insults *(48)*. An important point is, that in rat models of global ischemia the c-FOS protein *(30)* could be detected in CA1 pyramidal neurons by immunohistochemical methods.

The CA1-specific expression in dying hippocampal neurons has also been shown consistently for *hsp70 (49)* in all forebrain ischemia models, and in the rat also the HSP70 protein was found in CA1 neurons *(50)*. Although delayed in onset of induction as compared to *hsp70*, the same late mRNA accumulation in dying CA1 neurons was found for *grp78 (25)*. GRP78, the glucose regulated protein *(78)* is an endoplasmatic protein and belongs to the 70-kDa stress protein family. This late expression of both stress proteins could indicate prolonged cellular stress such as occurrence of misfolded or denatured proteins in both the cytoplasm and the endoplasmic reticulum. However, it is also noteworthy that *hsp70* is known to be expressed during cell cycle *(51)* and that the activation of a cell death program may be a consequence of conflicting growth regulatory signals leading to an unsuccessful attempt to proceed through the cell cycle *(52)*. In line with this notion is that MAP2c, an isoform of MAP2 (microtubule associated protein 2), which is normally expressed during embryogenesis, is re-expressed specifically in dying CA1 neurons after forebrain ischemia *(53)*. A recent study reports the specific activation of *p53* and *p21* (WAF1) in the CA1 sector following transient forebrain ischemia in the rat *(54)*. Expression of *p53* is known to be involved in the initiation of programmed cell death in various cell types *(55,56)*. In addition, a late CA1-specific accumulation of the cathepsins B, H, L has been shown in the gerbil *(57)*. Cathepsins are cysteine proteases involved in autophagy, a process known to greatly increase in severely injured cells *(58)*.

2.2. Is There Evidence for the Ischemia-Induced Activation of a Death Program in Delayed Neuronal Death?

The CA1-specific neuronal expression of genes at time points rather shortly preceding cell death in rat models as described above, indeed is similar to events in certain cell types in which a death program has clearly been identified. An expression of c-*fos* and c-*jun* accompanying death of neurons is found in sympathetic neurons undergoing programmed cell death because of NGF deprival *(16)*. Continuous c-*fos* expression has also been described for programmed cell death of neurons during development of the CNS *(59)*. Dimers of c-FOS and c-JUN, or homodimers of c-JUN act as transcription factors which can activate other genes, and it may be that "bad" proteins are induced by these factors. If this is true, however, after ischemia additional regulatory mechanisms must be postulated, since both proteins products are also formed in the CA3 sector at 6–24 h after forebrain ischemia *(30)*. Another immediate–early gene expressed in sympathetic neurons undergoing PCD because of NGF deprival, *mkp-1*, was found in the

CA1 sector following ischemia. The CA1 specific expression at 2 d after ischemia, however, was not observed *(29)*. An expression of *hsp70* shortly before cell death has also been reported for epithelial cells of the rat prostate gland, which die by programmed cell death following castration and subsequent depletion of male hormones *(19)*. However, also contrasting findings have been made. The secondary peak of c-*fos* expression has not been found in gerbil models *(60)*. In addition, in the gerbil accumulation of immunoreactivity for HSP70 and c-FOS was not observed *(34,61)*. Some genes that have been proposed to be markers for cell death programs were not found to be activated in dying CA1 neurons *(62,63)*. *CyclinD1* is induced in neurons undergoing PCD because of nerve growth factor withdrawal and, therefore, has been proposed to be a marker for PCD *(18)*. Following forebrain ischemia, however, this gene is expressed in proliferating microglia rather than in dying neurons *(62)*. *Sgp-2* mRNA is expressed in prostate gland cells during PCD *(20)*. Following transient forebrain ischemia, however, *sgp-2* mRNA was induced exclusively in reactive astrocytes *(63)*.

Besides investigation on gene expression profiles to assess whether delayed neuronal death includes an active death program, several other studies have been performed to address this question. A recent study demonstrated that CA1 neurons express high levels of the death promoting protein Bax, whereas the levels of the PCD-blocking protein Bcl-2 are relatively low *(64)*. This may indicate that CA1 neurons are prone to undergo programmed cell death. In addition, a further up-regulation of Bax levels was found in CA1 neurons in a rat cardiac arrest model *(64)*. Compatible with occurrence of PCD are some studies reporting a beneficial effect of infusion of the protein synthesis inhibitor cycloheximide on the survival of CA1 neurons following transient forebrain ischemia *(65–67)*. These findings implicate that the formation of a protein would be necessary for the progression of delayed neuronal death in the hippocampus, a suggestion that has also been made for PCD in neurons upon NGF withdrawal *(24)*. However, one study showed that cycloheximide infusion induced prolonged hypothermia in vivo *(68)* and it is well known that hypothermia can protect CA1 neurons following forebrain ischemia *(69,70)*. Therefore, although the in vivo effects of cycloheximide are compatible with the hypothesis of the activation of PCD, the significance of these results is debatable. Further evidence for occurrence of PCD has been provided by the demonstration of DNA fragmentation in the hippocampal CA1 sector after ischemia *(57,71–73)*. The activation of endonucleases cleaving DNA at internucleosomal sites is one of the hallmarks of most PCD paradigms *(24)*. Another indication for PCD in CA1 neurons in the rat after transient forebrain ischemia is that CA1 neurons can be rescued by high doses of brain-derived neurotrophic factor (BDNF) *(74)*. Although the precise roles of BDNF in the CNS are not completely understood, it appears to be an important factor for survival of neurons within the central nervous system (CNS) *(75)*. Since it is closely related to NGF, it is reasonable to assume that survival promoting effects of BDNF are exerted via similar pathways when compared with those of NGF in peripheral neurons.

In summary, accumulating evidence suggests that an active death program may be involved in delayed neuronal death in the hippocampus. For rat models, a role for newly formed proteins appears possible, although this has not yet been proven.

2.3. Genomic Responses vs General Protein Synthesis Inhibition

Cerebral ischemia is well known to cause prolonged disturbances of protein synthesis *(76,77)*. This functional disturbance develops within a few min of reperfusion following ischemia and is probably owing to inactivation of initiation factors *(78)*. In cell populations surviving the ischemic insult, protein synthesis recovers within several hours, although this time-course is much slower than the recovery of energy metabolism *(2)*. This response is a typical cellular stress response, and is also found following heat shock in most cell types *(9,79)*. Accordingly, in neuronal cell cultures, which were sublethally injured by heat shock, the protein synthesis capacity recovered with a very similar time-course compared to the in vivo situation following severe forebrain ischemia *(80)*. We have suggested previously that a recovery of protein synthesis is closely related to cellular survival following ischemia *(81)*. This notion is supported by our recent finding, that preventing the recovery of protein synthesis by cycloheximide decreased neuronal survival following otherwise sublethal heat shock in cell culture (Vogel et al., submitted). In selectively vulnerable brain regions, protein synthesis never recovers after transient forebrain ischemia and, therefore, delayed neuronal death may be related to this irreversible suppression *(82–84)*. It is conceivable that critical proteins that may be damaged in the reperfusion phase after ischemia have to be replenished to allow long-term survival of the cells. For example, it is well known that glutamine synthetase is inactivated by oxidation after ischemia *(85)*, whereas the cytoskeletal protein spectrin is cleaved *(86)*, presumably by calcium activated proteases such as calpain *(87)*.

The severely impaired protein synthesis capacity of the brain following forebrain ischemia, most probably is the reason why gene products of immediate early genes and heat shock protein encoding genes were not found by immunohistochemical methods within the first hours after the insults, although the mRNA induction started essentially immediately with reperfusion. Occurrence of immunohistochemically detectable proteins corresponds to the recovery of protein synthesis. For example c-*fos* mRNA was not significantly translated into protein up to 1 h of recovery following 30 min forebrain ischemia *(30)*. At all later time points examined, however, the c-*fos* mRNA expression patterns corresponded to the occurrence c-FOS immunoreactivity, and this was also true for the hippocampal CA1 sector. Therefore, despite a persisting overall protein synthesis inhibition in the hippocampal CA1 sector it may be that certain genes such as c-*fos* can be translated into proteins. The same observation has been reported for the heat shock protein *hsp70* in rat models of forebrain ischemia *(50,88)*. These observations correlate with a partial recovery of protein synthesis capacity in the CA1 sector in the rat *(84)*, and support the notion for a role of newly formed proteins in delayed neuronal death in the rat. It should be noted that in gerbil models of transient forebrain ischemia, a much more severe inhibition of protein synthesis has been observed *(83)*. Accordingly, several studies report on a failure to detect translation of mRNA to proteins in the CA1 sector of the gerbil hippocampus *(34,89,90)*. However, in more recent reports the increase of certain proteins has also been reported in the gerbil *(57)*.

In conclusion, the recovery of protein synthesis is an important process for long-term survival following ischemic insults. In addition, the permanent depression of protein synthesis is very likely to be fundamentally involved in the development of delayed neuronal degeneration in the hippocampal CA1 sector. However, the overall depression of protein synthesis does not exclude that the residual protein synthesis capacity allows the formation of some proteins, which may be formed preferentially because the respective mRNAs accumulate to high levels in the cells. It is conceivable that the persistent impairment of protein synthesis is one of the triggering event for the activation of a final death program *(24)*, which nevertheless may additionally depend also on newly formed specific proteins.

2.4. Ischemia-Induced Genomic Responses in Glia

As a general feature of both astrocytic and microglia gene induction, correlation to selective vulnerable brain regions is found almost exclusively once neuronal damage has occurred *(25)*. Concluding from these results a direct involvement of newly formed gene products synthesized by glial cells in selective vulnerability appears not very likely at present. Gene activations are more likely to be an important factor for persisting alterations of astrocyte and microglia phenotype in regions with neuronal damage, and include structural proteins, cell surface proteins, cytokines, and enzymes such as the inducible form of NO-Synthase *(62,63,91–93)*.

2.4.1. Gene Activation in Reactive Astrocytes Following Global Ischemia

It has been long known that astrocytes respond to transient forebrain ischemia mainly by hypertrophy that is accompanied by an increase of immunoreactivity for GFAP *(94–96)*. The increase of GFAP, which encodes an intermediate filament, in reactive astrocytes is thought to be important for glia scar formation *(97)*. Following ischemia, an increase can be detected immunocytochemically from 1 d after global ischemia onward. Ultrastructurally, morphological changes of astrocytes, however, can be observed as early as 3 h after 30 min of four-vessel occlusion *(98)*. An increase of *gfap* mRNA can also first observed 3–4 h after forebrain ischemia. This early increase of *gfap* mRNA was found to take place initially mainly in the hypothalamus (Yamashita et al., in preparation), but spread throughout the brain within 12 h. Another study, investigating *gfap* mRNA by Northern blot analysis found an increase also already at 4 h after 5 min forebrain ischemia in the gerbil *(99)*. Up to 1 d after the insult the increase of *gfap* mRNA and GFAP is found throughout the brain and is independent of occurrence of neuronal damage. At later survival periods (3 d and longer) astrocytes heavily expressing *gfap* mRNA and GFAP concentrate within and around regions with neuronal damage.

At survival times of 1 d and longer, *sgp-2* mRNA was increased in astrocytes in the same regions as *gfap* mRNA *(63)*. SGP-2 (= clusterin/apolipoprotein-J/TRPM-2) is a secreted multifunctional protein with proposed functions in complement regulation (cytolysis inhibition), cell membrane stabilization, or even cell death *(100)*, and it is not known at present what the pathophysiological role clusterin has in ischemia. Since SGP-2 immunoreactivity accumulates in dying neurons following status epilepticus, some researchers suspect an involvement in neuronal death, which would mean that reactive astrocytes secreting Sgp-2 would be directly involved in neuronal cell death *(101)*. In global ischemia, however, the cellular localization and the spatial pattern indicate an yet undefined role in astrogliosis *(63)*. Both hippocampal *gfap* and *sgp-2* expression accompany an astrocytic reaction preceding neuronal degeneration, but showing no topical correlation to selective vulnerability *(62,63)*. This widespread activation is probably directly related to effects of the ischemic conditions. The functional relevance is not clear at present.

From 2–3 d after the ischemic insult onward, the widespread astrocytic reaction decreases in brain areas not showing neuronal damage *(25,91)*. In contrast, astrogliosis is further enhanced in vulnerable regions, once neuronal death has occurred. This reaction then includes production of trophic factors such as basic fibroblast growth factor (bFGF), insulin-like growth factor 1 (IGF-1), but also the inducible form of nitric oxide-synthase (iNOS), amyloid precursor protein (APP) *(39,102–105)*, and additional structural proteins such as vimentin *(91)*. Both bFGF and IGF-1 are neuroprotective in vitro *(106)* and , therefore, it could be assumed that their production represents an attempt of the injured tissue to limit damage. However, induction of these factors are most pronounced or exclusively found at survival times when neuronal degeneration is already obvious *(102–104)*. In addition, IGF-1 infusion failed to rescue CA1 neurons in global ischemia, although the receptor density increased in the CA1 sector *(107)*. Moreover, induction of bFGF receptors was mainly found in reactive astrocytes. An autocrine function of bFGF for maintenance or regulation of astrocytic activation is therefore likely *(103)*. The pathophysiological role of astrocytically produced iNOS is unclear at present, but appears to be more likely involved in events following neuronal death and may have some role in repair processes *(108)*.

2.4.2. Gene Activation in Microglia

It is generally assumed that microglia play a harmful role in ischemia, since they have the capacity to release cytotoxic compounds *(109)*. This notion is supported by a recent study showing that following excitotoxic hippocampal lesioning, tissue plasminogen activator (tPA) is released by microglia and is responsible for tissue damage *(110)*. On the other hand, microglial activation may also play a beneficial role in ischemia, for example by releasing trophic factors, such as transforming growth factor-β1 (TGF-β1) *(see ref. 13)*. Microglial activation becomes apparent through immuno-histochemical markers well before cell death in the hippocampus *(92,111–115)*. Finsen et al. showed an early but transient microglial reaction almost throughout the entire hippocampus within the first 24 h after ischemia *(115)*. This transient reaction is indicated by rapid major histocompatibility complex class I (MHC I), but not class II, expression on microglia.

Once neuronal damage has occurred, the microglia reaction is further increased and includes prominent cell proliferation. Proliferation of GFAP-negative cells, which were most probably microglia, has been demonstrated by [3H]thymidine labeling 3 d after forebrain ischemia in the lateral cortex and the striatum *(96)*. In addition, by immune-electron microscopy, microglia undergoing mitosis could be observed within the hippocampus only in the CA1 sector and not before 3 d after ischemia *(113)*. Two recently described gene inductions in microglial are most probably involved in the transformation of microglia to a proliferating state following neuronal degeneration. We have shown that *cyclinD1* is expressed in microglia shortly before (1 and 2 d after ischemia in the striatum) or concomitantly (2 and 3 d after ischemia in the hippocampus) with the occurrence of proliferating microglial *(62)*. It is therefore likely that *cyclinD1* is involved in the microglia proliferation. It is well known that complexes of specific cyclins and cyclin-dependent kinases (CDK) are necessary to switch from the quiescent G0 state to the G1 state in the cell cycle. The phosphorylation of other cellular proteins by the cyclin-activated CDKs triggers the progression of the cell cycle *(116)*. In a recent study, disruption of the *cyclinD1* gene in mice resulted in neurological deficits, contrasting with an obviously normal histological appearance of the central nervous system *(117)*. Concluding from our results, this may be related to an altered microglia distribution and function in these animals. A role in microglial cell proliferation has also been proposed for the immediate-early gene *mkp-1*, that showed a transient mRNA increase most probably in microglial cells at 1 d recirculation in regions of the striatum showing severe neuronal damage *(29)*. *Mkp-1* induction has been previously found after mitogenic stimulation in fibroblasts *(118)*.

An important cytokine induced at the transcriptional level in microglia following global ischemia is transforming growth factor-β1 (TGF-β1) *(93)*. Although an early generalized expression of TGF-β1 mRNA may occur within the first 24 h following ischemia *(119)*, in our own experiments the onset of TGF-β1 mRNA synthesis was delayed compared to the rapidly occurring microglial reaction and was only pronounced from 3 d onward. The spatial distribution pattern correlated well with the distribution pattern of activated microglia as well with results on TGF-β1 mRNA expression in infant rats *(120)*. TGF-β1 induction could effect astrogliosis, and thereby influence gliotic scar formation in ischemia. For example, it is known that TGF-β1 can stimulate *sgp-2* expression in astrocytes *(121)*, which could explain the persisting *sgp-2* expression in astrocytes within the hippocampus at 7 d after ischemia. In addition, TGF-β1 is known to act as a classical wound healing factor by enhancing angiogenesis and extracellular matrix formation *(122)*. Similar to astrocytic IGF-I and bFGF synthesis following forebrain ischemia *(102,103)*, TGF-β1 synthesis by activated microglia may represent an endogenous CNS response limiting tissue damage in ischemia and possibly promoting repair mechanisms.

The complement receptor CR3 belongs to the B2 integrin family, which has one common β subunit and functions in adhesion-dependent inflammatory responses. In global ischemia, prominent infiltration of blood-borne cells into brain tissue has not been observed *(92)*, and the increase in immunoreactivity of CR3 (OX-42) was found on intrinsic microglia at 1 d after the insult *(92)*. The functional role of CR3 in microglia activation is not known at present. Interestingly, cr3- β-subunit mRNA expression was delayed as compared to the increase in CR3 immunoreactivity and restricted to regions in which neuronal death already had occurred *(62)*. This indicates a posttranscriptional mechanism for the early increase of CR3 immunoreactivity. In fact, it is well known that CR3 can be stored in intracellular pools, which may be not accessible in immunohistochemistry *(123,124)*.

2.4.3. The Amyloid Precursor Protein Is Induced
in Both Astrocytes and Microglia Following Transient Forebrain Ischemia

Amyloid precursor protein (APP) is an ubiquitously expressed protein that contains the insoluble βA4 amyloid, a main component of the senile plaque of Alzheimer's disease brains *(125)*. In ischemia, APP synthesis occurs both in reactive astrocytes and activated microglia, whereas the relative contributions of these two glial cell types vary with regard to the ischemia model and to the brain region *(105,126–129)*. Following 30 min of four-vessel occlusion, APP synthesis occurs predominantly in reactive GFAP-positive astrocytes in the hippocampal CA1 and CA4 sectors once neuronal degeneration has occurred *(105)*. In the parietal cortex, however, clusters of APP-immunoreactive microglia are observed *(105)*. In general, focal and global ischemia lead to a preferential upregulation of APP mRNA splice variants containing the Kunitz type inhibitor domain that is derived from reactive glia rather than neurons *(130,131)*. It thus seems possible that ischemia induces a glial APP synthesis and that persisting glial activation may contribute to amyloid formation particularly in the aged brain that may frequently suffer from ischemic periods. Amyloid formation may be additionally facilitated by ischemia-induced expression of Apolipoprotein E, which has been demonstrated in a gerbil model of transient forebrain ischemia *(132)*. The E4 isoform of this protein is thought to be a risk factor in Alzheimer's disease *(133)*.

3. GENOMIC RESPONSES IN FOCAL ISCHEMIA

Models to generate focal cerebral ischemia are intended to represent experimental approaches to investigate development of brain infarction relevant to human ischemic stroke *(134)*. However, there is much discussion in the field about which model of focal ischemia mimics most closely and representative the situation found in human stroke, and therefore would allow clinically relevant conclusions to be drawn. One concern to transient focal ischemia models is that these models may mix different pathophysiological parameters such as ischemic damage, reperfusion damage, and vascular damage and it may be difficult to clearly distinguish the contribution of each of these parameters. Nevertheless, such models are frequently used also for gene expression studies.

In general, the spectrum of genes investigated is similar to that in transient forebrain ischemia, including stress proteins (mainly *hsp70*), immediate early genes (c-*fos*, c-*jun*, *junB*, *zif268*), genes involved in programmed cell death *(p53, sgp2, bcl-2)*, glial markers (GFAP), cytokines (TGF-β, TNFα), amyloid precursor protein, and NO-Synthase. Since infarction results in leukocyte infiltration, and leukocyte adhesion is thought to be an important determinant of damage in transient focal ischemia, molecules involved in inflammatory processes are gaining increasing attention (IL-1, ELAM, CINC, MCP-1).

In contrast to global ischemia models, the extent of tissue damage is much more heterogeneous in focal ischemia models, depending on the precise location of occlusion of the middle cerebral artery (MCAO), the way of occluding the middle cerebral artery (electrocoagulation, thread occlusion), transient or permanent occlusion, and the animal strain used for the experiments. Consequently, the patterns of gene expression may differ quite dramatically among the different models. In addition, the precise time course of development of tissue damage is still a matter of discussion. Although it is usually assumed that most of the tissue damage occurs within the first few hours following transient or permanent MCAO, there is some evidence that tissue damage can progress for days *(135,136)*, although the proportion to the total lesion is not quite clear. This uncertainty on the time course of cell degeneration makes it difficult to interpret the role of any investigated gene or protein.

3.1. Transient Focal Ischemia

3.1.1. Stress Genes, BDNF, and Immediate Early Genes

Hsp70 is dramatically induced under stressful conditions in the brain, and it is assumed to be a marker for potentially injurious stress, which may or may not be followed by cell death *(137)*. If transient focal ischemia is limited to periods up to 30 min with subsequent reperfusion, the general genomic stress response is very similar to models of global transient forebrain ischemia. This may be exemplified by investigations on expression of the heat shock protein *hsp70* mRNA that is induced in all previously ischemic areas up to 24 h after the ischemic insult *(138)*, with a peak expression in the cortex at 8 h after ischemia *(139)*. Following 2 h thread occlusion, neuronal induction of HSP70 protein was never found in neurons residing in areas destined for infarction, but was found in sublethally injured neurons in the periphery of the predestined or actual infarct area from 9–96 h after ischemia, suggesting a protective role of HSP70 in these neurons *(140)*. A transient expression in cortical neurons outside the ischemic area after transient focal ischemia was also found for brain-derived neurotrophic factor (BDNF), which was accompanied by a bilateral increase of the respective receptor (trkB) and nerve growth factor (NGF) mRNAs *(141,142)*, suggesting a protective role for BDNF.

Similarly to other models of cerebral ischemia, induction of the immediate early gene c-*fos* was the most rapid genomic response following transient focal ischemia, and was found to be accompanied by *junB* induction *(143)*. In comparison to c-*fos*, onset of c-*jun* induction was delayed, but peak expression was found for both genes at 1 h of reperfusion, which later on declined *(144)*. A similar time-course was found for NGFI-A (or Krox-20), another immediate early gene encoding a transcription factor *(144,145)*. The functional consequences and the pathophysiological relevance of immediate early gene expression is unclear at present. By gel shift analysis, an increased DNA binding activity of the AP-1 transcription factor (a Fos/Jun heterodimer) was found, indicating functional consequences of the immediate early gene induction *(143)*. Although no direct evidence has been obtained so far, it appears very likely that immediate-early gene induction plays a role in alterations of the expression of other genes (secondary response genes).

3.1.2. Genes Associated with Programmed Cell Death

Besides immediate–early genes and stress proteins whose expression may be associated also with programmed cell death, two genes known to be involved in cell death in several instances have been investigated following transient focal ischemia, *p53* and *bcl-2*. Immunoreactivity of a mutant form of *p53* was found at 12 h after 2 h ischemia in the infarct area *(146)*. In addition, expression of *p53* mRNA was observed in both hemispheres peaking at 24 h after reperfusion *(146)*. In support for a role of *p53* in cell death after ischemia is the finding of reduced infarct sizes in permanent focal ischemia in *p53* knock-out mice *(147)*. However, even from this study, the role of *p53* is not very clear, since the mice with one intact and one knockout allele produced even smaller infarcts than animals with both alleles knocked out. Transient focal ischemia of 1 and 2 h duration induced BCL-2 in sublethally injured cells including glial cells, microglia, and neurons as assessed by immunohistochemistry *(148)*. BCL-2 is a protein known to protect against programmed cell death, and it therefore could be speculated that its induction represents an attempt to counteract activation of a death program in sublethally injured cells. Indeed, mice overexpressing BCL-2 developed smaller infarct, although these studies were done with permanent focal ischemia *(149)*. However, BCL-2 has also been proposed to act as a free radical scavenger *(150)* and the protective effect, therefore, may depend on this function.

3.1.3. Cytokines and Molecules Involved in Cell Attraction, Cell Adhesion and Tissue Infiltration

Inflammatory reactions involve infiltration of leukocytes, in which polymorphonuclear cells precede invasion of mononuclear cells and activation of macrophages. In transient focal ischemia, inflammatory reactions are thought to contribute directly to infarct development *(14)*. Only very recently molecules mediating theses events have attracted the attention of cerebral ischemia researchers, and, therefore, only few data from singular studies are available. By Northern blot analysis, it was demonstrated that induction of tumor necrosis factor-α (TNFα) and interleukin-1β (IL-1β) mRNAs occurred as early as 1 h after reperfusion following *(160)* min transient focal ischemia in spontaneously hypertensive (SHR) rats in the cortex. TNFα mRNA reached its peak at 3 h whereas IL-1β mRNA reached its peak at 6 h. Both cytokine mRNA levels remained elevated for up to 2 d after reperfusion *(152)*. Both molecules are known to induce the expression of adhesion molecules such as intercellular adhesion molecule-1 (ICAM-1), vascular cell adhesion molecule-1 (VCAM-1), endothelial cell adhesion molecule-1 (ELAM-1 = E-Selectin), and monocyte chemoattractant protein-1 (MCP-1), a chcmoattractant specific for monocytes *(151)*. Indeed, an upregulation of ICAM-1 mRNA, was found at 1 h of reperfusion in this model *(153)*. An early increase of ICAM-1 and P-Selectin immunoreactivity was also found after 2 h MCAO in baboons *(154)*. Supporting a role of ICAM-1 for infarct development in transient focal ischemia is the finding that intravenous application of anti-ICAM-1 antibodies or other antibodies potentially interfering with PMN adhesion following transient focal ischemia reduced the infarct volume *(155–157)*. In addition, it has been shown that the IL-1 receptor antagonist can ameliorate tissue damage *(158)*. The MCP-1 expression was delayed as compared to ICAM-1, starting at 3 h after ischemia and peaking between 12 h and 2 d *(159)*. This finding is in accordance with the observation that monocytes start to infiltrate the lesion area from 2 d after the insult onward *(160)*.

3.1.4. Amyloid Precursor Protein

One study reports that repeated reversible MCAO results in an increase of APP reactivity in astrocytic processes in perifocal regions and white matter tracts 3 d after the insult *(127)*. Dystrophic axons and neurons with accumulated APP were also found in the ipsilateral neocortex and hippocampus *(127)*. This study supports the notion that brain ischemia stimulates glial production of the APP protein.

3.2. Permanent Focal Ischemia

3.2.1. Stress Proteins and Immediate Early Genes

In permanent focal ischemia, *hsp70* mRNA and HSP70 immunoreactivity is mostly found in regions with decreased blood flow, i.e., in regions supplied by the MCA *(43)* up to 24 h after onset of MCAO. An exception are thread occlusion models, in which induction of *hsp70* (and immediate early genes) has also been found in the hippocampus and the thalamus, regions that are not affected by blood flow decreases *(161)* (Wiessner, unpublished). The infarcts in these models are very large and, in addition, may result in postocclusion hyperthermia (Yamashita et al., submitted). Therefore, *hsp70* induction in the mentioned regions may be related to secondary effects. Induction of the *hsp70* mRNA is an early event and found from 1 h after MCAO onward *(138)*. Increased mRNA levels for *hsp70* are generally found up to 24 h after onset of MCAO and decrease dramatically at later observation periods (Yamashita et al., in press). Interestingly, an increase of HSP70 immunoreactivity has been repeatedly described in the infarct core from 4–24 h after onset of MCA occlusion. However, whereas one study considered staining in the core region as unspecific *(137)*, other researches claim that immunoreactivity in the infarct core is owing to a specific staining of endothelial cells *(162)*. It should be noted that in thread occlusion a patchy expression of *hsp70* mRNA is found in the infarct core between 4–24 h after onset MCAO *(162)*, whereas this has not been observed in electrocoagulation models *(138)* (Yamashita et al., in press). This may indicate that although thread occlusion affects a larger area, blood flow cessation may be not as severe as in electrocoagulation. A consistent finding in all permanent ischemia models is that HSP70 immunoreactivity in neurons was observed in the peri-infarct area not before 24 h after MCAO *(137,162,* Vogel et al., in press). There are some notions, that in the immediate infarct periphery, astrocytes also express HSP70 in thread occlusion models *(163)*, although we did not observe that in an electrocoagulation model. The expression in noninfarcted cortical areas up to 24 h after MCAO is most pronounced in the outer layers and is associated with moderate selective neuronal damage in these regions *(164)*. The persistent expression of *hsp70* mRNA and HSP70 protein in cortical peri-infarct areas indicate ongoing cellular stress in these regions. One study reports the induction of several other stress proteins, i.e., *grp78*, *hsp27*, and *hsp47*. Induction of these mRNAs was delayed as compared to *hsp70 (165)*. For *grp78* this finding is in contrast to a study in transient focal ischemia in which a concomitant induction of *hsp70* and *grp78* was found *(166)*. In addition, *hsp27* was found to be induced throughout the affected hemisphere, i.e., including regions not affected by blood flow disturbances, indicating that heat shock proteins may be differentially induced.

Similar to global ischemia and transient focal ischemia, induction of immediate early genes is a prominent feature in the brains after permanent MCAO. As for global and transient focal ischemia, immediate early genes are differentially induced following permanent MCAO. Increased mRNA levels for c-*fos* have been found already 15 min after MCAO throughout the cortex of the affected hemisphere in neuronal nuclei, persisting up to 24 h *(138)*. Similar observation were made for *junB (161)*. c-FOS protein like immunoreactivity was found from 1 h after MCAO onward exclusively outside the ischemic core region throughout the injured hemisphere, peaking at 4 h after MCAO, and declining in regions remote from the infarct afterwards. In the immediate infarct vicinity, increased immunoreactivity persisted for up to 4 d *(167)*. As compared to c-*fos* and *junB*, onset of induction of c-*jun* and some other immediate-early genes was delayed both at the mRNA and protein level, but persisted for up to 24 h as well *(161)*. Similar results have been obtained in photothrombotic brain lesions *(168)*. In thread occlusion models for c-*fos*, a similar pattern as for *hsp70* was observed, i.e., induction was found also in the thalamus and the hippocampus and in the contralateral hemisphere *(161)*. Since pretreatment with MK-801 inhibited the widespread expression of c-*fos*, it has been suggested that spreading depression like peri-infarct depolarization are responsible for this expression pattern *(168)*. Interestingly, similar to c-*fos*, neuronal NO-Synthase (NOS-I) was already induced at the mRNA level 15 min following MCAO and peaking at 1 h and an association of NOS-I expression and survival of scattered neurons within lesioned areas has been suggested *(169)*.

It appears that from 1 h after permanent MCAO onward, a concomitant, neuronal expression of c-*fos*, c-*jun*, and *hsp70* is found in the peri-infarct region up to 1 d after onset of MCAO. In a recent study, we have directly compared these genes in adjacent brain sections and could verify this suggestion, and in addition show that this concomitant expression occurs in a region with preserved energy metabolism as assessed by ATP bioluminescence imaging and decreased blood flow (Wiessner et al., in preparation). This suggests that neurons in the peri-infarct penumbra zone suffer persisting stress, which most likely is caused by spreading depression like peri-infarct depolarizations *(170,171)*.

3.2.2. Cytokines and Molecules Involved in Cell Attraction, Cell Adhesion and Tissue Infiltration

The induction of several molecules involved in cell adhesion and tissue infiltration of blood-borne cells has been demonstrated in permanent focal ischemia. TNFα, IL-1β, and IL-6 mRNA were found in the ischemic hemisphere between 6 h and 5 d after MCAO, mainly in SHR rats *(172–174)*. One study reports an increase of IL-1β mRNA already at 15 min after MCAO in SHR rats 175. Between 6 and 12 h, TNFα immunore activity was found in nerve fibers within the ischemic area, whereas at 5 d, macrophages within the lesion area were immunoreactive *(173)*. Using Fischer rats, we could not detect TNFα mRNA early after MCAO, but only 7 d after the insult in lesioned areas in cells that were most likely macrophages (Gerken et al., in preparation). In contrast, we found an early increase of TGF-β1 at 6 and 12 h after MCAO in nonlesioned areas of the ischemic hemisphere (Gerken et al., in preparation), which has not been found in SHR rats *(176)*, and that may represent an protective attempt or anti-inflammatory response of the tissue. The cellular source of this early TGF-β1 mRNA expression is not clear at present. Several molecules that could mediate cell adhesion of circulating blood cells and that respond to growth factors such as TNFα and IL-1β were found to be increased at the mRNA level between 6 h and 2 d after MCAO in SHR rats. This included ICAM-1, endothelial-leukocyte adhesion molecule-1 (ELAM-1; E-Selectin), which is specifically found in endothelial cells, MCP-1, MIP-1α (macrophage inflammatory protein-1α) *(177)*, and cytokine-induced neutrophil attractant (CINC) *(153,159,178,179)*. Another molecule with chemoattractant activity for monocytes that is induced early in focal permanent ischemia in SHR rats is platelet-derived growth factor-B chain *(180)*. Despite the relative early onset of induction of these molecules, it appears that infiltration of polymorphonuclear cells and mononuclear cells is a rather late process in permanent MCAO and that this process does not contribute to development of the primary infarct. Accordingly, intravenous infusion of anti-ICAM-1 antibodies did not influence infarct size in permanent focal ischemia, whereas significantly decreasing the lesion area in a transient focal ischemia model *(155)*. The infiltrating cells may influence glia response and neurons directly by producing cytokines such as TGF-β1, IL-1 and TNFα, but may influence also the hemodynamic situation in the vicinity of the lesion area, for example, by induction of the inducible form of NO-synthase (iNOS) which has been found between 2 and 4 d after MCAO within the infarcted area in infiltrating polymorphonuclear cells *(181)*.

3.2.3. Amyloid Precursor Protein (APP)

An increase of APP mRNA, containing the Kunitz-type protease inhibitor domain, in permanent focal ischemia has been reported from 1 d after onset of MCAO onward 130. No attempt was made to localize the cell type, but concluding from the time course and the fact that the Kunitz-type proteinase inhibitor domain containing a form of the amyloid precursor proteins was induced, a glial origin appears likely. An immunohistochemical study in permanent MCAO reported an increase of APP immunoreactivity from 4 d after occlusion onward. The APP was found in the periphery of the infarct, although the cellular source was not clear *(128)*. No deposition of the β-A4 fragment was found in this study. Therefore, as in transient focal and transient forebrain ischemia, the increase of APP was a late event preceded by cellular damage.

3.3. Is There Evidence for Programmed Cell Death in Focal Ischemia?

Suggestions that an active death program may be involved in infarct development depend largely on four lines of evidence:

1. Several studies show that DNA degradation, as assessed by gel electrophoresis and *in situ* labeling procedures (TUNEL-staining), accompanies or follows development of cerebral infarction following both transient and permanent MCA-occlusion *(182–188)*.
2. One study reports a beneficial effect of continuous intraventricular cycloheximide infusion on infarct size 1 d after permanent MCAO *(189)*.
3. Studies on investigations of genes that are expressed in some cell types undergoing programmed cell death appear rather incomplete at present. The formation of any protein product depends on some preserved energy metabolism that is not given in the core region. Therefore, it appears rather unlikely that proteins contributing to a death program can be formed in cells within the core region of the developing infarct. However, the concomitant and persisting expression of c-*fos*, c-*jun*, and *hsp70* in the penumbra may indicate a similar situation as in dying CA1 neurons following transient forebrain ischemia in this region. In principle, it is conceivable that the ongoing metabolic stress in the penumbra may activate a stress induced death program.
4. In line with such a hypothesis is the finding that *p53* knockout mice develop smaller infarcts *(147)*. In addition, BCL-2 overproducing mice develop smaller infarcts *(149)*. In each of theses studies, however, underlying unspecific mechanisms, for example influences of the genetic manipulations on brain development cannot be ruled out at the moment.

In summary, at present the evidence for the involvement of an active program in cell death in the developing infarct appears rather sketchy and no conclusive assessment appears possible.

A type of cell death in which programmed cell death could be expected to be involved is the retrograde degeneration of neurons in thalamic region, which is probably owing to loss of synaptic input from infarcted cortical regions and loss of trophic support *(190)*. In fact, infusion of basic fibroblast growth factor (bFGF) inhibited degeneration of thalamic neurons following MCAO *(191)*. It is noteworthy that no study so far reported any correlation of expression of stress proteins and immediate early genes and this type of cell death. Obviously, this notion is contradictory to the suggestion in the global ischemia section, that a prolonged or secondary induction of c-*fos*, c-*jun*, and *hsp70* may indicate the activation of an active death program in the hippocampal CA1 sector following transient forebrain ischemia. However, it should also be noted that secondary degeneration in the thalamus has not been investigated in great detail, and the precise time course of occurrence of neuronal degeneration is not known. It occurs presumably between 3 and 7 d after onset of permanent focal ischemia. From our studies, it is clear that an astrocytic response to this lesion as assessed by *in situ* hybridization was not found at 3 d after MCAO, but after 7 d (Yamashita et al., in press). In addition, TUNEL staining revealed dead neurons in the thalamus at 7 d after MCAO, but not at 3 d (Wiessner, unpublished results). Therefore, the onset of cell degeneration must start somewhere in between. Since no study so far has investigated this period with higher temporal resolution, it may well be that the time-points of a secondary increase of c-*fos*, c-*jun*, and *hsp70* just have missed. This is an issue that clearly deserves further studies.

3.3. Glial Responses to Focal Ischemia

3.3.1. Astrocytic Responses

Following permanent MCAO, *in situ* hybridization revealed an early and widespread increase of *gfap* mRNA in nonischemic areas including the contralateral hemisphere starting between 3–6 h in regions remote from the ischemic area, and a delayed but persisting circumscribed expression in the peri-infarct border zone after 1 wk (Yamashita et al., in press). Only astrocytes adjacent to the lesion showed an increase in sulfated glycoprotein-2 mRNA (*see* Section 2.4.1.), which may improve membrane stability in these cells. Following photothrombosis, astrocytes lining the lesion area showed increased vimentin immunoreactivity, which was not observed in reactive astrocytes remote from the lesion *(192)*. It is noteworthy that there is obviously a very long delay between increase of *gfap* mRNA (6 h) and reports on increased GFAP immunoreactivity (2 d). Following transient focal ischemia, the increase of GFAP in regions remote from lesioned areas has not been described *(193,194)*. In these studies, an increase of GFAP immunoreactivity was found in regions adjacent to the infarct, in the same region where neuronal HSP70 induction was observed. However, HSP70 was not found in astrocytes, but exclusively in neurons. An induction of HSP70 in the border zone of the infarct has been described in one study in a thread occlusion model *(163)*. Whereas the persisting astrocytic activation in the infarct border zone is related to the formation of a glial scar, the functional role of the early widespread activation of astrocytes in not clear at present (Yamashita et al., in press). There are indications that it may be mediated by peri-infarct depolarizations *(192)*, although the spread to the contralateral hemisphere frequently found cannot be explained by this mechanism. Although several of the cytokines that have been reported to be induced following focal ischemia may be produced also by astrocytes, no study so far has attempted to demonstrate this. One study indicates that astrocytes may be involved in the production of macrophage inflammatory protein-1α *(177)*. From our results, a participation of astrocytes in the early expression of TGF-β1 mRNA in regions remote from the lesion area appears possible.

3.3.2. Microglial Responses

Two studies have shown that the microglial activation in both permanent and transient focal ischaemia is an early event that precedes the response previously observed in global ischaemia *(195,196)*. From 3 h after permanent MCAO onward, activated microglia were detected throughout the ipsilateral cortex using staining for activation markers such as MUC 101 and 102, CR3 complement receptor (OX42), MHC class II, and amyloid precursor protein *(196)*. The widespread ipsilateral microglia activation may be related to spreading depression-like waves induced from the infarct core since it is known that potassium-induced spreading depression waves can transiently activate microglia in the intact cortex *(197)*. From 24 h onward, microglial activation extended to the contralateral gyrus cinguli. Within the first 3 d of permanent focal ischemia, activated microglia did not express ED markers characteristic of hematogenous monocytes/macrophages *(196,198)*. From 3 d onward, a massive influx of ED1-positive, presumably blood-derived, macrophages took place at the site of infarction. In addition, ED1 was up-regulated on cells of microglial morphology at sites remote of the actual infarction. The macrophage/microglial response continued to increase for 7 d along with pronounced astrocytic reactions and was observed also in regions of secondary cell degeneration such as the thalamus *(198)*.

Regarding the genomic responses at the transcriptional level, not very much is known at the present. It appears likely that several of the cytokines found to be induced following both transient and permanent focal ischemia may be produced by activated microglia (*see* Sections 3.1.3. and 3.2.2.) However, in none of these studies attempts were made to clearly identify the cell type producing mRNA coding as for example for IL-1β, TNFα, and so on. This is clearly an issue that deserves further investigations.

4. NEW APPROACHES AND FUTURE PERSPECTIVES

So far, most studies on gene expression in models for ischemic brain insults were descriptive and, therefore, the functional relevance for the pathophysiological events and outcome has not yet been elucidated. In principle, methods to specifically interfere with the production of a specific protein are needed to clarify these issues. One possibility appears to be so-called antisense-oligonucleotides, which can block translation of specific mRNAs *(199)*. Another possibility is the use of knock-out mice or transgenic mice, in which specific genes have been deleted or specific proteins are overproduced, respectively *(200)*.

4.1. Modulation of Gene Expression Following Ischemia by Antisense Oligonucleotides

This technique is based on the (poorly understood) phenomenon that cells can take up short single-stranded nucleic acids (oligonucleotides), which subsequently can hybridize specifically to target mRNA molecules or DNA and inhibit transcription or translation *(199)*. Recently, two reports claimed successful application of this technique also in vivo in the CNS in the ischemia field. In one study, c-Fos protein formation in focal ischemia was successfully diminished *(201)*; however, the authors did not report on any effects of infarct development or neurological outcome. In another study,

NMDA receptor formation was impaired by antisense oligonucleotides that subsequently decreased infarct volume following permanent occlusion of the middle cerebral artery *(202)*. We have worked on the development of antisense oligonucleotides to block formation of the heat shock protein HSP70. Although we could impair formation of the protein in cell culture after heat shock in some experiments, these results were poorly reproducible. Difficulties to reproduce antisense-oligonucleotide mediated effects and unspecific effects appear to be general problems in the use of these technique. Methods to improve cellular uptake and stability of oligonucleotides may result in more general applications for ischemia research.

4.2. Use of Genetically Modified Animals in Ischemia Research

Few studies have been presented using genetically modified animals in cerebral ischemia research, and most of these were investigating effects on infarct development in focal cerebral ischemia. One reason is that such techniques have been developed quite recently. In addition, a technical problem for ischemia research is that targeted genomic disruptions so far can be accomplished only in mice, which owing to their small size only occasionally have been used for the development of cerebral ischemia models. Nevertheless, in principle mice models are available *(203)*, and it can be expected that in the near future an increasing number of studies in cerebral ischemia research using genetically modified animals will come up. A general problem in using knockout or transgenic mice is that these manipulations may have systemic effects by influencing development, function of various organs, and anatomy. An example for this notion may the BCL-2 overexpressing mouse *(149)*. Intracellular levels of BCL-2 are thought to be important for the regulation of programmed cell death, i.e., high levels of the protein inhibit execution of the program *(204)*. In BCl-2 overexpressing animals, a prominent decrease in infarct size was observed in permanent focal ischemia, and it has been suggested that this may indicate an involvement of programmed cell death in infarct developments *(149)*. However, the brains of these animals are anatomically clearly distinct from wild-type mice, i.e., they are larger due to failure of naturally occurring cell death during brain development. Certainly, this may influence hemodynamic parameters, which were not measured in the study. In addition, it is known that BCL-2 scavenges free radicals *(150)*, and, therefore the protective role of high intracellular amounts of BCL-2 may be attributed to this ability, rather than a specific effect on death programs. Another example showing that the use of genetically engineered animals does not necessarily produce unambiguously clear results, is the investigation of infarct sizes in *p53* knockout mice *(147)* that have a normally developed brains. The *p53* is thought to be a trigger for programmed cell death; it was therefore expected that, if programmed cell death may be in involved in infarct development, the disruption of the gene should result in smaller infarcts. This was indeed observed; however, surprisingly the heterozygous littermates, i.e., animals having one intact and one disrupted gene, developed even smaller infarcts than the animals with both genes disrupted *(147)*. The meaning of these results is at present unclear. Some other recent examples in which genetically modified animals have been used in ischemia research are superoxide-dismutase overproducing mice *(205)*, nitric oxide synthase knock-out mice *(206)*, and bFGF overexpressing mice *(207)*.

In principle, it would be desirable to have animals at hand that would allow researchers to conditionally disrupt genes in specific tissues at any chosen time-point. Indeed, methods showing routes to accomplish this goal have recently been reported *(208,209)*. It can be anticipated that in the near future animals allowing the conditional and tissue-specific modulation of gene activation will revolutionize research in the field of cerebral ischemia.

REFERENCES

1 Kleihues, P., Hossmann, K.-A., Pegg, A. E., Kobayashi, K., and Zimmermann, V., Resuscitation of the monkey brain after one hour complete ischemia. III. Indications of metabolic recovery, *Brain Res.,* **95** (1975) 61–73.

2 Hossmann, K.-A., Disturbances of protein synthesis and ischemic cell death, *Progr. Brain Res.,* **96** (1993) 161–177.

3 Dienel, G. A. and Cruz, N. F., Induction of brain ornithine decarboxylase during recovery from metabolic, mechanical, thermal, or chemical injury, *J. Neurochem.,* **42** (1984) 1053–1061.

4 Nowak, T. S., Jr., Synthesis of a stress protein following transient ischemia in the gerbil, *J. Neurochem.,* **45** (1985) 1635–1641.

5 Dienel, G. A., Cruz, N. F., and Rosenfeld, S. J., Temporal profiles of proteins responsive to transient ischemia, *J. Neurochem.,* **44** (1985) 600–610.

6 Kiessling, M., Dienel G. A., Jacewicz M., and Pulsinelli W. A., Protein synthesis in postischemic rat brain: a two-dimensional electrophoretic analysis, *J. Cereb. Blood Flow Metab.,* **6** (1986) 642–649.

7 Jacewicz, M., Kiessling, M., and Pulsinelli, W. A., Selective gene expression in focal cerebral ischemia, *J. Cereb. Blood Flow Metab.,* **6** (1986) 263–272.

8 Lindvall, O., Kokaia, Z., Bengzon, J., Elmer, E., and Kokaia M., Neurotrophins and brain insults, *Trends Neurosci.,* **17** (1994) 490–496.

9 Welch, W. J., Mammalian stress response: cell physiology, structure/function of stress proteins, and implications for medicine and disease, *Physiol. Rev.,* **72** (1992) 1063–1081.

10 Schreiber, S. S. and Baudry, M., Selective neuronal vulnerability in the hippocampus—a role for gene expression, *Trends Neurosci.,* **18** (1995) 446–451.

11 Morgan, J. I. and Curran, T., Stimulus-transcription coupling in the nervous system: involvement of the inducible proto-oncogenes fos and jun, *Annu. Rev. Neurosci.,* **14** (1991) 421–451.

12 Eddleston, M. and Mucke, L., Molecular profile of reactive astrocytes-implications for their role in neurologic disease, *Neuroscience,* **54** (1993) 15–36.

13 Gehrmann, J., Banati, R., Wiessner, C., Hossmann, K.-A., and Kreutzberg, G. W., Reactive microglia in ischemia: an early mediator of tissue damage?, *Neurpathol. Appl. Neurobiol.,* **24** (1995) 277–289.

14 Kochanek, P. M. and Hallenbeck, J. M., Polymorphonuclear leukocytes and monocytes/macrophages in the pathogenesis of cerebral ischemia and stroke, *Stroke,* **23** (1992) 1367–1379.

15 Wang, X. K. and Feuerstein, G. Z., Induced expression of adhesion molecules following focal brain ischemia, *J. Neurotrauma,* **12** (1995) 825–832.

16 Estus, S., Zaks, W. J., Freeman, R. S., Gruda, M., Bravo, R., and Johnson, E. M., Altered gene expression in neurons during programmed cell death–identification of c-jun as necessary for neuronal apoptosis, *J. Cell Biol.,* **127** (1994) 1717–1727.

17 Colotta, F., Polentarutti, N., Sironi, M., and Mantovani, A., Expression and involvement of c-fos and c-jun protooncogenes in programmed cell death induced by growth factor deprivation in lymphoid cell lines, *J. Biol. Chem.,* **267** (1992) 18,278–18,283.

18 Freeman, R. S., Estus, S., and Johnson, E. M. J., Analysis of cell-cycle-related gene expression in postmitotic neurons: selective induction of Cyclin D1 during programmed cell death, *Neuron,* **12** (1994) 343–355.

19 Buttyan, R., Zakeri, Z., Lockshin, R., and Wolgemuth, D., Cascade induction of c-fos, c-myc, and heat shock 70K transcripts during regression of the rat ventral prostate gland, *Mol. Endocrinol.,* **2** (1988) 650–657.

20 Buttyan, R., Olsson, C. A., Pintar, J., Chang, C., Bandyk, M., NG, P. Y., and Sawczuk, I. S., Induction of the TRPM-2 gene in cells undergoing programmed cell death, *Mol. Cell. Biol.,* **9** (1989) 3473–3481.

21 Oppenheim, R. W., Cell death during development of the nervous system, *Annu. Rev. Neurosci.,* **14** (1991) 453–501.

22 Siesjö, B. K., Historical overview. Calcium, ischemia, and death of brain cells, *Ann. NY Acad. Sci.,* **522** (1988) 638–661.

23 Siesjö, B. K., Lundgren, J., and Pahlmark, K., The role of free radicals in ishcemic brain damage: a hypothesis. In J. Krieglstein and H. Oberpichler (eds.) *Pharmacology of Cerebral Ischemia,* Wissenschaftliche Verlagsgesellschaft, Stuttgart, 1990, pp. 319–323.

24 Deckwerth, T. L. and Johnson, E. M. J., Temporal analysis of events associated with programmed cell death (apoptosis) of sympathetic neurons deprived of nerve growth factor, *J. Cell. Biol.,* **123** (1993) 1207–1222.

25 Wiessner, C., Neumann-Haefelin, T., Brink, I., Vogel, P., and Hossmann, K.-A., The genomic response of the rat brain following cerebral ischemia: generalized versus cell and region specific responses. In J. Krieglstein and H. Oberpichler-Schwenk (eds.) *Pharamcology of Cerebal Ischemia,* Wissenschaftliche Verlagsgesellschaft, Stuttgart, 1994, pp. 455–465.

26 Kirino, T., Tamura, A., and Sano, K., Selective vulnerability of the hippocampus to ischemia—reversible and irreversible types of ischemic cell damages, *Progr. Brain Res.,* **63** (1985) 39–55.

27 Horn, M. and Schlote W., Delayed neuronal death and delayed neuronal recovery in the human brain following global ischemia, *Acta Neuropathol.,* **85** (1992) 79–87.

28 Duchen, L. W., *General Pathology of the Neuron,* 4 ed., Edward Arnold, London, 1984.

29 Wiessner, C., Neumann-Haefelin, T., Vogel, P., Back, T., and Hossmann, K.-A., Transient forebrain ischemia induces an immediate-early gene encoding the mitogen-activated protein kinase phosphatase 3CH134 in the adult rat brain, *Neuroscience,* **64** (1995) 959–966.

30 Neumann-Haefelin, T., Wiessner, C., Vogel, P., Back, T., and Hossmann, K.-A., Differential expression of the immediate early genes c-fos, c-jun, junB, and NGFI-B in the rat brain following transient forebrain ischemia, *J. Cereb. Blood Flow Metab.,* **14** (1994) 206–216.

31 Wessel, T. C., Joh, T. H., and Volpe, B. T., In situ hybridization analysis of c-fos and c-jun expression in the rat brain following transient forebrain ischemia, *Brain Res.,* **567** (1991) 231–240.

32 Onodera, H., Kogure, K., Ono, Y., Igarashi, K., Kiyota, Y., and Nagaoka, A., Proto-oncogene c-fos is transiently induced in the rat cerebral cortex after forebrain ischemia, *Neurosci. Lett.,* **98** (1989) 101–104.

33 Kindy, M. S., Carney, J. P., Dempsey, R. J., and Carney, J. M., Ischemic induction of protooncogene expression in gerbil brain, J. Mol. Neurosci., **2** (1991) 217–228.

34 Kiessling, M., Stumm, G., Xie, Y., Herdegen, T., Aguzzi, A., Bravo, R., and Gass, P., Differential expression and translation of immediate early genes in the gerbil hippocampus after transient forebrain ischemia, *J. Cereb. Blood Flow Metab.,* **13** (1993) 914–924.

35 Nowak, T. S., Jr., Osborne, O. C., and Suga, S., Stress protein and proto-oncogene expression as indicators of neuronal pathophysiology after ischemia, *Prog. Brain Res.,* **96** (1993) 195–208.

36 Muller, M., Cleef, M., Rohn, G., Bonnekoh, P., Pajunen, A. E., Bernstein, H. G., and Paschen, W., Ornithine decarboxylase in reversible cerebral ischemia: an immunohistochemical study, *Acta Neuropathol.,* **83** (1991) 39–45.

37 Kawagoe, J.-I., Abe, K., Aoki, M., and Kogure, K., Induction of HSP90alpha heat shock mRNA after transient global ischemia in gerbil hippocampus, *Brain Res.,* **621** (1993) 121–125.

38 Paschen, W., Uto, A., Djuricic, B., and Schmitt, J., Hemeoxygenase expression after reversible ischemia of rat brain, *Neurosci. Lett.,* **180** (1994) 5–8.

39 Endoh, M., Pulsinelli, W. A., and Wagner, J. A., Transient global ischemia induces dynamic changes in the expression of bFGF and the FGF receptor, *Mol. Brain Res.,* **22** (1994) 76–88.

40 Lindvall, O., Ernfors, P., Bengzon, J., Kokaia, Z., Smith, M. L., Siesjo, B. K., and Persson, H., Differential regulation of mRNAs for nerve growth factor, brain-derived neurotrophic factor, and neurotrophin 3 in the adult rat brain following cerebral ischemia and hypoglycemic coma, *Proc. Natl. Acad. Sci. USA,* **89** (1992) 648–652.

41 Kitagawa, K., Matsumoto, M., Tagaya, M., Hata, R., Ueda, H., Niinobe, M., Handa, N., Fukunaga, R., Kimura, K., Mikoshiba, K., and Kamada, T., 'Ischemic tolerance' phenomenon found in the brain, *Brain Res.,* **528** (1990) 21–24.

42 Kirino, T., Tsujita, Y., and Tamura, A., Induced tolerance to ischemia in gerbil hippocampal neurons, *J. Cereb. Blood Flow Metab.,* **11** (1991) 299–307.

43 Nowak, T. S., Jr. and Jacewicz, M., The heat shock/stress response in focal cerebral ischemia, *Brain Pathol.,* **4** (1994) 67–76.

44 Nishi, S., Taki, W., Uemura, Y., Higashi, T., Kikuchi, H., Kudoh, H., Satoh, M., and Nagata, K., Ischemic tolerance due to the induction of HSP70 in a rat ischemic recirculation model, *Brain Res.,* **615** (1993) 281–288.

45 Sommer, C., Gass, P., and Kiessling, M., Selective c-jun expression in CA1 neurons of the gerbil hippocampus during and after acquisition of an ischemia-tolerant state, *Brain Pathol.,* **5** (1995) 135–144.

46 Abe, H., Xu, Z. K., and Nowak, T. S. J., Threshold depolarization intervals for ischemic tolerance versus expression of HSP72 and other ischemia-induced mRNAs, *J. Cereb. Blood Flow Metab.,* **15(Suppl. 1)** (1995) S67.

47 Jorgensen, M. B., Deckert, J., Wright, D. C., and Gehlert, D. R., Delayed c-fos protooncogene expression in the rat hippocampus induced by transient global cerebral ischemia: an in situ hybridization study, *Brain Res.,* **484** (1989) 393–398.

48 Dragunow, M., Yooung, D., Hughes, P., MacGibbon, G., Lawlor, P., Singleton, K., Sirimanne, E., Beilharz, E., and Gluckman, P., Is c-jun involved in nerve cell death follwoing status epilepticus and hypoxic-ischemia brain injury, *Mol. Brain Res.,* **18** (1993) 347–352.

49 Nowak, T. D. J., Prolonged expression of hsp70 mRNA as an early indicator of neuronal vulnerability following ischemia. In J. Krieglstein and H. Oberpichler (eds.) *Pharmacology of Cerebral Ischemia*, Wissenschaftliche Verlagsgesellschaft mbH, Stuttgart, 1990, pp. 117–121.

50 Tomioka, C., Nishioka, K., and Kogure, K., A comparison of induced heat-shock protein in neurons destined to survive and those destined to die after transient ischemia in rats, *Brain Res.,* **612** (1993) 216–220.

51 Milarski, K. and Morimoto, R. I., Expression of human HSP70 during the synthestic phase of teh cell cycle, *Proc. Natl. Acad. Sci. USA,* **83** (1986) 9517–9521.

52 Colombel, M., Olsson, C. A., Ng, P. Y., and Buttyan, R., Hormone-regulated apoptosis results from reentry of differentiated prostate cells into a defective cell cycle, *Cancer Res.,* **52** (1992) 4313–4319.

53 Saito, N., Kawai, K., and Nowak, T. S., Jr., Reexpression of developmentally regulated MAP2c mRNA after ischemia: colocalization with hsp72 mRNA in vulnerable neurons, *J. Cereb. Blood Flow Metab.,* **15** (1995) 205–215.

54 Stubberöd, P., Tomasevic, G., Kamme, F., and Wieloch, T., Expression of the tumor supressor p53 and WAF1 in the postischemic rat hippocampus, *Abstr. Soc. Neurosci.,* **20** (1994) 616.
55 Clarke, A. R., Purdie, C. A., Harrison, D. J., Morris, R. G., Bird, C. C., Hooper, M. L., and Wyllie, A. H., Thymocyte apoptosis induced by p53-dependent and independent pathways, *Nature,* **362** (1993) 849–852.
56 Lowe, S. W., Schmitt, E. M., Smith, S. W., Osborne, B. A., and Jacks, T., p53 is required for radiation-induced apoptosis in mouse thymocytes, *Nature,* **362** (1993) 847–849.
57 Nitatori, T., Sato, N., Waguri, S., Karasawa, Y., Araki, H., Shibanai, K., Kominami, E., and Uchiyama, Y., Delayed neuronal death in the CA1 pyramidal cell layer of the gerbil hippocampus following transient ischemia is apoptosis, *J. Neurosci.,* **15** (1995) 1001–1011.
58 Glaumann, H., Ericsson, L. E., and Marzella, L., Mechanisms of intralysosomal degradation with special reference to autophagocytosis and heterophagocytosis of cell organelles, *Int. Rev. Cytol.,* **73** (1981) 149–182.
59 Smeyne, R. J., Vendrell, M., Hayward, M., Baker, S. J., Miao, G. G., Schilling, K., Robertson, L. M., Curran, T., and Morgan, J. I., Continous c-fos expression precedes programmed cell death in vivo, *Nature,* **363** (1993) 166–169.
60 Ikeda, J., Nakajima, T., Osborne, O. C., Mies, G., and Nowak, T. S., Jr., Coexpression of c-fos and hsp70 mRNAs in gerbil brain after ischemia: induction threshold, distribution and time course evaluated by in situ hybridization, *Mol. Brain Res.,* **26** (1994) 249–258.
61 Vass, K., Welch, W. J., Nowak, T. S., Jr., Localization of 70 kDa stress protein induction in gerbil brain after ischemia, *Acta Neuropathol. (Berlin),* **77** (1989) 128–135.
62 Wiessner, C., Brink, I., Lorenz, P., Neumann-Haefelin, T., Vogel, P., and Yamashita, K., Cyclin D1 messenger RNA is induced in microglia rather than neurons following transient forebrain ischemia, *Neuroscience,* **72** (1996) 947–958.
63 Wiessner, C., Back, T., Bonnekoh, P., Kohn, K., Gehrmann, J., and Hossmann, K.-A., Sulfated glycoprotein-2 mRNA in the rat brain following transient forebrain ischemia, *Mol. Brain Res.,* **20** (1993) 345–352.
64 Krajewski, S., Mai, J. K., Krajewska, M., Sikorska, M., Mossakowski, M. J., and Redd, J. C., Up-regulation of bax protein-levels in neurons following cerebral ischemia, *J. Neurosci.,* **15** (1995) 6364–6376.
65 Shigeno, T., Yamasaki, Y., Kato, G., Kusaka, K., Mima, T., Takakura, K., Graham, D. I., and Furukawa, S., Reduction of delayed neuronal death by inhibition of protein synthesis, *Neurosci. Lett.,* **120** (1990) 117–119.
66 Goto, K., Ishige, A., Sekiguchi, K., Iizuka, S., Sugimoto, A., Yuzurihara, M., Aburada, M., Hosoya, E., and Kogure, K., Effects of cycloheximide on delayed neuronal death in rat hippocampus, *Brain Res.,* **534** (1990) 299–302.
67 Tortosa, A., Rivera, R., and Ferrer, I., Dose-related effects of cycloheximide on delayed neuronal death in the gerbil hippocampus after bilateral transitory forebrain ischemia, *J. Neurol. Sci.,* **121** (1994) 10–17.
68 Kiessling, M., Xie, Y., Ullrich, B., and Thilmann, R., Are the neuroprotective effects of the protein synthesis inhibitor cycloheximide due to prevention of apoptosis, *J. Cereb. Blood Flow Metab.,* **11(Suppl. 2)** (1991) S357.
69 Minamisawa, H., Nordström, C. H., Smith, M. L., and Siesjö, B. K., The influence of mild body and brain hypothermia on ischemic brain damage, *J. Cereb. Blood Flow Metab.,* **10** (1990) 817–822.
70 Busto, R., Dietrich, W. D., Globus, M. Y., and Ginsberg, M. D., Postischemic moderate hypothermia inhibits CA1 hippocampal ischemic neuronal injury, *Neurosci. Lett.,* **101** (1989) 299–304.
71 Ferrer, I., Tortosa, A., Macaya, A., Sierra, A., Moreno, D., Munell, F., Blanco, R., and Squier, W., Evidence of nuclear DNA fragmentation following hypoxia-ischemia in the infant rat brain, and transient forebrain ischemia in the adult gerbil, *Brain Pathol.,* **4** (1994) 115–122.

72 Heron, A., Pollard, H., Dessi, F., Moreau, J., Lasbennes, F., Ben-Ari, Y., and Charriaut-Marlangue, C. R., Regional variability in DNA fragmentation after global ischemia evidenced by combined histological and gel electrophoresis observations in rat brain, *J. Neurochem.,* **61** (1993) 1973–1976.

73 MacManus, J. P., Hill, I. E., Preston, E., Rasquinha, I., Walker, T., and Buchan, A. M., Differences in DNA fragmetnation following transient cerebral or decapitation ischemia in rats, *J. Cereb. Blood Flow Metab.,* **15** (1995) 728–737.

74 Beck, T., Lindholm, D., Castren, E., and Wree, A., Brain-derived neurotrophic factor protects against ischemic cell damage in rat hippocampus, *J. Cereb. Blood Flow Metab.,* **14** (1994) 689–692.

75 Ghosh, A., Carnahan, J., and Greenberg, M. E., Requirement for BDNF in activity-dependent survival of cortical neurons, *Science,* **263** (1994) 1618–1623.

76 Kleihues, P. and Hossmann K.-A., Protein synthesis in the cat brain after prolonged cerebral ischemia, *Brain Res.,* **35** (1971) 409–418.

77 Dienel, G. A., Pulsinelli, W. A., and Duffy, T. E., Regional protein synthesis in rat brain following acute hemispheric ischemia, *J. Neurochem.,* **35** (1980) 1216–1226.

78 Cooper, H. K., Zalwska, T., Kawakami, S., Hossmann, K.-A., and Kleinhues, P., The effect of ischemia and recirculation on protein synthesis in the rat brain, *J. Neurochem.,* **28** (1977) 929–934.

79 Lindquist, S., The heat-shock response, *Ann. Rev. Biochem.,* **55** (1986) 1151–1191.

80 Vogel, P., Wiessner, C., Dux, E., Uto, A., and Hossmann, K.-A., Heat shock in neuronal cell cultures: a simple model for studying mechansims of ischemic injury?, *Eur. J. Neurosci.,* **(Suppl. 7)** (1994) 16.01.

81 Hossmann, K.-A., Widmann, R., Wiessner, C., Duc, E., Djuricic, B., and Röhn, G., Protein synthesis after global ischemia and selective vulnerability. In J. Krieglstein and H. Oberpichler-Schwenk (eds.) *Pharmacology of Cerebral Ischemia*, Wissenschaftliche Verlagsgesellschaft, Stuttgart, 1992, pp. 289–299.

82 Bodsch, W., Takahashi, K., Barbier, A., Grosse, Ophoff, B., and Hossmann, K.-A., Cerebral protein synthesis and ischemia, *Progr. Brain Res.,* **63** (1985) 197–210.

83 Thilmann, R., Xie, Y., Kleihues, P., and Kiessling, M., Persistent inhibition of protein synthesis precedes delayed neuronal death in postischemic gerbil hippocampus, *Acta Neuropathol.,* **71** (1986) 88–93.

84 Widmann, R., Kuroiwa, T., Bonnekoh, P., and Hossmann, K.-A., [14C]leucine incorporation into brain proteins in gerbils after transient ischemia: relationship to selective vulnerability of hippocampus, *J. Neurochem.,* **56** (1991) 789–796.

85 Oliver, C. N., Starke-Reed, P. E., Stadtman, E. R., Lin, G. J., Carney, J. M., and Floyd, R. A., Oxidative damage to brain proteins, loss of glutamine synthetase activity, and production of free radicals during ischemial reperfusion-induced injury to gerbil brain, *Proc. Natl. Acad. Sci. USA,* **87** (1990) 5144–5147.

86 Seubert, P., Lee, K., and Lynch, G., Ischemia triggers NMDA receptor-linked cytoskeletal proteolysis in hippocampus, *Brain Res.,* **492** (1989) 366–370.

87 Siman, R., Noszek, J. C., and Kegerise, C., Calpain I activation is specifically related to excitatory amino acid induction of hippocampal damage, *J. Neurosci.,* **9** (1989) 1579–1590.

88 Wiessner, C., Vogel, P., Neumann-Haefelin, T., and Hossmann, K.-A., Molecular correlates of delayed neuronal death following transient forebrain ischemia, *Acta Neurochir.,* **66(Suppl.)** (1996) 1–7.

89 Uemura, Y., Konwall, N. W., and Beal, M. F., Global ischemia induces NMDA receptor-mediated c-fos expression in neurons resistant to injury in gerbil hippocampus, *Brain Res.,* **542** (1991) 343–447.

90 Nowak, T. S., Jr., Ikeda, J., and Nakajima, T., 70-kDa heat shock protein and c-fos gene expression after transient ischemia, *Stroke,* **21** (1990) III107–111.

91 Schmidt- Kastner, R. and Szymas, J., Immunohistochemistry of glial fibrillary acidic protein, vimentin and S-100 protein for study of astrocytes in hippocampus of rat, *J. Chem. Neuroanatomy, 3* (1990) 179–192.

92 Gehrmann, J., Bonnekoh, P., Miyazawa, T., Hossmann, K.-A., and Kreutzberg, G. W., Immunocytochemical study of an early microglial activation in ischemia, *J. Cereb. Blood Flow Metab., 12* (1992) 257–269.

93 Wiessner, C., Gehrmann, J., Lindholm, D., Topper, R., Kreutzberg, G. W., and Hossmann, K.-A., Expression of transforming growth factor-beta 1 and interleukin-1 beta mRNA in rat brain following transient forebrain ischemia, *Acta Neuropathol., 86* (1993) 439–446.

94 Schmidt-Kastner, R., Szymas, J., and Hossmann, K.-A., Immunohistochemical study of glial reaction and serum-protein extravasation in relation to neuronal damage in rat hippocampus after ischemia, *Neuroscience, 38* (1990) 527–540.

95 Petito, C. K. and Halaby, I. A., Relationship between ischemia and ischemic neuronal necrosis to astrocyte expression of glial fibrillary acidic protein, *Int. J. Dev. Neurosci., 11* (1993) 239–247.

96 Petito, C. K., Morgello, S., Felix, J. C., and Lesser, M. L., The two patterns of reactive astrocytosis in postischemic rat brain, *J. Cereb. Blood Flow Metab., 10* (1990) 850–859.

97 Eng, L. F. and Ghirnikar, R. S., GFAP and astrogliosis, *Brain Pathol., 4* (1994) 229–237.

98 Petito, C. K. and Babiak, T., Early proliferative changes in astrocytes in postischemic non-infarcted rat brain, *Ann. Neurol., 11* (1982) 510–518.

99 Kindy, M. S., Bhat, A. N., and Bhat, N. R., Transient ischemia stimulates glial fibrillary acid protein and vimentin gene expression in the gerbil neocortex, striatum and hippocampus, *Mol. Brain Res., 13* (1992) 199–206.

100 May, P. C. and Finch C. E., Sulfated glycoprotein 2: new realtionships of this multifunctional protein to neurodegeneration, *Trends Neurol. Sci., 15* (1992) 391–396.

101 Dragunow, M., Preston, K., Dodd J., Young, D., Lawlor, P., and Christie, D., Clusterin accumulates in dying neurons following status epilepticus, *Mol. Brain Res., 32* (1995) 279–290.

102 Takami, K., Iwane, M., Kiyota, Y., Miyamoto, M., Tsukuda, R., and Shiosaka, S., Increase of basic fibroblast growth factor immunoreactivity and its mRNA level in rat brain following transient forebrain ischemia, *Exp. Brain Res., 90* (1992) 1–10.

103 Takami, K., Kiyota, Y., Iwane, M., Miyamoto, M., Tsukuda, R., Igarashi, K., Shino, A., Wanaka, A., Shiosaka, S., and Tohyama, M., Upregulation of fibroblast growth factor-receptor messenger RNA expression in rat brain following transient forebrain ischemia, *Exp. Brain Res., 97* (1993) 185–194.

104 Gluckmann, P., Klempt, N., and Guan, J., A role for IGF-1 in the rescue of neurons following hypoxic-ischemic injury, *Biochem. Biophys. Res. Commun., 31* (1992) 593–599.

105 Banati, R., Gehrmann, J., Wiessner, C., Hossmann, K.-A., and Kreutzberg, G. W., Glial expression of the β-amyloid precursor protein (APP) in global ischemia, *J. Cereb. Blood Flow Metab., 15* (1995) 647–654.

106 Unsicker, K., Prehn, J. H. M., Backhaub, C., Krieglstein K., Otto D., and Krieglstein J., Growth factors and their implications for the injured brain: regulation and neuroprotective effects of FGFs and TGF-βs in ischemic and other lesion paradigms. In J. Krieglstein and H. Oberpichler-Schwenk (eds.) *Pharmacology of Cerebral Ischemia*, Wissenschaftliche Verlagsgesellschaft mbH, Stuttgart, 1992, pp. 229–238.

107 Bergstedt, K. and Wieloch T., Changes in insulin-like growth factor 1 receptor density after transient cerebral ischemia in the rat. Lack of protection against ischemic brain damage following injection of insulin-like growth factor 1, *J. Cereb. Blood Flow Metab., 13* (1993) 895–898.

108 Endoh, M., Maiese, K., and Wagner, J. A., Expression of the inducible form of nitric oxide synthase by reactive astrocytes after transient global ischemia, *Brain Res., 651* (1994) 92–100.

109 Banati, R. B., Gehrmann, J., Schubert, P., and Kreutzberg, G. W., Cytotoxicity of microglia, *Glia,* **7** (1993) 111–118.

110 Tsirka, S. E., Gualandris, A., Amaral, D. G., and Strickland, S., Excitotoxin-induced neuronal degeneration and seizure are mediated by tissue plasminogen factor, *Nature,* **377** (1995) 340–344.

111 Morioka, T., Kalehua, A. N., and Streit, W. J., Progressive expression of immuno-molecules on microglial cells in rat dorsal hippocampus following transient forebrain ischemia, *Acta Neuropathol.,* **83** (1992) 149–157.

112 Morioka, T., Kalehua, A. N., and Streit, W. J., The microglial reaction in the rat dorsal hippocampus following trasnient forebrain ischemia, *J. Cereb. Blood Flow Metab.,* **11** (1991) 966–973.

113 Gehrmann, J., Bonnekoh, P., Miyazawa, T., Oschlies, U., Dux, E., Hossmann, K.-A., and Kreutzberg, G. W., The microglial reaction in the rat hippocmapus following global ischemia: immuno-electron microscopy, *Acta Neuropathol.,* **84** (1992) 588–595.

114 Jorgensen, M. B., Finsen, B. R., Jensen, M. B., Castellano, B., Diemer, N. H., and Zimmer, J., Microglial and astroglial reactions to ischemic and kainic acid-induced lesions of the adult rat hippocampus, *Exp. Neurol.,* **120** (1993) 70–88.

115 Finsen, B. R., Jorgensen, M. B., Diemer, N. H., and Zimmer, J., Microglial MHC antigen expression after ischemic and kainic acid lesions of the adult rat hippocampus, *Glia,* **7** (1993) 41–49.

116 Sherr, C. J., G1 phase progression: cycling on cue, *Cell,* **79** (1994) 551–555.

117 Sicinski, P., Donaher, J. L., Parker, S. B., Li, T., Fazeli, A., Gardner, H., Haslam, S. Z., Bronson, R. T., Elledge, S. J., and Weinberg, R. A., Cyclin D1 provides a link between development and oncogenesis in the retina and breast, *Cell,* **82** (1995) 621–630.

118 Lau, L. F. and Nathans, D., Identification of a set of genes expressed during the G0/G1 transition of cultured mouse cells, *EMBO J.,* **4** (1985) 3145–3151.

119 Lehrmann, E., Kiefer, R., Finsen, B., Diemer, N. H., Zimmer, J., and Hartung, H. P., Cytokines in cerebral ischemia: expression of transforming growth factor beta-1 (TGF-beta 1) mRNA in the postischemic adult rat hippocampus, *Exp. Neurol.,* **131** (1995) 114–123.

120 Klempt, N. D., Sirimanne, E., Gunn, A. J., Klempt, M., Singh, K., Williams, C., and Gluckman, P. D., Hypoxia-ischemia induces transforming growth factor beta 1 mRNA in the infant rat brain, *Mol. Brain Res.,* **13** (1992) 93–101.

121 Morgan, T. E., Laping N. J., Rozovsky I., Oda T., Hogan T. H., Finch C. E., and Pasinetti G. M., Clusterin expression by astrocytes is influenced by transforming growth factor beta 1 and heterotypic cell interactions, *J Neuroimmunol.,* **58** (1995) 101–110.

122 Barnard, J. A., Bascom, C. C., Lyons, M. M., Sipes, N. J., and Moses, H. L., Transforming growth factor β in control of epidermal proliferation, *Am. J. Med. Sci.,* **296** (1988) 159–163.

123 Bainton, D. F., Miller, L. J., Kishimoto, T. K., and Springer, T. A., Leucocyte adhesion receptors are stored in peroxidase-negative granules of human neutrophiles, *J. Exp. Med.,* **166** (1987) 1641–1653.

124 O'Shea, J. J., Brown, E. J., Seligmann, B. E., Metcalf, J. A., Frank, M. M., and Gallin, J. I., Evidence for distinct intracellular pools of receptors for C3b and C3bi in human neutrophils, *J. Immunol.,* **134** (1985) 2580–2586.

125 Beyreuther, K. and Masters, C. L., Amyloid precursor protein (APP) and βA4 amyloid in the etiology of Alzheimer's disease: precursor-product relationships in the derangement of neuronal function, *Brain Pathol.,* **1** (1991) 241–251.

126 Abe, K., Tanzi, R. E., St. George-Hyslop, P. H., and Kogure, K., Induction of amyloid precursor protein (APP) gene after middle artery occlusion of rats, *J.Cereb. Blood Flow Metab.,* **11(Suppl. 2)** (1991) S351.

127 Kalaria, R. N., Bhatti, S. U., Palatinsky, E. A., Pennington, D. H., Shelton, E. R., Chan, H. W., Perry, G., and Lust, W. D., Accumulation of the beta amyloid precursor protein at sites of ischemic injury in rat brain, *Neuroreport,* **4** (1993) 211–214.

128 Stephenson, D. T., Rash, K., and Clemens, J. A., Amyloid precursor protein accumulates in regions of neurodegeneration following focal cerebral ischemia in the rat, *Brain Res.,* **593** (1992) 128–135.

129 Wakita, H., Tomimoto, H., Akiguchi, I., Ohnishi, K., Nakamura, S., and Kimura, J., Regional accumulation of amyloid beta/A4 protein precursor in the gerbil brain following transient cerebral ischemia, *Neurosci. Lett.,* **146** (1992) 135–138.

130 Abe, K., Tanzi, R. E., and Kogure, K., Selective induction of Kunitz-type protease inhibitor domain-containing amyloid precursor protein mRNA after persistent focal ischemia in rat cerebral cortex, *Neurosci. Lett.,* **125** (1991) 172–174.

131 Xie, Y. X., Herget T., Hallmayer J., Starzinski-Powitz A., and Hossmann K.-A., Determination of RNA content in postischemic gerbil brain by in situ hybridization, *Metab. Brain Dis.,* **4** (1989) 239–251.

132 Hall, E. D., Ostveen, J. A., Dunn, E., and Carter, D. B., Increased amyloid protein precursor and apolipoprotein E immunoreactivity in the selectively vulnerable hippocampus following transient forebrain ischemia, *Exp. Neurol.,* **135** (1995) 17–27.

133 Corder, E. H., Saunders, A. M., Strittmatter, W. J., Schmechel, D. E., Gaskell, P. C., Small, G. W., Roses, A. D., Haines, J. L., and Pericak-Vance, M. A., Gene dose of apolipoprotein E type 4 allele and the risk of Alzheimer's disease in late onset families, *Science,* **261** (1993) 921–923.

134 Hossmann, K.-A., Animal models of cerebral ischemia. 1. Review of literature, *Cereborvasc Dis.,* **1(Suppl. 1)** (1991) 2–15.

135 Dereski, M. O., Chopp, M., Knight, R. A., Rodolosi, L. C., and Garcia, J. H., The heteroenous temporal evolution of focal ischemic neuronal damage in the rat, *Acta Neuropathol.,* **85** (1993) 327–333.

136 Bartus, R. T., Dean, R. L., Cavanaugh, K., Eveleth, D., Carriero, D. L., and Lynch, G., Time-related neuronal changes following middle cerebral artery occlusion: implications for therapeutic intervention and the role of calpain, *J. Cereb. Blood Flow Metab.,* **15** (1995) 969–979.

137 Gonzalez, M. F., Shiraishi, K., Hisanaga, K., Sagar, S. M., Mandabach, M., and Sharp, F. R., Heat shock proteins as markers of neuronal injury, *Mol. Brain Res.,* **6** (1989) 93–100.

138 Welsh, F. A., Moyer, D. J., and Harris, V. A., Regional expression of heat shock protein-70 mRNA and c-fos mRNA following focal ischemia in rat brain, *J. Cereb. Blood Flow Metab.,* **12** (1992) 204–212.

139 Abe, K., Kawagoe, J., Araki, T., Aoki, M., and Kogure, K., Differential expression of heat shock protein 70 gene between the cortex and caudate after transient focal cerebral ischaemia in rats, *Neurol. Res.,* **14** (1992) 381–385.

140 Li, Y., Chopp, M., Zhang, Z. G., Zhang, R. L., and Garcia, J. H., Neuronal survival is associated with 72 kDa heat shock protein expression after transient middle artery oclusion, *J. Neurol. Sci.,* **120** (1993) 187–194.

141 Kokaia, Z., Zhao, Q., Kokaia, M., Elmer, E., Metsis, M., Smith, M. L., Siesjo, B. K., and Lindvall, O., Regulation of brain-derived neurotrophic factor gene expression after transient middle cerebral artery occlusion with and without brain damage, *Exp. Neurol.,* **136** (1995) 73–88.

142 Hsu, C. Y., An, G., Liu, J. S., Xue, J. J., He, Y. Y., and Lin, T. N., Expression of immediate early gene and growth factor mRNAs in a focal cerebral ischemia model in the rat, *Stroke,* **24** (1993) 178–188.

143 An, G., Lin, T. N., Liu, J. S., Xue, J. J., He, Y. Y., and Hsu, C. Y., Expression of c-fos and c-jun family genes after focal cerebral ischemia, *Ann. Neurol.,* **33** (1993) 457–464.

144 An, G., Lin, T. N., Liu, J. S., and Hsu, C. Y., Induction of Krox-20 expression after focal cerebral ischemia, *Biochem. Biophys. Res. Commun.,* **188** (1992) 1104–1110.

145 Abe, K., Kawagoe, J., Sato, S., Sahara, M., and Kogure, K., Induction of the 'zinc finger' gene after transient focal ischemia in rat cerebral cortex, *Neurosci. Lett.,* **123** (1991) 248–250.

146 Lı, Y., Chopp, M., Zhang, Z. G., Zalog, C., Niewenhuis, L., and Gautam, S., p53-immunoreactive protein and p53 mRNA expression after transient middle cerebral artery occlusion in rats, *Stroke,* **25** (1994) 849–855.

147 Crumrine, R. C., Thomas, A. L., and Morgan, P. F., Attenuation of p53 expression protects against focal ischemic damage in transgenic mice, *J. Cereb. Blood Flow Metab.,* **14** (1994) 887–891.

148 Chen, J., Graham, S. H., Chan, P. H., Lan, J. Q., Zhou, R. L., and Simon, R. P., Bcl-2 is expressed in neurons that survive focal ischemia in the rat, *Neuroreport,* **6** (1995) 394–398.

149 Martinou, J. C., Dubois-Dauphin, M., Staple, J. K., Rodriguez, I., Frankowski, H., Missotten, M., Albertini, P., Talabot, D., Catsicas, S., Pietra, C., and Huarte, J., Overexpression of BCL-2 in transgenic mice protects neurons from naturally occurring cell death and experimental ischemia, *Neuron,* **13** (1994) 1017–1030.

150 Hockenberry, D. M., Oltvai, Z. N., Yin, X.-M., Milliman, C. L., and Korsmeyer, S. J., Bcl-2 functions in an antioxidant pathway to prevent apoptosis, *Cell,* **75** (1993) 241–251.

151 Wang, X., Yue, T.-L., Barone, F. C., White, R. F., Gagnon, R. C., and Feuerstein, G. Z., Concomitant cortical expression of TNFalpha and IL-1β mRNAs follows early response gene expression in transient fical ischemia, *Mol. Chem. Neuropathol.,* **23** (1994) 103–114.

152 Merril, J. E. and Miller Jonakait, G., Interactions of the nervous and immune systems in development, normal brain homeostasis, and disease, FASEB J., **9** (1995) 611–618.

153 Wang, X., Siren, A.-L., Liu, Y., Yue T.-L., Barone, F. C., and Feuerstein, G. Z., Upregulation of intercellular adhesion molecule 1 (ICAM-1) on brain microvasvular endothelial cells in rat ischemic cortex, *Mol. Brain Res.,* **26** (1994) 61–68.

154 Okada, Y., Copeland, B. R., Mori E., Tung, M. M., Thomas, W. S., and del Zoppo, G. J., P-selectin and intercellular adhesion molecule-1 expression after focal brain ischemia and reperfusion, *Stroke,* **25** (1994) 202–211.

155 Zhang, R. L., Chopp, M., Jiang, N., Tang, W. X., Prostak, J., Manning, A. M., and Anderson, D. C., Anti-intercellular adhesion molecule-1 antibody reduces ischemic cell damage after transient but not permanent middle cerebral artery occlusion in the Wistar rat, *Stroke,* **26** (1995) 1438–1443.

156 Matsuo, Y., Onodera, H., Shiga, Y., Shozuhara, H., Ninomya, M., Kihara, T., Tamatami, T., Miyasaka M., and Kogure K., Role of adhesion molecules in brain injury after transient middle cerebral artery occlusion in the rat, *Brain Res.,* **656** (1994) 344–352.

157 Chen, H., Chopp, M., Zhang, R. L., Bodzin, G., Chen, Q., Rusche, J. R., and Todd, R. F., III, Anti-CD11b monoclonal antibody reduces ischemic cell damage after transient focal cerebral ischemia, *Ann. Neurol.,* **35** (1994) 458–463.

158 Betz, A. L., Yang, G. Y., and Davidson, B. L., Attenuation of stroke size in rats using an adenoviral vector to induce overexpression of interleukin-1 receptor antagonist in brain, *J. Cereb. Blood Flow Metab.,* **15** (1995) 547–551.

159 Wang, X., Yue, T. L., Barone, F. C., and Feuerstein, G. Z., Mònocyte chemoattractant protein-1 messenger RNA expression in rat ischemic cortex, *Stroke,* **26** (1995) 661–665.

160 Clark, R. K., Lee, E. V., Fish, C. J., White, R. F., Price, W. J., Jonak, Z. L., Feuerstein, G. Z., and Barone, F. C., Development of tissue damage, inflammation and resolution following stroke: an immunohistochemical and quantitative study, *Brain Res. Bull.,* **31** (1993) 565–572.

161 Kinouchi, H., Sharp, F. R., Chan, P. H., Koistinaho, J., Sagar, S. M., and Yoshimoto, T., Induction of c-fos, junB, c-jun, and hsp70 mRNA in cortex, thalamus, basal ganglia, and hippocampus following middle cerebral artery occlusion, *J. Cereb. Blood Flow Metab.,* **14** (1994) 808–817.

162 Kinouchi, H., Sharp, F. R., Koistinaho, J., Hicks, K., Kamii, H., and Chan, P. H., Induction of heat shock hsp70 mRNA and HSP70 kDa protein in neurons in the 'penumbra' following focal cerebral ischemia in the rat, *Brain Res.,* **619** (1993) 334–338.

163 Sharp, F. R., Lowenstein, D., Simon, R., and Hisanaga, K., Heat shock protein hsp72 induction in cortical and striatal astrocytes and neurons following infarction, *J. Cereb. Blood Flow Metab.,* **11** (1991) 621–627.

164 Iizuka, H., Sakatani, K., and Young, W., Selective cortical neuronal damage after middle cerebral artery occlusion in rats, *Stroke,* **20** (1989) 1516–1523.

165 Higashi, T., Takechi, H., Uemura, Y., Kikuchi, H., Nagata, K., Differential induction of mRNA species encoding several cases of stress proteins following focal cerebral ischemia in rats, *Brain Res.,* **650** (1994) 239–248.

166 Wang, S., Longo, F. M., Chen, J., Butman, M., Graham, S., Haglid, K. G., and Sharp, F. R., Induction of glucose regulated protein (grp78) and inducible heat shock protein (hsp70) mRNAs in rat brain after kainic acid seizures and focal ischemia, *Neurochem. Int.,* **23** (1993) 575–582.

167 Uemura, Y., Kowall, N. W., and Moskowitz, M. A., Focal ischemia in rats causes time-dependent expression of c-fos protein immunoreactivity in widespread regions of ipsilatcral cortex, *Brain Res.,* **552** (1991) 99–105.

168 Gass, P., Spranger, M., Herdegen, T., Bravo, R., Kock, P., Hacke, W., and Kiessling, M., Induction of FOS and JUN proteins after focal ischemia in the rat: differential effect of the N-methyl-D-aspartate receptor antagonist MK-801, *Acta Neuropathol.,* **84** (1992) 545–553.

169 Zhang, Z. G., Chopp, M., Gautam, S., Zaloga, C., Zhang, R. L., Schmidt, H. H., Pollock, J. S., and Forstermann, U., Upregulation of neuronal nitric oxide synthase and mRNA, and selective sparing of nitric oxide synthase-containing neurons after focal cerebral ischemia in rat, *Brain Res.,* **654** (1994) 85–95.

170 Back, T., Kohno, K., and Hossmann, K.-A., Cortical negative DC deflections following middle cerebral artery occlusion and KCL-induced spreading depression: effect on blood flow, tissue oxigenation and electroencephalogram, *J. Cereb. Blood Flow Metab.,* **14** (1993) 12–19.

171 Mies, G., Iijima, T., and Hossmann, K.-A., Correlation between periinfarct shifts and ischaemic neuronal damge in rat, *Neuroreport,* **4** (1993) 709–711.

172 Liu, T., McDonnell, P. C., Young, P. R., White, R. F., Siren, A. L., Hallenbeck, J. M., Barone, F. C., and Feurestein, G. Z., Interleukin-1 beta mRNA expression in ischemic rat cortex, *Stroke,* **24** (1993) 1746–1750.

173 Liu, T., Clark, R. K., McDonnell, P. C., Young, P. R., White, R. F., Barone, F. C., and Feuerstein, G. Z., Tumor necrosis factor-alpha expression in ischemic neurons, *Stroke,* **25** (1994) 1481–1488.

174 Wang, X., Yue, T. L., Young, P. R., Barone, F. C., and Feuerstein, G. Z., Expression of interleukin-6, c-fos, and zif268 mRNAs in rat ischemic cortex, *J. Cereb. Blood Flow Metab.,* **15** (1995) 166–171.

175 Buttini, M., Sauter, A., and Boddeke, H. W. G. M., Induction of interleukin-1β mRNA after focal cerebral ischemia in the rat, *Mol. Brain Res.,* **23** (1994) 126–134.

176 Wang, X., Yue, T.-L., White, R. F., Barone, F. C., and Feuerstein, G. Z., Transforming growth factor-β1 exhibits delayed gene expression folowing focal cerebral ischemia, *Brain Res. Bull.,* **36** (1995) 607–609.

177 Kim, J. S., Gautam, S. C., Chopp, M., Zaloga, C., Jones, M. L., Ward, P. A., and Welch, K. M., Expression of monocyte chemoattractant protein-1 and macrophage inflammatory protein-1 after focal cerebral ischemia in the rat, J. Neuroimmunol., **56** (1995) 127–134.

178 Wang, X., Yue, T. L., Barone, F. C., and Feuerstein, G. Z., Demonstration of increased endothelial-leukocyte adhesion molecule-1 mRNA expression in rat ischemic cortex, *Stroke,* **26** (1995) 1665–1668.

179 Liu, T., Young, P. R., McDonnell, P. C., White, R. F., Barone, F. C., and Feuerstein, G. Z., Cytokine-induced neutrophil chemoattractant mRNA expressed in cerebral ischemia, *Neurosci. Lett.,* **164** (1993) 125–128.

180 Iihara, K., Sasahara, M., Hashimoto, N., Uemura, Y., Kikuchi, H., and Hazama, F., Ischemia induces the expression of the platelet-derived growth factor-B chain in neurons and brain macrophages in vivo, *J. Cereb. Blood Flow Metab.,* **14** (1994) 818–824.

181 Iadecola, C., Zhang, F., Xu, S., Casey, R., and Ross, M. E., Inducible nitric oxide synthase gene expression in brain following cerebral ischemia, *J. Cereb. Blood Flow Metab.,* **15** (1995) 378–384.

182 Li, Y., Chopp, M., Jiang, N., Zhang, Z. G., and Zaloga, C., Induction of Dna fragmentation after 10 to 120 minutes of focal cerebral ischemia in rats, *Stroke,* **26** (1995) 1252–1257.

183 Li, Y., Chopp, M., Jiang, N., and Zaloga, C., In situ detection of DNA fragmetnation after focal cerebral ischemia in mice, *Mol. Brain Res.,* **28** (1995) 164–168.

184 Li, Y., Chopp, M., Jiang, N., Yao, F., and Zaloga, C., Temporal profile of in situ DNA fragmentation after trasnient middle cerebral artery occlusion in the rat, *J. Cereb. Blood Flow Metab.,* **15** (1995) 389–397.

185 MacManus, J. P., Hill, I. E., Huang, Z. G., Rasquinha, I., Xue, D., and Buchan, A. M., DNA damage consistent with apoptosis in transient focal ischaemic neocortex, *Neuroreport,* **5** (1994) 493–496.

186 Tominaga, T., Kure, S., Narisawa, K., and Yoshimoto, T., Endonuclease activation following focal ischmic injury in the rat brain, *Brain Res.,* **608** (1993) 21–26.

187 Charriaut-Marlangue, C., Margaill, I., Plotkine, M., and Ben-Ari, Y., Early endonuclease activation following reversible focal ischemia in the rat brain, *J. Cereb. Blood Flow Metab.,* **15** (1995) 385–388.

188 Linnik, M. D., Miller, J. A., Sprinklecavallo, J., Mason, P. J., Thompson, F. Y., Montgomery, L. R., and Schroeder, K. K., Apoptotic DNA fragmentation in the rat cerebral cortex induced by permanent middle cerebral artery occlusion, *Mol. Brain Res.,* **32** (1995) 116–124.

189 Linnik, M. D., Zobrist, R. H., and Hatfield, M. D., Evidence supporting a role for programmed cell death in focal cerebral ischemia in rats, *Stroke,* **24** (1993) 2002–2009.

190 Iizuka, H., Sakatani, K., and Young, W., Neural damage in the rat thalamus after cortical infarcts, *Stroke,* **21** (1990) 790–794.

191 Yamada, K., Kinoshita, A., Kohmura, E., Sakaguchi, T., Taguchi, J., Kataoka, K., and Hayakawa, T., Basic fibroblast growth factor prevents thalamic degeneration after cortical infarction, *J. Cereb. Blood Flow Metab.,* **11** (1991) 472–478.

192 Schroeter, M., Schiene, K., Kraemer, M., Hagemann, G., Weigel, H., Eysel, U. T., Witte, O. W., and Stoll, G., Astroglial responses in photochemically induced focal ischemia of the rat cortex, Exp. *Brain Res.,* **106** (1995) 1–6.

193 Li, Y., Chopp, M., Zhang, Z. G., and Zhang, R. L., Expression of glial fibrillary acidic protein in areas of focal cerebral ischemia accompanies neuronal expression of 72-kDa heat shock protein, *J. Neurol. Sci.,* **128** (1995) 134–412.

194 Chen, H., Chopp, M., Schultz, L., Bodzin, G., and Garcia, J. H., Sequential neuronal and astrocytic changes after transient middle cerebral artery occlusion in the rat, *J. Neurol. Sci.,* **118** (1993) 109–116.

195 Korematsu, K., Goto, S., Nagahiro, S., and Ushio, Y., Microglial response to transient focal cerebral ischemia: an immunocytochemical study on the rat cerebral cortex using anti-phosphotyrosine antibody, *J. Cereb. Blood Flow Metab.,* **14** (1994) 825–830.

196 Gehrmann, J., Yamashita, K., Back, T., Kreutzberg, G. W., and Wiessner, C., The microglial reaction in focal ischemia: an early generalized response not attenuable by post-ischaemic pharmacological intervention, *J. Cereb. Blood Flow Metab.,* **15(Suppl. 1)** (1995) S353.

197 Gehrmann, J., Mies, G., Bonnekoh, P., Banati, R., Iijima, T., Kreutzberg, G. W., and Hossmann, K.-A., Microglial reaction in the rat cerebral cortex induced by cortical spreading depression, *Brain Pathol.,* **3** (1993) 11–17.

198 Morioka, T., Kalehua, A. N., and Streit, W. J., Characterization of microglial reaction after middle cerebral artery occlusion in rat brain, *J. Comp. Neurol.,* **327** (1993) 123–132.

199 Cohen, J. S., Oligonucleotides as therapeutic agents, *Pharmac. Ther.,* **52** (1991) 211–225.

200 Aguzzi, A., Brandner, S., Sure, U., Ruedi, D., and Isenmann, S., Transgenic and knock-out mice: models of neurological disease, *Brain Pathol.,* **4** (1994) 3–20.

201 Liu, P. K., Salminen, A., He, Y. Y., Jiang, M. H., Xue, J. J., Liu, J. S., and Hsu, C. Y., Suppression of ischemia-induced fos expression and AP-1 activity by an antisense oligodeoxynucleotide to c-fos mRNA, *Ann. Neurol.,* **36** (1994) 566–576.

202 Wahlestedt, C., Golanov, E., Yamamoto, S., Yee, F., Ericson, H., Yoo, H., Inturrisi, C. E., and Reis, D. E., Antisense oligodeoxynucleotides to NMDA-R1 receptor channel protect cortical neurons from excitotoxicity and reduce focal ischemic infarction, *Nature,* **363** (1993) 260–263.

203 Barone, F. C., Knudsen, D. J., Nelson, A. H., Feuerstein, G. Z., and Willette, R. N., Mouse strain differences in susceptibility to cerebral ischemia are related to cerebral vascular anatomy, *J. Cereb. Blood Flow Metab.,* **13** (1993) 683–692.

204 Garcia, I., Martinou, I., Tsujimoto, Y., and Martinou, J.-C., Prevention of programmed cell death of sympathetic neurons by the bcl-2 proto-oncogene, *Science,* **258** (1992) 302–304.

205 Chan, P. H., Epstein, C. J., Kinouchi, H., Kamii, H., Imaizumi, S., Yang, G., Chen, S. F., Gafni, J., and Carlson, E., SOD-1 transgenic mice as a model for studies of neuroprotection in stroke and brain trauma, *Ann. New York Acad. Sci.,* **738** (1994) 93–103.

206 Huang, Z., Huang, P. L., Panahian, N., Dalkara, T., Fishman, M. C., and Moskowitz, M. A., Effects of cerebral ischemia in mice deficient in neuronal nitric oxide synthase, *Science,* **265** (1994) 1883–1885.

207 MacMillan, V., Judge, D., Wiseman, A., Settles, D., Swain, J., and Davis, J., Mice expressing a bovine basic fibroblast growth factor transgene in the brain show increased resistance to hypoxemic-ischemic cerebral damage, *Stroke,* **24** (1993) 1735–1739.

208 Gu, H., Marth, J. D., Orban, P. C., Mossmann, H., and Rajewsky, K., Deletion of a DNA polymerase β gene segment in T cells using cell type specific gene targeting, *Science,* **265** (1994) 103–106.

209 Gossen, M., Freundlieb, S., G. B., Müller, G., Hillen, W., and Bujard, H., Transcriptional activation by tetracyclines in mammalian cells, *Science,* **268** (1995) 1766–1769.

9

Hormonal Modulators of Cerebral Ischemia

Laura J. McIntosh and Robert M. Sapolsky

1. INTRODUCTION

Hormones are modulators of cellular physiology throughout the body, exerting both short-term and long-term effects on fundamental cell processes such as gene transcription, energy homeostasis, and signaling cascades. Because of the integral nature of their actions, the presence of hormones both before and after a metabolic crisis like cerebral ischemia (CI) can greatly affect the ability of hormone-sensitive tissues to respond to the crisis.

The loss of oxygen and glucose that defines CI arises most commonly from an interruption or reduction in cerebral blood flow. Neurons are dependent on a constant supply of nutrients, since they lack the glycogen reserves common in peripheral tissues; thus the first effect of CI is energy depletion. The ATP levels fall and energy dependent processes shut down. This produces a massive release of excitatory amino acids (EAAs) such as glutamate from axon terminals within the area of reduced blood flow. The resulting prolonged stimulation of glutamate receptors creates an inflow of sodium and calcium ions, leading to elevation of intracellular calcium concentrations and initiation of the calcium second messenger cascade. Heavy stimulation of the calcium cascade activates a variety of pathways, resulting in membrane dysfunction (lipid peroxidation, disruption of signalling pathways), cytoskeletal failure, free radical production, and enzyme overactivation (kinases, lipases, endonucleases, proteases), and can ultimately lead to cell death (for review, *see* refs. *1* and *2*).

The metabolic insult is worst at the focal ischemic core, which is irreversibly injured during total blood deprivation. The penumbral area may be salvaged depending on postischemic conditions and the amount of oxygen and glucose available from collateral sources. The importance of outside factors in susceptibility to CI is shown by the delay in cytopathological measures of neuron death until 20 h or more after the CI, the ability of specific pharmacological agents to alter the extent of damage, and by evidence suggesting that one mild ischemic event can alter neuronal physiology in such a way as to be protective against a second, ordinarily lethal ischemic event. Evidence also suggests both intrinsic and extrinsic factors are important for the selective vulnerability of specific cell types to CI (i.e., CA1 layer cells in the hippocampus) *(2)*.

Hormones that affect neuronal metabolism are logical modulators of CI. Glucocorticoids, insulin, thyroid hormone, testosterone, sympathetic catecholamines, and

From: Clinical Pharmacology of Cerebral Ischemia *Edited by: G. J. Ter Horst and J. Korf Humana Press Inc., Totowa, NJ*

corticotrophin-releasing factor all have mechanisms of action that have been shown to affect important physiological responses during or after CI. These hormones collectively help control glucose transport, cellular oxidation rate, membrane depolarization parameters, neurotransmitter release and reuptake, and body temperature. Obviously, these are complex functions and interactions between hormones add considerable dimensionality to the physiological story.

An extensive body of evidence now exists indicating that hormones can modulate or produce disease states, including neurodegeneration. Further, moderate levels below the threshold for outright cellular damage can still result in subtle metabolic derangements that become apparent in the presence of a second concurrent insult, such as ischemia. We will discuss the evidence indicating that a number of hormones have modulatory effects on the brain, in some instances aggravating injury arising from CI, and in others, decreasing it. We will concentrate primarily, but not exclusively, on glucocorticoids.

2. GLUCOCORTICOIDS AND CEREBRAL ISCHEMIA

The perception of stress, either physical or psychological, induces secretion of hormones from the hypothalamus, beginning a cascade that results within minutes in release of glucocorticoids (GCs) from the adrenals. GCs mobilize energy to cardiovascular, pulmonary, sensory, and muscle tissues to maximize the individual's ability to respond to the stressor. To provide this energy boost from a finite amount of available glucose, GCs suppress metabolism in the systems whose functions are not crucial in an emergency, such as digestion, reproduction, growth, and immune response. The GCs also block glucose uptake and storage in these temporarily dispensable systems, and promote energy release from body stores. Although the actions of GCs are highly effective in helping the individual respond to an immediate crisis, these extreme metabolic rearrangements are the basis for pathology in all GC-responsive tissues after prolonged exposure or exceptionally high GC concentrations. High GC levels owing to chronic stress, Cushing's syndrome, or exogenous administration (such as dexamethasone, prednisone, or hydrocortisone), can cause or worsen myopathy, fatigue, hypertension, ulcers, depression, decreased immune responses, adult-onset diabetes, and neuron death *(3,4)*.

This latter phenomenon of GC neurotoxicity has now been well documented. Not only are GCs capable of killing some neurons outright, GCs also predispose neurons to damage during concomitant insults, including CI. We will discuss the evidence for direct neuronal damage by GCs, the known mechanisms by which GCs act on neurons, and how these mechanisms work synergistically with the metabolic derangements of CI to exacerbate neuronal degeneration.

Although several brain regions are susceptible to GC-mediated damage, sustained exposure to GCs is particularly damaging to the brain region fundamental to learning and memory, the hippocampus. Hippocampal degeneration from elevated GC concentrations is seen in rodents *(5–11)*, primates *(12)*, and possibly humans *(13,14)*. The high concentration of corticosteroid receptors in the hippocampus is thought to account for this regional vulnerability *(15)*. Physiologically high levels of GCs achieved by behavioral means can also be shown to produce damage in rodent and primate models *(16–20)*. Conversely, long-term lowering of GC levels appear to alleviate the neuron loss and cognitive impairments typical of aging *(21–23)*. The GCs, like all good things in life, are best in moderation *(24)*.

GC levels must be exceptionally high and of long duration to kill neurons—not a common occurrence. Therefore most research effort has been devoted to understanding the effects of GCs under more physiological circumstances. At elevated levels, but below the threshold for direct cell death, GCs still have deleterious effects upon the nervous system. This is seen in GC exacerbation of neurodegeneration following hypoxia/ischemia, hypoglycemic shock, or seizure *(25)*. The presence of GCs does not appear to alter most cellular parameters under resting conditions, but when the neurons are challenged by a concurrent insult such as CI, metabolic derangements become evident and lead to significantly increased damage. Possible mechanisms for this neuronal endangerment will be discussed in the following paragraphs.

The GCs increase the magnitude of events integral to the EAA cascade triggered during CI. Glutamate released from the presynaptic neuron, resulting from either stimulation or energy failure, accumulates in the synaptic cleft after ischemia *(26–29)*. GCs appear to increase and prolong the concentration of glutamate by inhibiting uptake into the glia *(30)*. This results in strong stimulation of the EAA receptors, some of which are calcium channels, and cytosolic calcium levels rise *(1)*. Although high levels of calcium ions are seen in the postsynaptic neuron, it is not clear what proportion comes through the EAA receptor calcium channels and what amount is mobilized from cellular stores. However, these abnormally high cytosolic free calcium levels are maintained by GC-mediated inhibition of calcium efflux or sequestration at both the calcium ATPase and the Ca^{++}/Na^+ exchanger. Since calcium ions initiate a complex cascade of events, including enzyme activation and oxygen radical production, the effects are seen in membrane impairment *(31)*, DNA and protein degradation, and disruption of the cytoskeleton *(32)*. GCs also have a demonstrated capacity to exacerbate breakdown of the cytoskeletal proteins spectrin *(33,34)* and tau *(35)*. (*see* ref. *25* for a full review of GC involvement in EAA cascade.)

The most directly damaging consequence of too much calcium is thought to be the increase in reactive oxygen species (ROS) such as superoxide and hydroxyl radical, which can impair protein function, fragment lipid membranes, and form DNA adducts *(36)*. In primary hippocampal cultures, physiological levels of GCs can exacerbate the toxicity of an oxygen radical generator, and can raise levels of ROS in the cultures by approx 10% *(37)*, a magnitude that may translate directly into increased cellular damage *(38)*. The GCs also decrease tissue activity of the antioxidant enzymes superoxide dismutase, glutathione peroxidase, and catalase in a brain region-specific manner (data unpublished). Since excess ROS are one of the final pathways capable of causing lesions that can kill the cell, these experiments hinting at GC enhancement of ROS production, and decrease of defense capacity, point to mechanisms by which GCs may exacerbate neuron death.

The effects of GCs on the EAA cascade are compounded by GC disruption of neuronal energy state. All neurological insults known to be exacerbated by GCs are ultimately energetic crises. Hypoxia/ischemia, seizure, and hypoglycemia, made worse by the presence of GCs *(4)*, either disrupt energy availability (ischemia and hypoglycemia) or place excessively high energy demands on the cells (ischemia and seizure). In both cases, energy stores such as ATP and phosphocreatine decline, and the rate of glucose uptake can become rate limiting in the energy production cycles *(39)*. Under these circumstances, tissues try to compensate by up-regulating glucose transport *(40,41)*. One theory to explain GC neuroendangerment focuses on interference with these compensatory processes, thereby making the energy crisis worse.

The GCs affect neuronal energy flux by decreasing glucose uptake *(42–44)*. Within 6–12 h after an increase in GC levels, glucose transporters are translocated from the plasma membrane to intracellular storage sites *(45)*. Low transporter levels are maintained over the long-term by GC inhibition of glucose transporter gene transcription *(46)*. The GC-mediated inhibition of glucose transport into hippocampal neurons results in approximately a 25% decrease in uptake. Measurements of ATP levels indicate that resting levels of this fundamental energy currency is maintained, even in the presence of high GC levels *(47,48)*. Likewise, the energy-dependent functions of the housekeeping enzymes appear to be preserved *(15,49)*. However, neuronal vulnerability to GCs is demonstrable in the accelerated decline in ATP concentrations in both glia *(48)* and neurons in culture *(47)* during oxygen and glucose deprivation. The energy dependency of the GC-mediated pathology is seen in the amelioration of damage by energy supplementation *(50–52)*.

However, it is interesting to note that glucose supplementation directly before ischemia appears to *increase* damage, possibly by increasing substrates for the anaerobic pathways that produce lactic acid. Only postischemic energy supplementation, when oxygen is available, lessens damage *(53)*.

Precipitate energy declines are likely to be an important component in GC neuroendangerment. Neurons must maintain costly ion gradients for signaling and neurotransmitter reuptake, which can require more than half of the energy generated by glutamatergic neurons *(54)*. Neurons are more likely to survive ischemic insults when their metabolic rates are lowered *(55)*, suggesting that some ischemic damage may be owing to shifting from aerobic metabolism to anaerobic processes in order to cope with energy demand. Without the high demand from ion fluxes, the low energy vegetative processes (as low as 5% of total cell energy requirements *[54]*) necessary for viability can be maintained for a considerable period of time. The severity of nearly every step of the EAA cascade is dependent on the amount of available energy. Glutamate is removed from the synapse through exchangers dependent on ionic gradients. Glutamate release is increased during energy failure because membrane depolarization allows nonvesicular neurotransmitter efflux. NMDA receptors depend on membrane potential for their magnesium blockade, without which there is increased calcium influx. Also, the costly repair of damage produced by calcium excess would be impaired by a lack of energy *(39,56)*. Therefore, the ability of GCs to decrease glucose uptake and exacerbate the decline in ATP levels during necrotic insults to neurons and glia is likely to explain how they worsen various steps in the EAA/calcium/ROS cascade. As proof that these synergies are based at least in part on energy availability, energy supplementation decreases the effect of GCs on EAA accumulation, intracellular free calcium, and the resultant degenerative events *(33,57,58)*.

It should be emphasized that not all insults to the brain are exacerbated by the presence of GCs, and some controversies exist. First, although all regions of the brain have corticosteroid receptors, it is the regions with higher concentrations of the Type II receptor that are most vulnerable to damage *(17,59)*. Synthetic Type II ligands such as methylprednisolone, RU28362, and dexamethasone reproduce the same pattern of toxicity *(17,60–63)*, whereas Type II antagonists block GC-mediated degeneration *(64)*. Excitotoxic damage in the hypothalamus and cerebellum is not increased by GCs, but the hippocampus, cortex, and striatum are vulnerable *(25)*. Second, raising body temperature significantly increases hypoxic-ischemic damage, and GCs can elevate temperature. Morse and Davis *(65)* suggest GC endangerment is largely owing to the temperature effect, but while the temperature may be an important component, other studies that control body temperature still see a GC augmentation of hypoxic-ischemic damage *(18,66,67)*. Third, GC exacerbation of ischemic injury appears to be relevant to adult brain, but not to neonatal brain *(68)*. Studies by neonatologists indicate that GCs may decrease ischemic damage in vivo *(69–72)* although primary cell cultures derived from fetal brains demonstrate GC toxicity *(17,44,51,52)*. Fourth, the GC exacerbation of ischemic injury appears to be most consistently demonstrated for global ischemia; in contrast, for focal ischemia due to middle cerebral artery occlusion, megadose administration of GCs has demonstrated both protection *(73)* and exacerbation of the damage *(74)*.

2.1. Clinical Relevance

As noted, a number of studies now suggest that the broadly deleterious effects of GCs apply to the primate and human hippocampus, as well as to the that of the rodent. Should that prove to be the case, a number of populations of individuals might be at risk for some of the adverse interactions between GCs and CI. One group would be individuals exposed to chronically elevated GC levels independent of their CI. This would include individuals prescribed high-dose regimens for any of a variety of maladies unrelated to CI. In 1990 for example, 16 million prescriptions for GCs were written, largely to control immune diseases or inflammation *(75)*. Another, if less dramatic example would be aged individuals. Not only do older individuals tend to have higher basal levels of circulating GCs, but they also have a decreased ability to moderate their physiological response to stressors *(76)*. Coupled with the increasing likelihood of older individuals experiencing cardiovascular accidents that produce CI, this makes the elderly population potentially vulnerable to the adverse effects of elevated GC exposure around the time of CI.

Probably of greater concern is the administration of high-dose GC concentrations in the immediate aftermath of CI. This is, of course, typically done to control poststroke edema. A number of authorities have been less than sanguine about the efficacy of GCs to control the edema under those circumstances, owing to the fact that the anti-inflammatory effects of GCs are most relevant to vascular edema, whereas poststroke edema is typically cellular in nature. The findings regarding GC neuroendangerment emphasizes even further the advantages of using nonsteroidal anti-inflammatory compounds in those situations.

Finally, disease is itself a stressful event, and GC concentrations are elevated for up to 2 wk after an episode of CI *(67,77–79)*. Increased GC concentrations in plasma of CI patients and experimental animals have been correlated with increased mortality and decreased functional recovery *(62,65,78–80)*. Moreover, longitudinal data indicate that stroke is associated with major depression within 2–3 yr *(81,82)*, that appears to correlate with hypercortisolism *(83)* and reactivity to the dexamethasone suppression test within 1 wk after a CI event *(78,84)*. Whereas these studies are correlative, the experimental literature suggests that the considerable GC stress-response at the time of CI might contribute to the damage in these individuals. In support of this possibility, adrenalectomy or pharmacologic blockade of GC synthesis decreases the extent of hippocampal, cortical, or striatal damage after ischemia *(18,65,74,85)*. These abnormalities of regulatory control, and release of GCs both during and after CI, appear to support the conclusions reached in laboratory studies—that GCs can be detrimental to functional and cognitive recovery both short and long-term.

The interaction between GCs and ischemic injury, reviewed here at length, is not the only example of a hormone modulating the severity of ischemic damage to the brain. The remainder of the chapter is devoted to reviewing the smaller, but in some cases still compelling literatures suggesting these hormone/insult interactions.

3. CORTICOTROPIN-RELEASING FACTOR AS NEUROENDANGERING

Corticotropin-releasing factor (CRF, also often known as CRH, for corticotropin-releasing hormone) was the first hypothalamic factor whose existence was inferred from physiological evidence. Its actual characterization, however, waited decades, in part because of the complexity of its structure *(86)*. As has been the case following the discovery of many of the hypothalamic peptides, CRF has turned out to play a variety of neurotransmitter or neuromodulatory roles within the CNS, in addition to its primary neuroendocrine role of stimulating the release of ACTH from the pituitary. Most broadly, CRF mediates the activation of numerous projections to the sympathetic nervous system. This is quite striking, in that this peptide thus serves a central role in integrating the adrenocortical and autonomic branches of the stress–response *(87)*. As part of the visceral and behavioral effects of CRF within the brain, the peptide is a potent excitatory agent, causing electrophysiological excitation and even epileptiform activity in the hippocampus, cortex, and other structures *(88–91)*.

It is the marked excitatory potential of CRF that makes it relevant to ischemic neuronal injury. As emphasized in the section on GCs, all necrotic injuries have strong energetic components, and a substantial part of the protective effects of hypothermia and barbiturate treatment is the fact that they decrease energy expenditure in necrotically-endangered neurons—put colloquially, a neuron is more likely to survive if it works less following an insult. This logic prompted tests of whether CRF antagonists might be neuroprotective following necrotic insults. To date, it has been shown that such antagonists protect the hippocampus against global ischemia and increase postischemic EEG recovery *(92)*, as well as protect the cortex and basal ganglia against middle cerebral artery occlusion *(93,94)*. Furthermore, such antagonists decrease glutamatergic damage to the striatum *(93)* and hippocampus (in prep.). The magnitude of the protection is quite striking, in that antagonists reduce damage by approx 50%.

These findings suggest that CRF adds to neuronal injury during necrotic insults. In support of this, focal ischemia causes a 2.5-fold increase in CRF mRNA in the ischemic cortex *(94)*. How might CRF worsen injury? Insofar as neuroendocrine CRF ultimately causes GC secretion, this may be the route for neuroendangerment. However, one of these studies presented circumstantial evidence that this was not the route of CRF's action in this case *(93)*; this needs to be demonstrated more explicitly. Other peripheral factors could come into play—given the role of CRF in mediating sympathetic arousal, it can raise body temperature and, if released prior to the insult, can also raise preischemic circulating glucose concentrations, both potential routes for aggravating injury. However, control data suggest that the CRF antagonists were neuroprotective without changing these peripheral parameters *(93)*. A critical means of eliminating such peripheral mechanisms of CRF action would be to show that the peptide worsens ischemic injury in vitro. This was tested in a single study by the same group that made the initial in vivo observation concerning CRF being deleterious. The outcome of this in vitro study, however, was highly confusing. Hippocampal slices were made anoxic, with the endpoint being electrophysiological recovery in the postischemic period. In agreement with the in vivo literature, a CRF antagonist facilitated recovery of the slices; however, unexpectedly, CRF itself did the same *(95)*. In our opinion, the authors labored valiantly but ultimately unsuccessfully to explain this paradox; the relevance of this report to the in vivo picture remains unsettled.

Insofar as the in vivo studies suggest that CRF contributes to ischemic damage, it is most likely to occur via the excitatory effects of the peptide. This will bias ischemic tissue towards the greater accumulation of synaptic excitatory amino acids, and of postsynaptic calcium. This must be tested more directly.

In addition to exploring those issues, future studies should focus on two additional facets. First, none of the cited studies involved the administration of the CRF antagonists only in the period *after* the insult; in contrast, studies involved administration both pre- and post. Postinsult efficacy is, of course, a prerequisite for any eventual therapeutic use of any such compounds. Second, newer generations of antagonists must be developed that can be delivered to the brain more readily than via the ventricles, and with minimal peripheral side-effects.

4. ESTROGENS AS NEUROPROTECTIVE

One of the most exciting and provocative findings in Alzheimer's disease research in recent years has been the fact that postmenopausal estrogen replacement therapy very markedly reduces the risk of the disease *(96,97)*. This has generated the question of whether estrogen can protect against necrotic insults as well. This question is prompted by the demonstration that the amyloid β-peptide, which appears to play a role in the neurodegeneration seen in Alzheimer's disease, damages neurons through calcium mobilization and oxygen radical generation. The convergence of such toxicity with that provoked by excitatory amino acids is made more explicit by there being a toxic synergy between the amyloid peptide and glutamatergic excitotoxins *98–101*.

In support of these speculations, a recent and careful report has shown that estrogen protects cultured hippocampal neurons against glutamate, hypoglycemia, and a potent pro-oxidant *(99)*. Significant, although lesser protection was afforded by progesterone, whereas other steroids such as testosterone, aldosterone and vitamin D had no effect; in support of the GC endangerment story, the authors observed that corticosterone significantly worsened the toxicity of these insults. The estrogenic protection was quite potent, causing about a fivefold increase in the LD50 for glutamate.

What mechanisms might account for this protection? The first possibility is that estrogen acts as an antioxidant, specifically intercalating into membranes and stabilizing them against peroxidative attack. All steroids have this potential, independent of their classical receptor-mediated, genomic effects, and estrogen is one of the more potent steroidal antioxidants *(102–104)*. In support of this scenario, the estrogenic protection shown by Goodman and colleagues required estrogen concentrations quite a bit higher than the K_d of the estrogen receptor; more importantly, the effect occurred even in the presence of protein synthesis inhibitors, indicating a nontraditional steroid effect.

As a second mechanism, estrogen blunts glutamate-induced calcium mobilization in postsynaptic neurons *(102)*. This could be owing to a direct, receptor-independent effect of estrogen on calcium dynamics. Alternatively, it may be secondary to estrogen's antioxidant effects, as speculated by the authors. Oxygen radicals can increase cytosolic calcium concentrations in neurons through a number of mechanisms *(105)*. However, Goodman and coworkers did not show that estrogen reduces oxygen radical levels, but rather reduces peroxidative lipid damage. Although there has not yet been a direct demonstration that such lipid damage elevates cytosolic calcium concentrations, the authors emphasize, reasonably, that an intact and functional membrane would be an important prerequisite for calcium homeostasis.

Finally, estrogen has long been known to promote neuronal sprouting and arborization, and to play a dynamic role in the remodeling of the brain *(106)*. As a likely mediating mechanism, estrogen induces a variety of neurotrophins, and will potentiate their induction following a necrotic insult *(107,108)*. The protective potential of neurotrophins following such insults has been demonstrated *(109)* and could be a route by which estrogen protects neurons. However, such estrogenic actions are receptor- and genomically-mediated, in contrast to the protection shown in the present study. Thus, one can tentatively conclude that the protection by estrogen uncovered by Goodman and colleagues most likely arose from its antioxidant actions and ability to reduce calcium mobilization.

5. TESTOSTERONE AS NEUROPROTECTIVE

A somewhat peculiar footnote appeared recently concerning GC neuroendangerment. The deleterious effects of GCs become more pronounced in aged brains; for example, a regime of repeated stressors that fail to cause damage to the hippocampus in young rats will do so in aged subjects *(16)*. Prompted by this, Mizoguchi and colleagues *17)* reported that testosterone administration protected male rats from stress-induced degeneration in the hippocampus. The authors speculated that the declining concentrations of androgens with age would thus be the explanation for why GC neurotoxicity becomes more pronounced in older animals.

This interesting idea has not held up, however, in our opinion. First, GC neurotoxicity and neuroendangerment occur in the female hippocampus as well as the male *(23)*. Given the neuroprotective potential of estrogen, one could argue by extension that the declining levels of estrogen in the aged female rat enhances hippocampal vulnerability, much as posited for testosterone and the male; this has not been directly demonstrated. As a second problem, there seems to be no particularly compelling cellular mechanism of action by which testosterone might serve this protective role, and a number of in vitro studies have failed to note protective effects of testosterone against necrotic insults *(102,110)*. Third, the picture of dramatic declines in testosterone concentrations with age is somewhat a myth in gerontology, and free testosterone fractions decline only mildly from middle-age on *(111)*. Finally, a recent study failed to replicate the testosterone-dependency of GC neurotoxicity *(112)*. Therefore, we tentatively feel that any role played by testosterone in modulating ischemic injury is likely to be a small one, at best.

6. INSULIN AS NEUROPROTECTIVE

The ability of preischemic hyperglycemia to worsen neurological outcome, once one of the most challenging paradoxes in neurology, has come to be a cornerstone of our understanding of ischemia and the possible role of anaerobic acidosis in causing ischemic injury. The glycemic literature that emerged in the 1980s generated the obvious prediction that preischemic administration of insulin, insofar as it would lower glucose concentrations at the time of the ischemia, should be neuroprotective. A spate of papers since then has indeed demonstrated insulin's neuroprotective potential against global ischemic models *(113–118)* as well as against middle cerebral artery occlusion *(119)*. In these studies, lowering glucose concentrations to approximately the 3–5 m*M* range proved quite protective, as assessed by the size or severity of the lesion produced, the survival rate, the incidence of postischemic seizures, neurologic deficit scores in general, and cognitive function; in some instances *(118)*, ischemic animals administered insulin were statistically indistinguishable from control animals without ischemia.

As with most facets of physiology and pharmacology, an inverse-U function occurs, in that higher insulin concentrations, in which resulting glucose concentrations were <3 m*M*, exacerbated ischemic damage *(114,115)*. This could readily be interpreted with a modification of the prevailing wisdom at the time. A conclusion from the hyperglycemia paradox was that, in the case of ischemia, energy could be deleterious, if it was the wrong kind of energy (i.e., of a type that promoted anaerobic acidosis). The lesson of the higher-dose insulin studies was that energy was still essential to a neuron, and the energetic consequences of severe hypoglycemia could more than offset the protective effects.

As more understanding emerged concerning the effects of insulin during ischemia, it became apparent that its neuroprotective actions were, in fact, independent of its glycemic effects. This came around the same time that there was some waning of enthusiasm for the notion that ischemic injury was exclusively, or even predominantly mediated by anaerobic proton production *(110)*. As the first piece of evidence that insulin's actions did not require preischemic hypoglycemia, it was discovered that the hormone was protective even if administered during the postischemic reperfusion period *(115,116,118)*. And as equally compelling evidence, the protective effects of preischemic insulin administration still occurred, to at least as great of an extent, even if the hypoglycemia was artificially reversed *(113,117)*. These unexpected findings suggested that insulin's neuroprotective effects arose from its actions within the central nervous system.

Insulin receptors can be found throughout the nervous system, and the peptide has a variety of neurotransmitter and/or neuromodulatory roles *(120)*. Thus, the emphasis in the field shifted to understanding the relevance of those roles to the neuroprotection. The strongest hint is insulin's generally inhibitory effects on neuronal electrophysiology *(121,122)*, as well as its ability to inhibit cerebral metabolism *(123)*. As the most likely mechanism underlying electrophysiologic inhibition, insulin inhibits astrocyte uptake of GABA *(124)*, and causes a marked increase in extracellular GABA concentrations in the hippocampus during ischemia *(125)*; both of these effects are independent of its glycemic actions. An increase in GABAergic tone during a necrotic insult is a well-documented mechanism for neuroprotection, given GABA's capacity to inhibit excitatory amino acid release and suppress seizure activity. As one possible confound, the same microdialysis study that reported that insulin enhanced extracellular GABA accumulation during ischemia failed to note an insulin effect on glutamate accumulation *(125)*.

A number of other central actions of insulin might be relevant to its neuroprotection. The peptide stimulates Na^+/K^+-ATPase activity *(126)*, which is likely to decrease neuronal excitability and to enhance postischemic ionic homeostasis. In addition, insulin enhances norepinephrine release and inhibits its reuptake *(127,128)*. Both effects will enhance noradrenergic tone, a demonstrated route of protection during ischemia *(129)*; as one complication, however, these insulin effects are not particularly robust in brain regions vulnerable to ischemia. Finally, insulin-like growth factors can be neuroprotective *(130)*; whether insulin itself can act in a neurotrophic manner during necrotic insults remains to be demonstrated.

Thus, a number of possible mechanisms could give rise to insulin's direct protective effects in the ischemic brain. As a final complication, whereas insulin protects against both global ischemia and infarct, the protection in the latter case appears to be entirely a function of its glycemic effects *(119)*, in that insulin administration to artificially maintained normoglycemic rats was not neuroprotective. This singular observation is strengthened by the fact that authors were the same who originally demonstrated the hypoglycemia-independent effects of insulin on global ischemia. This underlines the heterogeneity of mechanisms of ischemic injury, and the validity of dichotomizing between global and focal ischemia *(131)*. These general findings regarding insulin and global ischemia may reflect even more broadly salutary effects of this hormone in the endangered brain; for example, a recent paper reports that insulin improves cognitive performance in Alzheimer's patients, independent of its glycemic effects *(132)*.

7. SYMPATHETIC CATECHOLAMINES AND ISCHEMIC INJURY

As just discussed, the discovery that preischemic hyperglycemia worsens neurologic outcome prompted the examination of insulin's neuroprotective potential. A similar line of reasoning led to the examination of the effects of sympathetic catecholamines. As classically "counter-regulatory" hormones, epinephrine and norepinephrine antagonize insulin's actions, raising circulating glucose concentrations by stimulating glycogenolysis and gluconeogenesis. Moreover, given the cerebrovascular effects of catecholamines, they not only cause hyperglycemia but increase cerebral perfusion rates and thus promote the delivery of glucose to the brain. This generates the simple prediction that elevated preischemic catecholamine concentrations should add to the resulting damage, and pharmacologic blunting of the sympathetic system preischemia should be neuroprotective. This has now been shown in a number of models of incomplete brain ischemia *(133–139)*.

More recent studies have shown that although preischemic hyperglycemia does have the potential to worsen ischemic damage, postischemic glucose supplementation can, in fact, be quite protective *(110)*. This prompted the demonstration that administration of catecholamines in the postischemic reperfusion period *decreases* damage *(66)*. (It should be noted that this finding superficially resembles that of Blomqvist and colleagues *[129]*, who reported that activation of the noradrenergic locus cereleus projection to the forebrain also decreased ischemic damage. However, it must be recalled that the circulating catecholamines do not penetrate the blood–brain barrier, and any of their effects must arise from their peripheral actions.) Whereas only a single study has shown postischemic catecholamines to be protective, this strikes us as a particularly credible report, coming from the laboratory of Siesjo and colleagues in Lund. The authors noted how ischemic insults stimulate tremendous sympathetic output *(140)*, and speculated that the mobilization of this stress-response during ischemia can be viewed as a defensive adaptation on the part of the body. This is in sharp contrast to the fact that the stressor of ischemia also stimulates CRF activation and GC release, both of which, as noted, add to damage. This seeming contradiction will be discussed in Section 9.

8. HYPOTHYROIDISM AS A ROUTE OF NEUROPROTECTION

Thyroid hormones are among the most versatile endocrine messengers in the body, in terms of the extent of tissues over which they exert effects, and the broad range of cellular events they can influence. A number of studies have demonstrated that ischemic damage to the liver and kidney is reduced in animals surgically or chemically rendered hypothyroid *(141,142)*. Given the general picture of postischemic tissue as being energetically fragile, it is not surprising that elimination of a hormone that markedly increases metabolism should protect these organs. In line with that, these studies indicated that hypothyroidism also diminished oxygen radical generation in these organs.

These observations suggested an obvious extension to the brain, insofar as hypothyroid individuals also have a suppression of cerebral metabolism *(143)*. In support of this, a recent report shows that hypothyroidism in the gerbil also protects the ischemic forebrain, decreasing damage anywhere from 50 to 100% in a number of brain regions *(144)*. Importantly, the authors show that this protection is not secondary to hypothermia, a likely if uninformative route of protection.

This protection by hypothyroidism most plausibly arises from the decrease in metabolism, and would be commensurate with the protective effects of barbiturates and hypothermia. As a proximal outcome, ischemic tissue in hypothyroid animals, with their lower metabolic rate, would likely produce less oxygen radicals. However, in contrast to the literature regarding ischemia in peripheral organs, hypothyroidism did not change indices of oxygen radical generation or levels of some key antioxidants in the postischemic gerbils (*see* ref. *144*). Perhaps related to this, thyroid hormones do not exert identical effects on oxidative metabolism in neural and nonneural tissues *(145)*. Regardless of whether there is a specific change in the profiles of reactive oxygen species, the decreased metabolic rate will leave neurons with greater energy stores to control the fluxes of glutamate and calcium. In support of this, a subsequent microdialysis study from the same group indicated that hypothyroid animals had less accumulation of extracellular glutamate during ischemia than control gerbils *(144)*. As an additional and indirect peripheral mechanism, hypothyroidism causes hypoglycemia, and enhanced insulin release *(143)* as reviewed above, both of these effects can protect ischemic brain tissue.

This small literature suggests that thyroid hormones, most probably by enhancing both peripheral and CNS metabolism, will exacerbate ischemic brain damage. Numerous investigators have emphasized the increasing vulnerability of the aging brain to necrotic insults, and have sought to understand this. Interestingly, one might predict that the decreasing thyroid tone with old age *(146)* should be a source of protection for the aged brain. This remains to be examined.

9. CONCLUSIONS

In this review, we have considered the capacity of GCs and, more briefly, a number of other hormones to modulate ischemic neuronal injury. Broadly, current findings suggest that GCs, CRF, preischemic catecholamines, and thyroid hormones can worsen damage, whereas protection is afforded by estrogen, insulin, and postischemic catecholamines.

Several themes are worth emphasizing. One is how in some cases, the modulating effect of a hormone on ischemic injury does not arise from the traditional peripheral action of that hormone, but rather from more recently-discovered direct actions within the brain. Examples of this would include the glycemia-independent actions of insulin in decreasing damage, and the direct effects of GCs in increasing damage.

A second theme is the varied mechanisms of action by which hormones influence ischemic outcome, commensurate with the broad range of hormone actions in general. Some patterns that emerge are that hormones can modulate ischemic damage insofar as they change circulating glucose concentrations (sympathetic catecholamines, insulin in the case of focal infarct), directly alter the excitability and energetic profile of endangered neurons (CRF, insulin, GCs), or alter facets of oxygen radical production and oxidative damage (estrogen, GCs). Directly related to the issue of altering circulating glucose concentrations is the question of *when* the particular hormones are exerting their effects, as preischemic and postischemic hyperglycemia can have opposing effects on ischemic injury.

Another issue to be considered is the role of "stress" in modulating ischemic injury. Both physical and psychological stressors will mobilize CRF release within the CNS, and increase circulating catecholamine and GC concentrations. If these all occur prior to the ischemic insult, all are likely to worsen the outcome. However, if this collective stress–response is mobilized *only* at the time of the insult (and ischemia, as noted, triggers a considerable activation of all these branches of the stress-response), the GC and CRF components are likely to add to injury, whereas the sympathetic component is likely to decrease damage. Which components will prevail? As is often the case in physiology, the only way to tell the relative weights of these facets is to combine them into a whole; it remains to be tested whether the integrated, multiendocrine stress–response, if mobilized only at the time of an ischemic insult, adds to or lessens the resulting damage. This seems an important study to carry out.

This issue of the effects of stress-induced secretion of these hormones brings up another issue. The hope in a review such as this is to eventually derive useful therapies with the knowledge of how various endocrine factors modulate ischemic injury. The most common way in which such therapies are conceptualized is to detect something protective that the body does at such times and to improve on it—for example, to develop insulin analogs that are even more neuroactive than insulin itself is, and to use that therapeutically. The possibility that the endogenous stress-response can add to ischemic damage raises the specter of interventions that prevent the body from having a typical and maladaptive response to ischemia. Thus, the use of CRF antagonists or GC synthesis inhibitors would represent a more novel, but no less hopeful approach to therapy.

Finally, the ability to devote an entire chapter in an ischemia volume to endocrinology is a hopeful sign in and of itself; our knowledge of the basic cascade of neuron death following ischemia has now matured sufficiently that we can even consider the modulators of that cascade. Hopefully, this will pave the way for future interventions.

REFERENCES

1 Choi, D., Excitotoxic cell death, *J. Neurobiol.*, **23** (1992) 1261–1276.
2 Kuroiwa, T. and Okeda, R., Neuropathology of cerebral ischemia and hypoxia: recent advances in experimental studies on its pathogenesis, *Path. Internat.*, **44** (1994) 171–181.
3 Munck, A., Guyre, P., and Holbrook, N., Physiological actions of glucocorticoids in stress and their relation to pharmacological actions., *Endocrine Rev.*, **5** (1984) 25–43.
4 Sapolsky, R., *Stress, the Aging Brain, and the Mechanisms of Neuron Death*, The MIT Press, Cambridge, MA, 1992.
5 Aus der Muhlen, K. and Ockenfels, H., Morphologische veranderungen im diencephalon und telencephalon nach storngen des regelkreises adenohypophyse-ncbennicrenrinde III. Ergebnisse beim meerschweinchen nach verabreichung von cortison und hydrocortison, *Z. Zellforsch.*, **93** (1969) 126–140.
6 Levy, A., Dachir, S., Arbel, I., and Kadar, T., Aging, stress and cognitive function, *Ann. NY Acad. Sci.*, **717** (1994) 79–88.
7 Levy, A., Dachir, S., Kadar, T., Shukitt-Hale, B., Stillman, M., and Lieberman, H., Nimodipine, stress, and central cholinergic function, *Drugs Develop.*, **2** (1993) 223–233.
8 Magarinos, A. and McEwen, B., Blockade of glucocorticoid synthesis by cyanoketone prevents chronic stress-induced dendritic atrophy of hippocampal neurons in the rat, *Soc. Neurosci. Abs.*, **19** (1993) 168.
9 Sapolsky, R., Krey, L., and McEwen, B., Prolonged glucocorticoid exposure reduces hippocampal neuron number; implications for aging, *J. Neurosci.*, **5** (1985) 1221–1227.
10 Talmi, M., Carlier, E., and Soumireu-Mourat, B., Similar effects of aging and corticosterone treatment on mouse hippocampal function, *Neurobiol. Aging*, **14** (1993) 239–245.
11 Woolley, C., Gould, E., Frankfurt, M., and McEwen, B., Naturally occurring fluctuation in dendritic spine density on adult hippocampal pyramidal neurons, *J. Neurosci.*, **10** (1990) 4035–4039.
12 Sapolsky, R., Uno, H., Rebert, C., and Finch, C., Hippocampal damage associated with prolonged glucocorticoid exposure in primates, *J. Neurosci.*, **10** (1990) 2897–2903.
13 Axelson, D., Doraiswamy, P., McDonald, W., Boyko, O., Tupler, L., Patterson, C. N., Ellinwood, J., and Krishnan, K., Hypercortisolemia and hippocampal changes in depression, *Psychiatry Res.*, **47** (1993) 163–173.
14 Starkman, M., Gebarski, S., Berent, S., and Schteingart, D., Hippocampal formation volume, memory dysfunction, and cortisol levels in patients with Cushing's syndrome, *Biol. Psychiat.*, **32** (1992) 756–764.
15 McEwen, B., de Kloet, R., and Rostene, W., Adrenal steroid receptors and actions in the nervous system, *Physiol. Rev.*, **66** (1986) 1121–1187.

16 Kerr, D., Campbell, L., Applegate, M., Brodish, A., and Landfield, P., Chronic stress-induced acceleration of electrophysiologic and morphometric biomarkers of hippocampal aging, *J. Neurosci.*, **11** (1991) 1316–1324.

17 Mizoguchi, K., Tatsuhide Kunishita, De-Hua Chui, and Takeshi Tabira, Stress induces neuronal death in the hippocampus of castrated rats, *Neurosci. Lett.*, **138** (1992) 157.

18 Sapolsky, R. and Pulsinelli, W., Glucocorticoids potentiate ischemic injury to neurons: Therapeutic implications, *Science*, **229** (1985) 1397–1401.

19 Uno, H., Flugge, G., Thieme, C., Johren, O., and Fuchs, E., Degeneration of the hippocampal pyramidal neurons in the socially stressed tree shrew, *Soc. Neurosci. Abs.*, **17** (1991) 52.20.

20 Uno, H., Tarara, R., Else, J., Suleman, M., and Sapolsky, R., Hippocampal damage associated with prolonged and fatal stress in primates, *J. Neurosci.*, **9** (1989) 1705–1712.

21 Landfield, P., Baskin, R., and Pitler, T., Brain-aging correlates; retardation by hormonal-pharmacological treatments, *Science*, **214** (1981) 581–585.

22 Meaney, M., Aitken, D., Bhatnager, S., Berkel, C. v., and Sapolsky, R., Effects of neonatal handling on age-related impairments associated with the hippocampus, *Science*, **239** (1988) 766–769.

23 Meaney, M., Aitken, D., and Sapolsky, R., Postnatal handling attenuates neuroendocrine, anatomical and cognitive dysfunctions associated with aging in female rats, *Neurobiol. Aging*, **12** (1991) 31–40.

24 Pundits, (1990s).

25 Sapolsky, R., Glucocorticoids, stress and exacerbation of excitotoxic neuron death, *Sem. Neurosci.*, **6** (1994) 323–331.

26 Bagley, J. and Moghaddam, B., Rapid sampling of extracellular glutamate in the prefrontal cortex and hippocampus in response to repeated stress: effect of diazepam, *Soc. Neurosci. Ab.*, **21** (1995) 463.

27 Lowy, M., Wittenberg, L., and Novotney, S., Adrenalectomy attenuates kainic acid-induced spectrin proteolysis and heat shock protein 70 induction in hippocampus and cortex, *J. Neurochem.*, **63** (1994) 886–894.

28 Moghaddam, B., Stress preferentially increases extraneuronal levels of excitatory amino acids in the prefrontal cortex: Comparison to hippocampus and basal ganglia, *J. Neurochem.*, **60** (1993) 1650–1656.

29 Moghaddam, B., Boliano, M., Stein-Behrens, B., and Sapolsky, R., Glucocorticoids mediate the stress-induced accumulation of extracellular glutamate, *Brain Res.*, **655** (1994) 251–256.

30 Chou, Y., Lin, W., and Sapolsky, R., Glucocorticoids increase extracellular [3H]D-aspartate overflow in hippocampal cultures during cyanide-induced ischemia, *Brain Res.*, **654** (1994) 8–15.

31 Dugan, L. and Choi, D., Excitotoxicity, free radicals, and cell membrane changes, *Ann. Neurol.*, **35** (1994) S17–21.

32 Orrenius, S., Burkitt, M., Kass, G., Dypbukt, J., and Nicotera, P., Calcium ions and oxidative cell injury, *Ann. Neurol.*, **32** (1992) S33–S42.

33 Elliott, E., Mattson, M., Vanderklish, P., Lynch, G., Chang, I., and Sapolsky, R., Corticosterone exacerbates kainate-induced alterations in hippocampal tau immunoreactivity and spectrin proteolysis in vivo, *J. Neurochem.*, **61** (1993) 57–64.

34 Lowy, M., Gault, L., and Yamamoto, B., Adrenalectomy attenuates stress-induced elevations in extracellular glutamate concentrations in the hippocampus, *J. Neurochem.*, **61** (1993) 1957–1960.

35 Stein-Behrens, B., Mattson, M., Chang, I., Yeh, M., and Sapolsky, R., Stress exacerbates neuron loss and cytoskeletal pathology in the hippocampus, *J. Neurosci.*, **14** (1994) 5373–5379.

36 Dykens, J., Isolated cerebral and cerebellar mitochondria produce free radicals when exposed to elevated Ca^{++} and Na^+: Implications for neurodegeneration, *J. Neurochem.,* **63** (1994) 584–591.

37 McIntosh, L. and Sapolsky, R., Glucocorticoids increase oxidative parameters in hippocampal and cortical neuronal cultures exposed to adriamycin, *Exp. Neurol.,* (1996), in press.

38 Sohal, R., and Dubey, A., Mitochondrial oxidative damage, hydrogen peroxide release, and aging, *Free Rad. Med.,* **16** (1994) 621–626.

39 Auer, R. and Siesjo, B., Biological differences between ischemia, hypoglycemia, and epilepsy, *Ann. Neurol.,* **24** (1988) 699–723.

40 Lee, W. and Bondy, C., Ischemic injury induces brain glucose transporter gene expression, *Endocrinology,* **133** (1993) 2540–2544.

41 McDougal, D., Carter, J., Pusateri, M., Manchester, J., and Lowry, O., Glucose metabolism assessed with 2-deoxyglucose and the effect of glutamate in subdivisions of rat hippocampal slices, *J. Neurochem.,* **59** (1992) 1915–1921.

42 Horner, H., and Sapolsky, D. P. R., Glucocorticoids inhibit glucose transport in cultured hippocampal neurons and glia, *Neuroendocrinology,* **52** (1990) 57–62.

43 Kadekaro, M., Masanori, I., and Gross, P., Local cerebral glucose utilization is increased in acutely adrenalectomized rats, *Neuroendocrinology,* **47** (1988) 329–335.

44 Virgin, C., Ha, T., Packan, D., Yang, S., Horner, H., and Sapolsky, R., Glucocorticoids inhibit glucose transport and glutamate uptake in hippocampal astrocytes: Implications for glucocorticoid neurotoxicity, *J. Neurochem.,* **57** (1991) 1422–1428.

45 Horner, H., Munck, A., and Lienhard, G., Dexamethasone causes translocation of glucose transporters from the plasma membrane to an intracellular site in human fibroblasts, *J. Biol. Chem.,* **262** (1987) 17,696–17,703.

46 Garvey, W., Huecksteadt, T., Lima, F., and Birnbaum, M., Expression of a glucose transporter gene cloned from brain in cellular models of insulin resistance: Dexamethasone decreases transporter mRNA in primary cultured adipocytes, *Mol. Endrocrinol.,* **3** (1989) 1132–1141.

47 Lawrence, M. and Sapolsky, R., Glucocorticoids accelerate ATP loss in cultured hippocampal neurons, *Brain Res.,* **646** (1994) 303–308.

48 Tombaugh, G. and Sapolsky, R., Corticosterone accelerates hypoxia-induced ATP loss in cultured hippocampal astrocytes, *Brain Res.,* **588** (1992) 154–159.

49 Bennett, M., Lehman, J., Mlady, G., and Rose, G., Synergy between corticosterone and sodium azide in producing a place learning deficit in the Morris water maze, *Soc. Neurosci. Abs.,* **19** (1993) 188.

50 Sapolsky, R., Glucocorticoid toxicity in the hippocampus: Reversal by supplementation with brain fuels, *J. Neurosci.,* **6** (1986) 2240–2246.

51 Sapolsky, R., Packan, D., and Vale, W., Glucocorticoid toxicity in the hippocampus: in vitro demonstration, *Brain Res.,* **453** (1988) 367–373.

52 Tombaugh, G., Yang, S., Swanson, R., and Sapolsky, R., Glucocorticoids exacerbate hypoxic and hypoglycemic hippocampal injury in vitro: Biochemical correlates and a role for astrocytes, *J. Neurochem.,* **59** (1992) 137–145.

53 Tombaugh, G. and Sapolsky, R., Evolving concepts about the role of acidosis in ischemic neuropathology, *J. Neurochem.,* **61** (1993a) 793–807.

54 Ames, A. and Li, Y., Energy requirements of glutamatergic pathways in rabbit retina, *J. Neurosci.,* **12** (1992) 4234–4242.

55 Ames, A., Maynard, K. I., and Kaplan, S., Protection against CNS ischemia by temporary interruption of function-related processes of neurons, *J. Cereb. Blood Flow Metab.,* **15** (1995) 433.

56 Beal, M., Does impairment of energy metabolism result in excitotoxic neuronal death in neurodegenerative illnesses?, *Ann. Neurol.,* **31** (1992) 119–130.

57 Elliott, E. and Sapolsky, R., Corticosterone impairs hippocampal neuronal calcium regulation: possible mediating mechanisms, *Brain Res.,* **602** (1993) 84–90.

58 Stein-Behrens, B., Elliott, E., Miller, C., Schilling, J., Newcombe, R., and Sapolsky, R., Glucocorticoids exacerbate kainic acid-induced extracellular accumulation of excitatory amino acids in the rat hippocampus, *J. Neurochem.,* **58** (1992) 1730–1735.

59 Packan, D. and Sapolsky, R., Glucocorticoid endangerment of the hippocampus: Tissue, steroid and receptor specificity, *Neuroendocrinology,* **51** (1990) 613–620.

60 Brooke, S., Miller, J., and Sapolsky, R., Exacerbation of gp 120 effects on rat primary neuronal cultures by cortocosterone, *Soc. Neurosci. Ab.,* **20** (1994) 1049.

61 Hall, E., *Steroids and neuronal destruction or stabilization*, Wiley, 1990, pp. 206–215.

62 Koide, T., Wieloch, T., and Siesjo, B., Chronic dexamethasone pretreatment aggravates ischemic neuronal necrosis, *J. Cereb. Blood Flow Metab.,* **6** (1986a) 395–408.

63 Uhler, T., Frim, D., Pakzaban, P., and Isacson, O., The effects of megadose methylprednisolone and U-78517F on toxicity mediated by glutamate receptors in the rat neostriatum, *Neurosurgery,* **34** (1994) 122–128.

64 Talmi, M., Carlier, E., Bengelloun, W., and Soumireu-Mourat, B., Chronic RU 486 treatment reduces age-related alterations of mouse hippocampal function, *Neurobiol. Aging,* **17** (1996) 9–14.

65 Morse, J., and Davis, J., Regulation of ischemic hippocampal damage in the gerbil: Adrenalectomy alters the rate of CA1 cell disappearance, *Exptl. Neurol.,* **110** (1990) 86–94.

66 Koide, T., Wieloch, T., and Siesjo, B., Circulating catecholamines modulate ischemic brain damage, *J. Cereb. Blood Flow Metab.,* **6** (1986b) 559–565.

67 Miller, G. and Davis, J., Post-ischemic surge in corticosteroids aggravates ischemic damage to gerbil CA1 pyramidal cells, *Soc. Neurosci. Abs.,* **138** (1991) 756.

68 Tuor, U., Chumas, P., and Bigio, M. R. D., Prevention of hypoxic-ischemic damage with dexamethasone is dependent on age and not influenced by fasting, *Exptl. Neurol.,* **132** (1995) 116–122.

69 Barks, J., Post, M., and Tuor, U., Dexamethasone prevents hypoxic-ischemic brain damage in the neonatal rat, *Ped. Res.,* **29** (1991) 558–563.

70 Chumas, P., Bigio, M. D., Drake, J., and Tuor, U., A comparison of the protective effect of dexamethasone to other potential prophylactic agents in a neonatal rat model of cerebral hypoxia-ischemia, *J. Neurosurg.,* **79** (1993) 414–420.

71 Tuor, U., Simone, C., Arellano, R., Tanswell, K., and Post, M., Glucocorticoid prevention of neonatal hypoxic-ischemia damage: role of hyperglycemia and antioxidant enzymes, *Brain Res.,* **604** (1993) 165–172.

72 Tuor, U., Simone, C., Barks, J., and Post, M., Dexamethasone prevents cerebral infarction without affecting cerebral blood flow in neonatal rats, *Stroke,* **24** (1993) 452–457.

73 de Courten-Myers, G., Kleinholz, M., Wagner, K., Xi, G., and Myers, R., Efficacious experimental stroke treatment with high-dose methylprednisolone, *Stroke,* **25** (1994) 487–492.

74 Smith-Swintosky, V., Pettigrew, L., Sapolsky, R., Phares, C., Craddock, S., Brooke, S., and Mattson, M., Metyrapone, an inhibitor of glucocorticoid production, reduces brain injury induced by focal and global ischemia and seizures, *J. Cereb. Blood Flow Metab.,* (1996), in press.

75 Eufemio, M., Advances in the therapy of osteoporosis—part V. Steroid-induced osteoporosis., *Ger. Med. Today,* **9** (1990) 41–56.

76 Seeman, T. and Robbins, R., Ageing and the hypothalamic-pituitary-adrenal response to challenge in humans, *Endocrine Rev.,* **15** (1994) 233–260.

77 Oka, M., Effect of cerebral vascular accident on the level of 17-hydroxycorticosteroids in plasma, *Acta Med. Scand.,* **156** (1956) 221–226.

78 Olsson, T., Urinary free cortisol excretion shortly after ischaemic stroke, *J Int Med,* **228** (1990) 177–181.

79 Olsson, T., Marklund, N., Gustafson, Y., and Näsman, B., Abnormalities at different levels of the hypothalamic-pituitary-adrenocortical axis early after stroke, *Stroke,* **23** (1992) 1573–1576.

80 Feibel, J., Hardy, P., and Campbell, R., Prognostic value of the stress response following stroke, *JAMA,* **238** (1977) 1374–1376.

81 Parikh, R., Lipsey, J., Robinson, R., and Price, T., Two-year longitudinal study of post-stroke mood disorders: Dynamic changes in correlates of depression at one and two years, *Stroke,* **18** (1987) 579–584.

82 Robinson, R., Starr, L., Lipsey, J., Rao, K., and Price, T., A two-year longitudinal study of post-stroke mood disorders: Dynamic changes in associated variables over the first six months of follow-up, *Stroke,* **15** (1984) 510–517.

83 Åström, M., Olsson, T., and Asplund, K., Different linkage of depression to hypercortisolism early versus late after stroke. A 3-year longitudinal study, *Stroke,* **24** (1993) 52–57.

84 Carroll, B., Feinberg, M., Greden, J., Tarika, J., Albala, A., Haskett, R., James, N., Kronfol, Z., Lohr, N., Steiner, M., Vigne, J. D., and Young, E., A specific laboratory test for the diagnosis of melancholia. Standardization, validation and clinical utility., *Arch. Gen. Psych.,* **38** (1981) 15–22.

85 Stein, B. and Sapolsky, R., Chemical adrenalectomy reduces hippocampal damage induced by kainic acid, *Brain Res.,* **473** (1988) 175–181.

86 Vale, W., Speiss, J., Rivier, C., and Rivier, J., Characterization of a 41 residue ovine hypothalamic peptide that stimulates the secretion of corticotrophin and b endorphin, *J. Neurosci.,* **9** (1981) 1705–1712.

87 Johnson, E., Kamilaris, T., Chrousos, G., and Gold, P., Mechanisms of stress: A dynamic overview of hormonal and behavioral homeostasis, *Neurosci. Biobehav. Rev.,* **16** (1992) 115–130.

88 Aldenhoff, J., Groul, D., Rivier, J., Vale, W., and Siggins, G., CRF decreases postburst hyperpolarizations and excites hippocampal neurones, *Science,* **221** (1983) 875–877.

89 Ehlers, C., Henrikson, S., Wang, M., Rivier, J., Vale, W., and Bloom, F., Corticotropin releasing factor produces increases in brain excitability and convulsive seizures in rats, *Brain Res.,* **278** (1983) 332.

90 Marrosu, F., Fratta, W., Carcangiu, P., Giaheddu, M., and Gessa, G., Localised epileptiform activity induced by murine CRF in rats, *Epilepsia,* **29** (1988) 369–373.

91 Siggins, G., Groul, D., Aldenhoff, J., and Pittman, Q., Electrophysiological actions of corticotropin releasing factor in the central nervous system, *Fed. Proc.,* **44** (1985) 237–247.

92 Lyons, M., Anderson, R., and Meyer, F., Corticotropin releasing factor antagonist reduces ischemic hippocampal neuronal injury, *Brain Res.,* **545** (1991) 339–342.

93 Strijbos, P., Relton, J., and Rothwell, N., Corticotropin-releasing factor antagonist inhibits neuronal damage induced by focal cerebral ischemia or activation of NMDA receptors in the rat brain, *Brain Res.,* **656** (1994) 405–408.

94 Wong, M., Loddick, S., Bongiorno, P., Gold, P., Souza, E. D., Licinio, J., and Rothwell, N., Is corticotrophin releasing hormone neurotoxic during focal cerebral ischemia?, *Soc. Neurosci. Abs.,* **21** (1995) 230.

95 Fox, M., Anderson, R., and Meyer, F., Neuroprotection by corticotropin releasing factor during hypoxia in the rat, *Stroke,* **24** (1993) 1072–1076.

96 Henderson, V., Paganini-Hill, A., Emanuel, C., Dunn, M., and Buckwalter, J., Estrogen replacement therapy in older women. Comparisons between Alzheimer's disease cases and nondemented control subjects, *Arch. Neurol.,* **51** (1994) 896–900.

97 Simpkins, J., Singh, M., and Bishop, J., The potential role for estrogen replacement therapy in the treatment of the cognitive decline and neurodegeneration associated with Alzheimer's disease, *Neurobiol. Aging,* **15** (1994) S195–S197.

  98 Behl, C., Davis, J., Lesley, R., and Schubert, D., Hydrogen peroxide mediates amyloid beta protein toxicity, *Cell,* **77** (1994) 817–827.

  99 Goodman, Y., Steiner, M., Steiner, S., and Mattson, M., Nordihydroguaiaretic acid protects hippocampal neurons against amyloid β-peptide toxicity, and attenuates free radical and calcium accumulation, *Brain Res.,* **654** (1994) 171–176.

 100 Hensley, K., Carney, J., Mattson, M., Aksenova, M., Harris, M., Wu, J., Floyd, R., and Butterfield, D., A model for β-amyloid aggregation and neurotoxicity based on free radical generation by the peptide: relevance to Alzheimer disease, *Proc. Natl. Acad. Sci. USA,* **91** (1994) 3270–3274.

 101 Mattson, M., Cheng, B., Davis, D., Bryant, K., Liederburg, I., and Rydel, R., β-amyloid peptides destabilize calcium homeostasis and render human cortical neurons vulnerable to excitotoxicity, *J. Neurosci.,* **12** (1992) 376–389.

 102 Goodman, Y., Bruce, A., Cheng, B., and Mattson, M., Estrogens attenuate and corticosterone exacerbates excitotoxicity, oxidative injury, and amyloid b-peptide toxicity in hippocampal neurons, *J. Neurochem.,* **66** (1996) 1836–1844.

 103 Parthasarathy, S., Morales, A., and Murphy, A., Antioxidant: a new role for RU-486 and related compounds, *J. Clin. Invest.,* **94** (1994) 1990–1995.

 104 Subbiah, M., Kessel, B., Agrawal, M., Rajan, R., Abplanalp, W., and Rymaszewski, Z., Antioxidant potential of specific estrogens on lipid peroxidation, *J. Clin. Endocrinol. Metab.* **77** (1993) 1095–1097.

 105 Araki, N., Greenberg, J., Uematsu, D., Sladky, J., and Reivich, M., Effect of superoxide dismutase on intracellular calcium in stroke, *J. Cereb. Blood Flow Metab.,* **12** (1992) 43–52.

 106 McEwen, B., Coirini, H., Westlind-Danielsson, A., Frankfurt, M., Gould, E., Schumacher, M., and Wooley, C., Steroid hormones as mediators of neural plasticity, *J. Steroid Biochem. Mol. Biol.,* **39** (1991) 223–232.

 107 Kito, S., Semba, J., Miyoshi, R., Shingo, A., and Shimizu, E., Effects of steroid hormones on growth factors in kainic acid-induced injured brain, *Soc. Neurosci. Ab.,* **21** (1995) 297.

 108 Singh, M., Meyer, E., and Simpkins, J., The effect of ovariectomy and estradiol replacement on brain-derived neurotrophic factor mRNA expression in cortical and hippocampal brain regions of female Sprague-Dawley rats, *Endocrinology,* **136** (1994) 2320–2324.

 109 Mattson, M., Lovell, M., Furukawa, K., and Markesbery, W., Neurotrophic factors attenuate glutamate-induced accumulation of peroxides, elevation of intracellular calcium concentration, and neurotoxicity and increase antioxidant enzyme activities in hippocampal neurons, *J. Neurochem.,* **65** (1995) 1740–1751.

 110 Tombaugh, G. and Sapolsky, R., Endocrine features of glucocorticoid endangerment in hippocampal astrocytes, *Neuroendocrinology,* **57** (1993b) 7–15.

 111 Harman, S. and Talb ert, G., Reproductive Aging. In C. Finch and E. Schneider (eds.) *Handbook of the Biology of Aging,* Van Nostrand Reinhold, New York, 1985, pp.

 112 Clark, A., Mitre, M., and Brinck-Johnsen, T., Anabolic-androgenic steroid and adrenal steroid effects on hippocampal plasticity, *Brain Res.,* **670** (1995) 64–69.

 113 Fukouka, S., Yeh, H., Mandybur, T., and Tew, J., Effect of insulin on acute experimental cerebral ischemia in gerbils, *Stroke,* **20** (1989) 396–399.

 114 LeMay, D., Gehua, L., Zelenock, G., and D'Alecy, L., Insulin administration protects neurologic function in cerebral ischemia in rats, *Stroke,* **19** (1988) 1411–1419.

 115 Voll, C. and Auer, R., The effect of postischemic blood glucose levels on ischemic brain damage in the rat, *Ann. Neurol.,* **24** (1988) 638–646.

 116 Voll, C., and Auer, R., Postischemic seizures and necrotizing ischemic brain damage: Neuroprotective effect of postischemic diazepam and insulin, *Neurology,* **41** (1991a) 423–428.

 117 Voll, C. and Auer, R., Insulin attenuates ischemic brain damage independent of its hypoglycemic effect, *J. Cereb. Blood Flow Metab.,* **11** (1991b) 1006–1014.

118 Voll, C., Whishaw, I., and Auer, R., Postischemic insulin reduces spatial learning deficit following transient forebrain ischemia in rats, *Stroke,* **20** (1989) 646–651.

119 Hamilton, M., Tranmer, B., and Auer, R., Insulin reduction of cerebral infarction due to transient focal ischemia, *J. Neurosurg.,* **82** (1995) 262–268.

120 Unger, J., Livingstone, J., and Moss, A., Insulin receptors in the central nervous system: Localization, signalling mechanisms and functional aspects, *Prog. Neurobiol.,* **36** (1991) 343–362.

121 Palovcik, R., Phillips, M., Kappy, M., and Raizada, M., Insulin inhibits pyramidal neurons in hippocampal slices, *Brain Res.,* **309** (1984) 187–191.

122 Saller, C. and Chiodo, L., Glucose suppresses basal firing in haloperidol-induced firing rate of central dopaminergic neurons, *Science,* **210** (1980) 1269–1271.

123 Grunstein, H., James, D., Storlien, L., Smythe, G., and Kraegen, E., Hyperinsulinemia suppresses glucose utilization in specific brain regions: In vivo studies using the euglycemic clamp in the rat, *Endocrinology,* **116** (1985) 604–610.

124 Bouhaddi, K., Thomopoulos, P., Fages, C., Khelil, M., and Tardy, M., Insulin effect on GABA uptake in astroglial primary cultures, *Neurochem. Res.,* **13** (1988) 1119–1124.

125 Shuaib, A., Ijaz, M., Waqar, T., Voll, C., Kanthan, R., Miyashita, H., and Liu, L., Insulin elevates hippocampal GABA levels during ischemia: This is independent of its hypoglycemic effect, *Neuroscience,* **67** (1995) 809–814.

126 Stahl, W., The Na, K-ATPase of nervous tissue, *Neurochem. Internat.,* **8** (1986) 449–476.

127 Masters, B., Shemer, J., Judkins, J., Clarke, D., Le Roith, D., and Raizada, M., Insulin receptors and insulin action in dissociated brain cells, *Brain Res.,* **417** (1987) 247–256.

128 McCaleb, M., Myers, R., Singer, G., and Willis, G., Hypothalamic norepinephrine in the rat during feeding and push-pill perfusion with glucose, 2-deoxyglucose, or insulin, *Am. J. Physiol.,* **236** (1979) R312–321.

129 Blomqvist, P., Lindvall, O., and Wieloch, T., Lesions of the locus coeruleus system aggravate ischemic damage in the rat brain, *Neurosci. Lett.,* **58** (1985) 353–358.

130 Gluckman, P., Klempt, N., and Gaun, J., A role of IGF-I in the rescue of CNS neurons following hypoxic-ischemic injury, *Biochem. Biophys. Res. Comm.,* **182** (1992) 593–599.

131 Choi, D., Cerebral hypoxia: Some new approaches and unanswered questions, *J. Neurosci.,* **10** (1990) 2493–2499.

132 Craft, S., Newcomer, J., Kanne, S., Dagogo-Jack, S., Cryer, P., Sheline, Y., Luby, J., Dagogo-Jack, A., and Alderson A., Memory improvement following induced hyperinsulinemia in Alzheimer's disease, *Neurobiol. Aging,* **17** (1996) 123–130.

133 Hoffman, W., Baughman, V., and Albrecht, R., Interaction of catecholamines and nitrous oxide ventilation during incomplete brain ischemia in rats, *Anesthesia Analgesia,* **77** (1993) 908–912.

134 Hoffman, W., Cheng, M., Thomas, C., Baughman, V., and Albrecht, R., Clonidine decreases plasma catecholamines and improves outcome from incomplete ischemia in the rat, *Anesthesia Analgesia,* **73** (1991) 460–464.

135 Hoffman, W., Kochs, E., Werner, C., Thomas, C., and Albrecht, R., Dexmeditomidine improves neurologic outcome from incomplete ischemia in the rat. Reversal by the apha 2-adrenergic antagonist atipamezole, *Anesthesiology,* **75** (1991b) 328–332.

136 Hoffman, W., Pelligrino, D., Werner, C., Kochs, E., Albrecht, R., and Schulte, J., Ketamine decreases plasma catecholamines and improves outcome from incomplete cerebral ischemia in rats, *Anesthesiology,* **76** (1992) 755–762.

137 Shu, C., Hoffman, W., Thomas, C., and Albrecht, R., Sympathetic activity enhances glucose-related ischemic injury in the rat, *Anesthesiology,* **78** (1993) 1120–1125.

138 Standefer, M. and Little, J., Improved neurological outcome in experimental focal ischemia treated with propranolol, *Neurosurgery,* **18** (1986) 136–140.

139 Werner, C., Hoffman, W., Thomas, C., Miletich, D., and Albrecht, R., Ganglionic block-
    ade improves neurologic outcome from incomplete ischemia in rats: Partial reversal by
    exogenous catecholamines, *Anesthesiology,* **73** (1990) 923–929.
140 Cechetto, D., Experimental cerebral ischemic lesions and autonomic and cardiac effects
    in cats and rats, *Stroke,* **24** (1993) I6–I9.
141 Patter, M., Hypothyroidism protects against free radical damage in ischemic acute renal
    failure, *Kidney Internat.,* **29** (1986) 1162–1166.
142 Swaroop, A., and Ramasarma, T., Heat exposure and hypothyroid conditions decrease
    hydrogen peroxide generation in liver mitochondria, *Biochem. J.,* **226** (1985) 403–408.
143 DeGroot, L., Larsen, P., and Refetoff, S., *The Thyroid and Its Diseases*, Wiley, New
    York, 1984.
144 Shuaib, A., Ijaz, S., Hemmings, S., Galazka, P., Ishaqzay, R., Liu, L., Ravindran, J., and
    Miyashita, H., Decreased glutamate release during hypothyroidism may contribute to pro-
    tection in cerebral ischemia, *Exptl. Neurol.,* **128** (1994) 260–265.
145 Martensson, J., Goodwin, C., and Blake, R., Mitochondrial glutathione in hypermetabolic
    rats following burn injury and thyroid hormone administration: evidence of a selective
    effect on brain glutathione by burn injury, *Metab. Clin. Exp.,* **41** (1992) 273–277.
146 Mariotti, S., Franceschi, C., Cossarizza, A., and Pinchera, A., The aging thyroid, *Endo-
    crine Rev.,* **16** (1985) 686–715.

10

Inflammatory Responses to Cerebral Ischemia

Implications for Stroke Treatment

John Sharkey, John S. Kelly, and Steven P. Butcher

1. BACKGROUND

In a recent editorial, Dr. Schmid-Schönbein *(1)* succinctly summarized the case for inflammation in the pathogenesis of cerebral ischemia;

> There is increasing evidence to suggest that reperfusion following an ischemic episode leads to an oxidative burst that serves to induce the expression of a number of proinflammatory genes, such as cytokines or leukocyte adhesion molecules. This in turn leads to an accumulation of leukocytes during reperfusion and the initiation of a proinflammatory event, with elevation of endothelial permeability, enhancement of oxygen free radical production, migration of leukocytes into the parenchyma, and expansion of tissue necrosis.

On reading this synopsis, a number of questions come to mind. Do these inflammatory responses only occur in reperfusion injury, or are they also part of a general pathophysiological response to cerebral ischemia? What is the time-course of the various components involved in the inflammatory response, including production and release of proinflammatory proteins, leukocyte adhesion and infiltration into the ischemic tissue (Fig. 1)? What is the contribution of the resident inflammatory cells (astrocytes and microglia)? Do these various cell targets and biochemical processes offer a useful avenue for therapeutic intervention?

2. INFLAMMATORY RESPONSES TO CEREBRAL ISCHEMIA

The brain is traditionally thought of as an immunologically privileged site since it lacks a lymphoid system, and antigen presenting dendritic cells *(2)*. However, there is increasing evidence that inflammatory responses play an important role in a variety of pathological conditions affecting the central nervous system (CNS), including multiple sclerosis, Behçet's syndrome, trauma, and cerebral ischemia. Ischemic brain injury not only causes endogenous parenchymal cell damage, but also induces inflammatory responses that include the infiltration and accumulation of polymorphonuclear leukocytes (PMNLs) and monocytes/macrophages, as well as the stimulation of microvascular proliferation *(3–7)*. In addition, astrocytes and microglia resident within the brain respond to ischemic insults and can rapidly precipitate a local inflammatory response. The importance of these cells, together with their circulating counterparts,

From: Clinical Pharmacology of Cerebral Ischemia *Edited by:* G. J. Ter Horst and J. Korf *Humana Press Inc., Totowa, NJ*

235

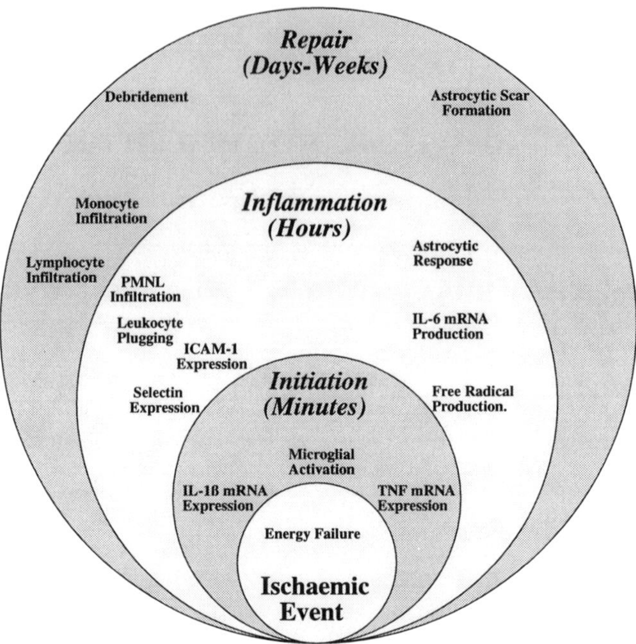

Fig. 1. Schematic overview of the inflammatory responses to focal cerebral ischemia. Within minutes of energy failure, proinflammatory cytokines (particularly TNFα and IL-1β are expressed within microglia. In the first hours after the insult, adhesion molecules are expressed and leukocytes accumulate and plug the microvasculature. Neutrophils transmigrate across the endothelium and infiltrate into the ischemic penumbra where they release free radicals and other cytotoxic products. The microglia retract their processes and begin to engulf tissue debris before releasing free radicals and NMDA-like neurotoxins. Subsequently, the microglia are joined by circulating monocytes/macrophages. Over the succeeding days and weeks, ischemic tissue is debrided and an astrocytic scar formed. In transient ischemia, these events may be accelerated *(see text)*.

the PMNLs and macrophages, to the pathophysiology of cerebral ischemia is emerging. Recent evidence also suggests that cytokines such as interleukin-1 (IL-1), interleukin-6 (IL-6), and tumor necrosis factor (TNFα, as well as the chemokines, interleukin-8 (IL-8), cytokine-induced neutrophil chemoattractant (CINC), and monocyte chemoattractant protein (MCP-1) are rapidly produced in the brain in response to ischemia. Such cytokines up-regulate leukocyte adhesion molecules (β_2 integrins, ICAM's, and the selectins) and activate inflammatory cells to produce toxic mediators.

Therapeutic strategies aimed at suppressing the inflammatory responses to cerebral ischemia offer a new and exciting avenue for the treatment of stroke. In this review we will examine: the immunological responses to cerebral ischemia, the contribution of resident and circulating inflammatory cells to ischemic injury, and strategies of neuroprotection involving suppression of inflammatory responses. We will also examine some recent insights into signal transduction processes involved in the pathogenesis of ischemic cell death gained from studies of the potent immunosuppressants, FK-506 and cyclosporin-A.

2.1. Circulating Leukocytes

The appearance of PMNLs and monocytes in ischemic tissue was originally believed to represent a pathophysiological response to pre-existing injury. However, since the early 1970s, increasing evidence suggests that activated PMNLs and monocytes accumulate in the microvasculature before transmigrating into the ischemic tissue where they may contribute to tissue damage by reducing local tissue perfusion, initiating thrombosis, and releasing free radicals, proteinases, and other cytotoxic substances *(6,8)*.

2.1.1. Impairment of Cerebral Tissue Perfusion

Defects in local tissue perfusion, the so called "no-reflow" phenomenon, has been reported in cardiac and brain models of ischemia/reperfusion (*see* refs. *5,6,9,11* for discussion). This has been attributed to physical blockade of the capillaries by leukocytes *(6,11)*. In the baboon, microvascular "no-reflow" was reported to occur in >60% of capillaries examined within 1 h of reperfusion following middle cerebral artery (MCA) occlusion. Further examination revealed the presence of single PMNLs obstructing individual capillaries, whereas clumps of granulocytes partially occluded the postcapillary venules *(9)*. The percentage of vessels exhibiting "no-reflow" was reduced in this model by prior administration of the anti-CD18 MAb IB41 *(12)*. More recently, Garcia and coworkers *(13)* examined the effects of permanent MCA occlusion on rat microvessels using light and electron microscopy. Leukocyte plugging of the microvasculature was noted, with PMNLs seen in close apposition to the endothelial surface within 1 h of occlusion followed after 4 h by monocytes.

2.1.2. Initiation of Thrombosis

The role of leukocytes in the initiation of thrombosis remains somewhat controversial. Epidemiological studies suggest that leukocytes may contribute to the initiation of stroke *(14)*. This view is supported by findings that the coagulation cascade is activated in stroke patients *(15)*, that activated monocytes and PMNLs exhibit potent clot-promoting activity in vitro and that leukocyte–platelet aggregates are found in cerebral thrombosis (for reviews, *see* refs. *6* and *16*). However, direct experimental data supporting a role for leukocytes in the initiation of thrombosis is lacking *(13)*, and activated leukocytes have little effect on cerebral blood flow in normal animals *(6)*. Conversely, since activated monocytes, and to a lesser extent PMNLs, support clot lysis by release of plasminogen activating factor and other lytic enzymes (including elastase), it has been argued that leukocytes may play an antithrombotic role in cerebral ischemia *(16,17)*. This view is supported by reports of increased levels of elastase *(18)*, and reduced procoagulant activity of PMNLs and monocytes, in patients following ischemic stroke *(16,17)*.

2.1.3. Secretion of Free Radicals and Toxins

In a recent study, Matsuo and coworkers *(19)* examined the time-course of free radical production during reperfusion following 1 h MCA occlusion in the rat. They found that ascorbate radical (AR) formation showed a biphasic profile, increasing immediately after reperfusion for approx 1 h before returning to baseline values until approx 24 h when a secondary increase was noted. Since neutrophil infiltration began some 6–12 h after reperfusion in this model, they argued that circulating neutrophils caused the initial surge in free radicals, whereas infiltrating neutrophils were responsible for the delayed increase. This hypothesis was supported by studies demonstrating that free radical surges could be attenuated by an antineutrophil antibody (RP3), superoxide dismutase or catalase *(19)*.

The time-course of PMNL accumulation after stroke in humans, and after permanent MCA occlusion in animal models, has been reviewed *(6)*. In both conditions, maximal PMNL infiltration into a "reactive zone" around the ischemic core occurred between 48–72 h, with further infiltration into the core by 7 d postocclusion. The PMNL accumulation is greater in animal models of transient focal cerebral ischemia with reperfusion than in permanent occlusion models *(93)*. Based on studies reporting PMNL infiltration in global ischemia only when the duration of the insult exceeded 40 min, Kochanek and Hallenbeck *(6)* concluded that PMNL infiltration is unlikely to be of clinical importance.

A recent electron microscopy study by Garcia and his associates *(13)* discussed the fate of PMNLs and monocytes following permanent MCA occlusion in rat. They observed that, whereas PMNLs were seen in close apposition to the endothelial cell surface after 1 h of ischemia, transmigration did not occur until at least 12 h following MCA occlusion. Similarly, monocytes were first seen in vessels 4 h postocclusion, but did not appear in the parenchyma until 8 h later. Widespread infiltration of monocytes did not occur until 72 h postocclusion, peaking after 7 d. A similar pattern of PMNL and monocyte infiltration was reported in spontaneously hypertensive rats (SHR) following permanent MCA occlusion *(20)*. However, when SHRs were subjected to transient MCA occlusion, infiltration occurred after 6 h for PMNLs and 1 d for monocytes. Contrarily, instead of increasing infarct size as seen in other strains of rat and in other tissues, reperfusion injury in the SHR model appears to reduce the volume of brain damage *(20)*.

2.1.4. Neutrophils

Neutrophils are initially the predominant leukocyte to enter the brain following ischemia followed by infiltration of lymphocytes and mononuclear phagocytes *(21,22)*. Neutrophils play an important role in ischemia–reperfusion injury by releasing a variety of cytotoxic products, including free radicals *(23–26)*. Activated neutrophils are a major source of free radicals associated with ischemia-reperfusion injury in heart *(27)* and lung *(28)*. Similarly, inhibition of neutrophil function has a significant protective effect in ischemia-reperfusion brain injury *(29–34)*. Neutropenia-induced protection against ischemic damage has been demonstrated in a variety of tissues including heart *(35–38)*, lung *(39,40)*, spinal cord *(41,42)*, and brain *(4,33,43–45)*. However, as noted by Chopp and coworkers *(30,46)*, the immunosuppression associated with antineutrophil therapy make this approach impracticable for clinical use. Consequently, other approaches, particularly those that interfere with leukocyte adhesion and infiltration, have become more popular.

2.1.5. Lymphocytes

It is generally believed that PMNLs and monocytes, but not lymphocytes, are involved in the pathogenesis of cerebral ischemia *(6)*. However, two recent studies suggest that T-lymphocytes may play a role. Morioka and coworkers *(47)* reported the infiltration of T-cells into ischemic tissue following MCA occlusion in the rat. A more extensive study by Jander and coworkers *(22)* examined the effects of photothrombotic lesions to the rat cortex upon lymphocyte infiltration. Using selective MAbs against lymphocyte surface antigens, infiltrating CD^{8+} cytotoxic T-cells and natural killer cells were observed between 1–2 d postlesion. These cells had circumscribed the infarct by d 3 and began to infiltrate the infarct by d 7. At each stage, the T-cell response preceded that of infiltrating macrophages. Considering the paucity of data, it is uncertain whether a pronounced T-cell response is a common feature of focal cerebral ischemia or is merely a consequence of irradiation-induced photoperoxidation of the endothelium.

2.1.6. Monocytes/Macrophages

The contribution played by blood-borne monocytes in ischemic brain injury is not well understood. In man, the infiltration of macrophages into ischemic tissue occurs some 48–72 h after the insult *(48,49)*. However, as noted recently *(49)*, clinical data for earlier time points is somewhat lacking. In animal models, infiltration of activated monocytes into the ischemic penumbra has been observed as early as 12 h postocclusion, with infiltration into the ischemic core occurring some 12–48 h later depending on the model *(39)*. Once these reactive monocytes reach the site of injury they appear to engulf tissue debris before releasing free radicals and NMDA-like neurotoxins *(51–53)*. Although the infiltration of monocytes and subsequent release of neurotoxic agents has been implicated in the progressive loss of motor neurons observed following moderate ischemia in the spinal cord *(41,54)*, it would appear unlikely that such a delayed event would contribute significantly to the evolution of infarction in focal cerebral ischemia.

2.2. Endogenous Cells

2.2.1. Microglia and Astroglia

Microglia behave as the resident macrophages of the brain, where they are suggested to play a critical role in the inflammatory response, phagocytosing injured or dying cells and scavenging debris from degenerating cells *(2,52)*. It has also been proposed that, like their circulating counterparts, microglia may injure cells by releasing cytokines, free radicals, and a neurotoxin that can be inhibited by NMDA receptor antagonists *(49,53,55)*. Monocytes and microglia are a source of nitric oxide (NO) *(56)*, and secrete a variety of cytokines including IL-1 and TNFα that stimulate both angiogenesis and glial proliferation *(2)*. In addition, microglia have been shown to secrete basic fibroblast growth factor, which has neurotrophic activity *(57)*.

The majority of resident microglia are quiescent, lacking phagocytic or secretory capacities *(58)*. However, within hours of an ischemic insult, their processes retract and the microglia assume a macrophage-like phenotype *(59)*. There is also evidence that the time-course of this transformation may be related to the type of ischemia. In the rat, microglial activation has been observed 24 h after permanent MCA occlusion *(47)*. In contrast, antiphosphotyrosine labeling of microglia is reported to increase within 3 h of reperfusion following transient MCA occlusion *(60)*. Differences in the time-course of microglial responses in permanent and transient ischemia models have been attributed to ischemic severity *(60)*. However, it is also likely that reperfusion damage, or even differences in staining methods, may contribute to the observed discrepancies. A systematic comparison of the microglial response in both temporary and permanent ischemia models would answer this question.

Whereas inhibition of monocyte infiltration, by combined treatment with colchicine and chloroquine, reduces ischemic injury and improves neurological outcome in rabbits subjected to spinal cord ischemia, this approach does not confer neuroprotection in models of focal cerebral ischemia *(31,41,52)*. Similarly, methyl prednisolone has been shown to reduce the expression of JE/MCP-1 mRNA (a potent and selective monocyte attractant) and to attenuate macrophage accumulation in rat brain following focal ischemic insult *(61)*. However, pharmacological inhibition of monocyte infiltration did not influence lesion size, neuronal damage or neurological outcome *(61)*. It has been argued that the rapidity and severity of the ischemia produced in the rat MCA occlusion model destroys both neurons and glia at the site of infarction thereby preventing the initiation of local inflammatory response. In contrast, the moderate ischemia produced in spinal cord models spares resident microglia, thereby facilitating inflammatory processes and consequent extension of the insult *(52,61)*.

Resident microglia have also been implicated in the phenomenon of ischemic preconditioning, in which a brief exposure to ischemia confers protection against subsequent ischemic episodes *(87)*. Kato and coworkers *(87)* have reported that a preconditioning insult of 3 min forebrain ischemia caused a selective activation of microglia in the CA1 subfield of hippocampus. In contrast, 6 min of forebrain ischemia was associated with a more intense activation of the microglia with retraction of the pseudopodia and subsequent damage to CA1 neurons. However, if the brain was subjected to preconditioning prior to the insult, CA1 cell damage induced by a 6-min ischemic episode was prevented and microglial activation returned towards control values by 7 d. Such findings suggest that inhibition of microglial activation may not represent a viable therapeutic intervention for the treatment of cerebral ischemia.

Astroglia have long been considered impediments to the regeneration of the injured brain *(2,63,64)*. However, most in vitro studies suggest that astroglia actually support neuronal growth and survival *(49,65)*. In one such study, Guilian and coworkers examined the effects of microglial and astroglial secretions upon the growth of cultured ciliary neurons. They found that microglia secrete small (<500 Dalton), heat-stable, nonproteinaceous neurotoxins with biological activity blocked by NMDA receptor antagonists. These factors were selectively neurotoxic and did not reduce oligodendroglia, astroglia, or Schwann cell number in vitro. Although these microglial neurotoxins have yet to be identified, they would appear to act via a mechanism similar to excitatory amino acids because APV and MK-801 were neuroprotective, but L-channel blockers and CNQX were not. High performance liquid chromatography revealed a sharply defined peak that was not copurified by quinolinate, glutamate, or aspartate. Interleukin 1α, 1β, and TNFα, lactic acid , proteases, and free radicals were also excluded. By contrast, the secretions from astroglia were proteinaceous (>10 kDa), promoted neuronal growth in culture, and attenuated microglial toxicity.

2.3. Proinflammatory Cytokines

2.3.1. Interleukin-1

To date, by far the best characterized cytokine response in cerebral ischemia involves IL-1. IL-1 is a potent inflammatory mediator that can stimulate astrocyte proliferation *(66,67)* and brain edema *(68)*, activate leukocyte infiltration *(6,69,70)*, and induce *de novo* synthesis of endothelial membrane adhesion molecules (ICAM-1 and E-selectin)

(71,72). IL-1 has also been found in the cerbrospinal fluid (CSF) of patients with traumatic brain injury *(73)* and is expressed in animal models of cerebral ischemia, neurodegeneration, and traumatic brain injury *(74–78)*. IL-1 describes three proteins, IL-1α, IL-1β, and IL-1 receptor antagonist (IL-1ra), which, although they show significant sequence homology, are derived from separate gene products. Both IL-1α and IL-1β are agonists for the Type I IL-1 receptor, whereas IL-1ra behaves as an endogenous receptor antagonist *(79)*. Autoradiographical and immunohistochemical studies indicate that the IL-1β isoform predominates within the CNS *(80,81)*, and is involved in the evolution of ischemia brain injury *(74–76,78,82)*.

IL-1β mRNA has been detected in the rat brain within 30 min of reperfusion following transient (10 min) forebrain ischemia using *in situ* hybridization techniques *(83)*. Interestingly, this early response appears to be a function of the ischemic insult rather than reperfusion since IL-1β mRNA can also be detected as early as 15 min following onset of permanent MCA occlusion *(74)*. In both studies, IL-1β mRNA appeared to be localized in microglia rather than astrocytes or neurons. These findings are consistent with the view that monocytes are the principle source of IL-1β in brain injury *(84)*. There is emerging evidence that the magnitude of the IL-1β response may be dependent on the ischemic model. Northern blot studies have shown that permanent MCA occlusion results in a 2–3-fold increase in IL-1β mRNA between 6 and 12 h postocclusion *(75)*. Increases in IL-1β of a similar magnitude have been reported using immunometric assays *(82)*. In contrast, Wang and coworkers *(78)* reported that IL-1β mRNA increased up to 29-fold following transient MCA occlusion. The observation that IL-1β mRNA is rapidly expressed does not necessarily mean that it is being released because IL-1β is regulated independently at the transcription and translation level *(85)*. Pro-IL-1β precursor protein requires cleavage at the Asp116-Ala117 site by interleukin-1 converting enzyme (ICE) to become biologically active. Whereas we do not know how quickly IL-1β is released in ischemic tissue, the requirement for ICE in the production of IL-1β may offer a method of neuroprotection.

Intracerebroventricular injections of recombinant human IL-1β enhanced edema formation, increased neutrophil infiltration, and increased infarct size in a dose-dependent manner in rats subjected to 60 min of MCA occlusion with reperfusion *(86)*. The edematous reaction, neutrophil infiltration, and infarct size were attenuated by anti-IL-1β neutralizing antibodies or by infusion of the putatively selective IL-1b blocker zinc protoporphyrin *(86)*. Edema induced by transient MCA occlusion is augmented by irradiation-induced leukopenia *(87)*. Since edema can also be significantly attenuated by infusions of the IL-1 neutralizing antibody, these authors suggested that IL-1β may act synergistically with other proinflammatory agents, such as leukotrienes and PAF, in the formation of edema *(87)*.

The IL-1ra has been shown to be neuroprotective in models of permanent MCA occlusion *(88–90)*. As little as 10 μg given directly into the ventricles 10 min before and 30 min after MCA occlusion were found to reduce ischemic damage by 50%. More recently, Betz and coworkers *(88)*, using a recombinant adenovirus vector system, found that rat brains that overexpress IL-1ra, by up to 50-fold over control levels, exhibit a 64% reduction in the volume of ischemic brain damage as a result of permanent MCA occlusion. Since these studies were performed in models of permanent occlusion, suppression of IL-1-mediated reperfusion injury could not account for the reduction in infarct size. Garcia and coworkers *(91)* also reported that IL-1ra significantly improved neurological score, reduced the number of PMNLs within the ischemic area, and attenuated ischemic brain damage produced by permanent MCA occlusion. They speculated that IL-1ra may prevent the infiltration of PMNLs by blocking the production of IL-8 in endothelial cells; IL-8 being necessary for the transendothelial migration of PMNLs *(91)*.

2.3.2. Other Cytokines

Tumor necrosis factor (TNFα and IL-6 may also play roles in the pathogenesis of ischemic brain injury. TNFα is primarily produced by activated macrophages in the periphery *(92)* and by microglia in the ischemic brain *(93–95)*. It is rapidly produced in ischemic tissue, with increases in mRNA *(93,94)* and protein *(93)* detectable within 1 h of MCA occlusion in the rat, peaking approx 3 h (mRNA) and 8–24 h (protein) following the insult. Uno and coworkers *(95)* have also reported rapid and prolonged increases in TNFα positive microglia within the hippocampal CA1 subfield of mice subjected to transient (10 min) forebrain ischemia *(95)*. Like other cytokines, its precise role in the pathophysiology of cerebral ischemia is unclear. TNFα is a potent proinflammatory cytokine that may contribute to ischemic injury through a variety of mechanisms. TNFα is toxic to oligodendrocytes *(77)*, stimulates astrocytic proliferation *(77)*, and has been implicated in the demyelination and gliosis associated with brain injury *(77,93)*. Alternatively, TNFα may exacerbate brain injury by stimulating the release of other cytokines (IL-1, IL-6, colony stimulating factor) *(96)*, by altering blood–brain barrier permeability *(77,97,98)*, by activating neutrophils to release free radicals *(77)*, or by direct damage to the endothelium *(99)*. TNFα also stimulates expression of leukocyte adhesion molecules *(100)* and iNOS, and enhances the action of endothelin-1 on isolated endothelial smooth muscle cells in vitro *(99)*. Direct microinjections of TNFα into rat cortex cause leukocyte accumulation within the microvasculature of the injected hemisphere in vivo *(94)*. Thus, whereas no neuroprotection studies have been performed to date, the inhibition of TNFα action may provide a useful avenue through which reduction of ischemic damage might be achieved.

Clinical studies suggest that interleukin-6 (IL-6) may be involved in the etiology of cerebral ischemia, with elevated levels of IL-6 recorded in the serum and CSF of stroke patients *(101,102)*. These authors reported that the magnitude of this increase is predictive of infarct size *(101,102)*. Similarly, animal studies have demonstrated increases in IL-6 production following stroke with rapid and prolonged increases in IL-6 mRNA expression reported following rat MCA occlusion *(103)*. At present, one can only speculate as to the role of IL-6 in stroke. IL-6 may act as a cytokine brake, attenuating the effects of IL-1 and TNFα by stimulating the production of their respective circulating antagonists, IL-1ra and TNFα receptor p55 *(104)*. Alternatively, IL-6 may exacerbate ischemic injury by stimulating gliosis and breakdown of the blood–brain barrier permeability since transgenic mice overexpressing IL-6 exhibit lifelong reactive gliosis *(104a)* and a disrupted blood–brain barrier *(105)*.

2.4. Cell Adhesion Molecules

The role of adhesion molecules in inflammation has been recently reviewed *(71,72,106)*. The mechanisms involved in leukocyte recruitment have been elucidated in the periphery, and it would not be unreasonable to anticipate that similar mecha-

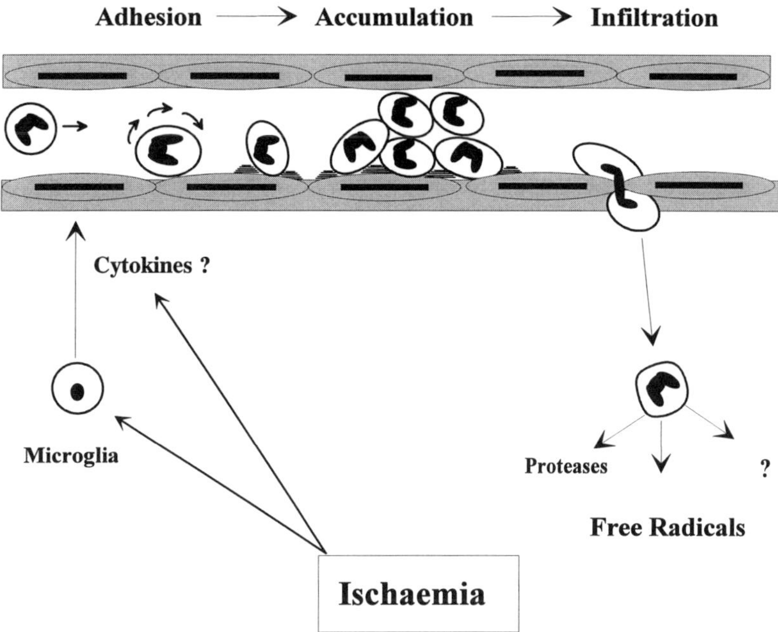

Fig. 2. A diagrammatic representation of the processes involved in leukocyte infiltration following an ischemic insult. In this simplified model, cytokines (possibly TNFα and IL-1β) released in response to the ischemic insult stimulate both the selectin-mediated binding and rolling behavior, and the β2 integrin and ICAM-1 mediated aggregation and transmigration of leukocytes across the endothelium. Neutrophils and monocytes then migrate to the ischemic area where they can release proteases, free radicals, and other cytotoxic products. Oxygen free radicals released during reperfusion can accelerate and accentuate these processes.

nisms occur within the CNS. In general, the first steps in the inflammatory response to tissue injury involve the adhesion of leukocytes to, and transmigration across, the vascular endothelium (Fig. 2). Both leukocytes (PMNLs and monocytes) and endothelial cells play an active role in this process by regulating the expression of molecules which mediate adhesion. Three major families of proteins play a role in leukocyte-endothelial interactions; β_2 integrins, intracellular adhesion molecules (ICAMs), and selectins.

2.4.1. β_2 Integrins

The β_2 or CD18 integrins are a family of heterodimeric adhesive proteins that share a common β chain (CD18) but differ with respect to their alpha chains (CD11a,b,c). Members of the β_2 integrins include lymphocyte function-associated antigen (LFA: CD11a/CD18), Mac-1 (CD11b/CD18), and p150.95 (CD11c/CD18). The LFA is expressed by all lineages of white blood cells while Mac-1 and p150.95 are largely restricted to macrophages, neutrophils and natural killer cells. Although β_2 integrins are thought to be involved in the migration of leukocytes across the endothelium, they have also been implicated in most leukocyte-adhesion functions. The Mac-1 is believed to mediate the binding of stimulated neutrophils to the endothelium and may be involved in the pathophysiology of ischemia/reperfusion injury in heart *(107)*, liver *(35)*, and brain (*see* Table 1).

Table 1
Effect of Antiadhesion Therapy
on the Outcome in Animal Models of Focal Ischemia

Species	Target	Ischaemia model	Outcome	Reference
Rat				
	β2 Integrins	*Intraluminal filament*		
	(CD18)	1 h MCAo, 24 h reperfusion	56-66%	*34*
		2 h MCAo, 1 wk reperfusion	51%	*192*
	LFA-1	*Intraluminal filament*		
	(CD11a)	1 h MCAo 24 h reperfusion	42-47%	*34*
	MAC-1	*Intraluminal filament*		
	(CD11b)	2 h MCAo 46 h reperfusion	43%	*46*
		2 h MCAo, 1 wk reperfusion	37–44%	*192*
		2 h MCAo 46 h reperfusion	28%	*30*
		2 h MCAo 46 h reperfusion	37%	*32*
		Permanent MCAo (48 h	No effect	*32*
		Permanent MCAo (48 h)	No effect	*30*
	ICAM-1	*Intraluminal filament*		
	(1A29)	1 h MCAo, 24 h reperfusion	66–83%	*34*
		2 h MCAo, 22 h reperfusion	41%	*193*
		2 h MCAo, 1 wk reperfusion	44%	*30*
		1 wk permanent MCAo	No effect	*30*
Rabbit				
	(CD18)	*Snare ligature*		
		8–40 min occlusion, 90 h reperfusion: spinal cord	Neurological Improvement	*31*
		Microspheres		
		Irreversible brain	No effect	*31*
		Balloon catheter		
		Reversible spinal cord (30 min)	No effect	*112*
	ICAM-1	Embolic	Neurological Improvement	*29*
Cat				
	(CD18)	*Surgical clip*	No Effect	
		90 min MCAo, 3 h reperfusion	(CBF)	
Baboon				
	(CD18)	*Surgical clip*		
		3 h MCAo , 1 h reperfusion	Reduced no. of occluded microvessels	*12*

Studies on rats were performed using the intraluminal filament approach, whereas a variety of approaches were used in the other species *(Italics)*. Outcome, according to these authors was defined variously as; % reduction in ischemic volume, alterations in cerebral blood flow (CBF), improvement in neurological score, or reduction in the number of occluded microvessels.

2.4.2. Intercellular Adhesion Molecules

The integrins bind to proteins on the surface of the endothelium that belong to the immunoglobulin superfamily (ICAM-1, ICAM-2, ICAM-3). ICAM-1 expression can be induced by exposure to the cytokines TNFα, IL-1, and interferon γ. ICAM-2 is constitutive and restricted to endothelial cells and monocytes. ICAM-3 is highly expressed by leukocytes. ICAM-1 has been implicated in neutrophil–endothelial interactions and transendothelial migration. The functional roles of ICAM-2 and ICAM-3 are not known.

2.4.3. Selectins

The selectins are the third family of adhesive proteins that mediate adhesion by binding to carbohydrate residues on glycoproteins and glycolipids. This family comprises three distinct molecules; E-selectin (ELAM-1), P-selectin (PADGEM or GMP-140), and L-selectin (LAM-1 or LECAM). E-selectin is expressed by stimulated endothelium and is primarily involved in neutrophil and monocyte adhesion to endothelial cells. P-selectin is stored in the α-granules of platelets and the Weibel-Palade bodies of the endothelial cells. Cell activation by proinflammatory agents (thrombin or histamine) results in a rapid and transient recruitment of P-selectin to the cell surface, where it is implicated in neutrophil rolling and local inflammation. L-selectin is constitutively expressed on the surface of various leukocytes, including monocytes, neutrophils, and a subset of lymphocytes. It was originally described as a lymphocyte homing receptor, but it has also been implicated in the adherence of neutrophils to cytokine-stimulated endothelial cells.

2.4.4. Cell Adhesion Molecules and Cerebral Ischemia

According to one general model of leukocyte-endothelial interactions *(71)*, the selectins mediate the initial binding and rolling behavior of leukocytes, with β_2 integrins and ICAM-1 involved in aggregation and transmigration across the endothelium (Fig. 2). These authors proposed that antiselectin treatments would block the initial binding events, whereas anti-β_2 integrin therapy would block the recruitment process, thereby leading to the eventual detachment of leukocytes *(71)*. The expression of adhesion molecules in response to cerebral ischemia has only recently been examined. To date there is evidence that ICAM-1, P-selectin, and E-selectin (ELAM-1) are expressed in ischemic brain tissue. Up-regulation of ICAM-1 has been demonstrated on microvessels following ischemia in rat *(22,33,78,018)*, baboon *(109)*, and human postmortem tissue *(110)*. Matsuo and coworkers *(33)* reported the presence of ICAM-1 positive immunoreactivity on capillary endothelial cells within 1 h of reperfusion following a 60-min period of MCA occlusion. Increases in ICAM-1 have been detected as early as 3 h following permanent surgical occlusion of the rat MCA *(78,108)*. This profile of induction was reported to be generally similar following transient MCA occlusion *(78,108)*. Interestingly, the magnitude of the response was approx twice that produced by permanent MCA occlusion, even though the authors used a protracted period of ischemia (160 min). In the study by Okada and coworkers *(109)*, a 3-h period of MCA occlusion in the baboon followed by reperfusion produced an up-regulation of both ICAM-1 and P-selectin. They reported that ICAM-immunoreactivity could not be detected 2 h after the initiation of ischemia, but peaked between 1 to 4 h following reperfusion, and returned to preischemic levels by 24 h postocclusion. In contrast, P-selectin immunoreactivity increased to its maximum level by 2 h of occlusion, a level that was maintained throughout the 24-h reperfusion period. The vascular distribution of ICAM-1 and P-selectin within precapillary arterioles and postcapillary venules is consistent with the favored site of PMNL adhesion and granulocyte transmigration during inflammation. Wang and coworkers *(108)* reported that E-selectin mRNA expression peaks in ischemic tissue some 12 h after permanent MCA occlusion in the rat and persists for up to 2 d after the onset of ischemia. However, since the measurements were made by Northern blotting, no information was available as to the exact localization of the E-selectin.

2.4.5. Anti-Adhesion Therapy

With two notable exceptions, the administration of anti-adhesion antibodies either before, or up to 4 h following initiation of the ischemic insult consistently reduces infarct size associated with transient MCA occlusion (Table 1). One reason for these two exceptions may be the choice of antibody. Takeshima and coworkers *(111)*, reported that the anti-CD18 MAb (60.3: 2 mg/kg), given 40–50 min into the insult had no effect on cerebral blood flow (CBF), somatosensory-evoked potential (SEP) amplitude, and infarct volume produced by 90 min of ischemia followed by 3 h reperfusion in the cat. Using the same concentration of this CD18 antibody, Forbes and coworkers *(112)* found no effect on cerebral blood flow or the volume of damage produced by a 90-min MCA occlusion followed by 3 h reperfusion in the rabbit. Although the 60.3 antibody has been shown to reduce ischemic-reperfusion damage in the intestine (*see* ref. *111*), it should be noted that in studies in which neuroprotection was reported, antibodies other than 60.3 were used. In contrast to the data concerning transient models of focal cerebral ischemia, several authors have reported that antiadhesion therapy (CD18, MAC-1, ICAM-1) is ineffective in permanent occlusion models (Table 1). The reason for this discrepancy is unclear, particularly since ICAM-1, β_2 integrins and selectins are expressed following both transient and permanent focal cerebral ischemia.

2.5. Immunosuppressants

2.5.1. Corticosteroids

Corticosteroids are potent anti-inflammatory agents with a long record of clinical usage. Consequently, these agents have been widely investigated as potential treatments for CNS trauma and ischemia. However, until relatively recently, the consensus of opinion from clinical studies and from animal work was that pre- or posttreatment with corticosteroids is ineffective for the treatment of cerebral ischemia in the adult. Indeed, a recommendation to that effect was published by the WHO in 1989. However, there has been increasing evidence that corticosteroids can protect against ischemic insult, particularly in neonates (*see* Chapter 9).

2.5.2. Cyclosporin-A, FK-506, and Rapamycin

The potent immunosuppressants, FK-506 and cyclosporin (CsA) are in clinical use for the treatment of allograft rejection following organ transplantation *(113)*. Recent reports detailing the anti-ischemic activity of FK-506 and CsA have also provided an insight into the biochemical pathways involved in the pathogenesis of cerebral ischemia. In order to examine the neuroprotective actions of these agents, it is important to understand the mechanisms by which they, and the related immunosuppressant, rapamycin, induce immunosuppression (Fig. 3). Stimulation of the T-cell receptor (TCR/CD3 complex) leads to a calcium-dependent, and IL-2-mediated, transition of T-cells from G0 to G1 and thence into the cell cycle (*see* refs. *114* and *115* for

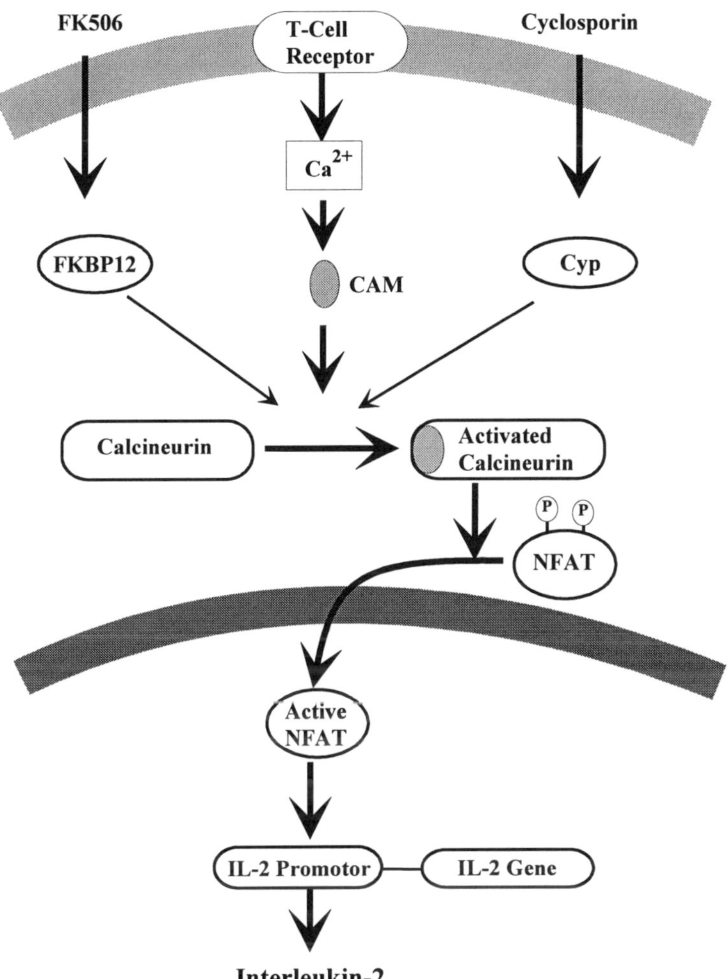

Fig. 3. Immunophilin-mediated inhibition of T-cell activation. Stimulation of the T-cell receptor complex results in an IP$_3$-mediated increase in intracellular calcium which activates, *inter alia*, the genes encoding IL-2 and its receptor. FK-506 and cyclosporin-A inhibit this process by binding to their respective immunophilins, FKBP12 and cyclophilin (Cyp), to inhibit the calcium-calmodulin-dependent protein phosphatase, calcineurin.

reviews). In recent years, the transduction pathway involved in T-cell activation has begun to be elucidated. Activation of the TCR/CD3 complex results in a phospholipase C/IP$_3$-mediated increase in intracellular calcium *(116)*, resulting in the activation of several genes including IL-2 and its receptor. The rate limiting enzyme in the pathway is the calcium/calmodulin-dependent protein phosphatase 2B (calcineurin) *(116)*, which dephosphorylates the transcription factor NF-ATc (nuclear factor of activated T-cells) *(117)*. Dephosphorylation of cytosolic NF-AT facilitates its translocation into the nucleus where, having combined with a newly synthesized nuclear subunit (NF-At$_n$), it forms transcriptionally competent NF-AT that binds to the promoter region of the IL-2 gene *(115,118)*.

FK-506 and CsA act by suppressing the calcium-dependent signal transduction pathway that promotes interleukin-2 gene transcription in helper T-cells *(114,115,119,120)*. FK-506, a macrolide lactone isolated in 1984 from cultures of the bacterium *Streptomyces tsukubaensis* (*see* ref. *121* for review), binds to a family of intracellular proteins termed FK-506 binding proteins (FKBPs), a subclass of the immunophilin protein family *(122,123)*. A complex of FK-506 and a 12 kDa immunophilin (FKBP12) inhibits the activity of the catalytic subunit of calcineurin *(116,119,124)*, thereby preventing activation of the IL-2 gene. Suppression of T-cell proliferation by FK-506 can be partly reversed by the administration of exogenous IL-2 *(121)*. FK-506 also inhibits transcription of several lymphokine genes up-regulated in activated CD^{4+} helper T-cells, including IL-2, IL-3, IL-4, granulocyte macrophage colony stimulating factor (GM-CSF), TNFα and interferon gamma (INFγ) *(121)*. CsA is a cyclic polypeptide derived from the fungus *Tolypocladium inflatum*, which has little structural resemblance to FK-506. However, despite these structural differences, CsA shows a remarkably similar immunosuppressant profile, although it is some 30–100-fold less potent in inhibiting T-cell proliferation responses *(125)* and preventing allograft rejection in vivo *(121)*. These similarities can be readily explained by findings that show that CsA binds to an 18 kDa immunophilin (cyclophilin), and inhibits the transcription of interleukin-2 via calcineurin and NFAT by the same mechanism as FK-506 *(114)*.

A third immunosuppressant, rapamycin, was isolated from the bacterium *Streptomyces gygrocopicus*, is structurally similar to FK-506, and competitively binds to FKBP-12 with similar affinity. However, unlike FK-506, the rapamycin-FKBP complex does not inhibit calcineurin, has no effect on the nuclear translocation of NFAT, and does not inhibit cytokine production. Instead, rapamycin appears to block more distal events in the T-cell activation cascade by blocking the cell cycle progression at the late G1 to S stage via a transduction pathway involving the recently described rapamycin—FKBP12 target proteins (RAFT or FRAP proteins: for reviews, *see* refs. *126,127*). Moreover, rapamycin can reverse the suppression of T-cell activation by FK-506 *(128)*.

2.5.2.1. NEUROPROTECTION

Recent studies suggest that both cyclosporin and FK-506 may be useful in the treatment of stroke. While CsA has little direct effect upon cerebral tissue perfusion *per se* *(129,130)*, it reduces brain edema and the volume of infarction produced by MCA occlusion *(131,132)*. Also, CsA has been shown to attenuate microglial activation, reduce tissue damage following chronic cerebral hypoperfusion *(133)* and to prevent the reduction in hippocampal muscarinic receptors observed 7 d after transient forebrain ischemia *(133,134)*. In 1994, we demonstrated that FK-506, at immunosuppressant doses (0.1–1 mg/kg), significantly attenuated the volume of ischemic damage produced by MCA occlusion in the rat *(135)*. Although similar doses of CsA were ineffective (Fig. 4), higher doses proved to be neuroprotective *(45,131)*. Conversely, rapamycin, which exhibits a similar affinity for FKBP-12 as FK-506, does not protect in the rat MCA occlusion model *(135)*. Although the precise cellular target or biochemical mechanism underlying these effects has still to be resolved, the finding that rapamycin pretreatment attenuated the neuroprotective action of FK-506 *(135)* confirmed the involvement of immunophilins and suggested that immunosuppression *per se* is not necessary for neuroprotective efficacy.

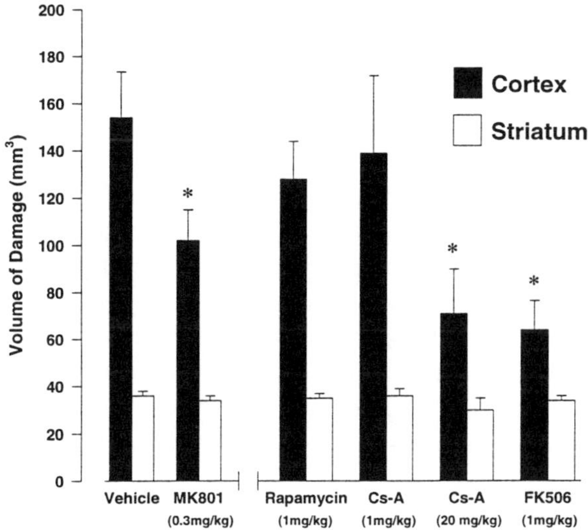

Fig. 4. Effects of the immunosuppressants rapamycin, cyclosporin-A, and FK-506 on the volume of ischemic brain damage produced by endothelin-induced MCA occlusion in the rat. The neuroprotective effect of noncompetitive NMDA receptor antagonist, MK-801 was examined in a parallel group of rats for comparison. All drugs were administered by iv infusion 1 min after vessel occlusion.

2.5.2.2. ANTI-INFLAMMATORY ACTION

Several lines of evidence point to an antiinflammatory mechanism as mediating the neuroprotective actions of CsA and FK-506 in animal models of stroke. In peripheral tissues, including liver *(136–138)* , heart *(120)*, kidney *(137,139)*, and intestine *(140)*, FK-506 and CsA have been shown to protect against ischemia-reperfusion damage. It has been suggested that pretreatment with FK-506 protects against ischemia-reperfusion injury in the liver by inhibiting the production of TNFα and IL-6 *(92,136)*. Alternatively, the cardioprotective effect of FK-506 may involve inhibition of super-oxide radical production in neutrophils *(120)*. The involvement of PMNL suppression in the neuroprotective actions of CsA has also been proposed *(132)*, although no direct evidence has been reported. However, there is evidence that attenuation of microglial responses may be involved in the neuroprotective actions of CsA *(133)*. Kondo and coworkers *(133)* reported that CsA significantly reduced the loss of hippocampal CA1 cells observed following transient forebrain ischemia in the gerbil. This CsA-mediated neuroprotection was accompanied by a significant reduction in the expression of HLA-DR class II antigen on microglia. Suppression of microglial activation was also noted by Wakita and coworkers *(141)*, who observed that microglia responses in white matter under conditions of chronic cerebral hypoperfusion, which results in tissue rarefaction, could be attenuated by CsA.

Recent studies in our laboratories suggest that suppression of microglial activation may not be the principle mechanism of FK-506-mediated neuroprotection. In an attempt to identify the contribution of microglia and astrocytes to the neuroprotective properties of FK-506, we examined three experimental conditions associated with glial activation; excitotoxic lesions of the striatum, deafferentation of facial motor neurons, and endothelin-induced MCA occlusion. Excitotoxic lesions of the striatum result in widespread neuronal loss associated with astrocyte hypertrophy and microglial activation. While FK-506 is reported to attenuate excitotoxic damage in cell culture models in vitro *(142)*, we found no evidence that FK-506 reduced the excitotoxic brain damage associated with intrastriatal injections of quinolinate, NMDA, or AMPA in vivo (Butcher and Sharkey, unpublished data). This model therefore allowed us to examine the effects of FK-506 on glial responses without the confounding effects of cellular neuroprotection. Interestingly, pretreatment with FK-506, at anti-ischemic doses, did not affect lesion-induced alterations in the number or distribution of GFAP-positive astrocytes or OX-42 labeled microglia (Butcher and coworkers; unpublished data). FK-506 is also reported to promote regrowth of the sciatic nerve in response to crush injury *(143)*, and to accelerate regrowth of the damaged facial nerve in young animals *(127)*. Transection of the facial nerve in adult rats results in atrophy of facial motor neurons and a transient loss of choline acetyltransferase (ChAT)-like immunoreactivity. In contrast to the neurotrophic effects observed in neonates, FK-506 had no effect on the phenotype of the adult facial motor neuron following transection, and did not affect lesion-induced astrocyte or microglial responses (Butcher and coworkers, unpublished data). Similarly, 3 days after MCA occlusion in rat, a ring of GFAP positive astrocytes and OX-42 labeled microglia were observed around the ischemic core. Whereas FK-506 treatment was associated with a significant reduction in the volume of brain damage, the astrocyte or microglial response was unaffected. However, the area of glial activation was reduced in accordance with the neuroprotective effect.

On balance, the available data suggest that whereas suppression of inflammatory cell activation may be an important mechanism in mediating the anti-ischemic activity of FK-506 and CsA in peripheral tissues, attenuation of glial responses to tissue injury is probably not their principle mechanism of action in the CNS. Obviously, such a view would be at variance with the attenuation of microglial activation reported following global ischemia by CsA *(133)*. However, the observations of Condo and coworkers *(133)* could equally be explained by a direct neuroprotective effect of CsA, which would reduce the volume of ischemic damage and consequently decrease the extent of the microglial response. Future studies will examine the involvement of leukocytes and cytokines in the neuroprotective actions of FK-506 and related compounds.

2.5.2.3. Direct Neuronal Actions

Although research on the immunophilins has primarily focused on their role in immunological responses *(121,144)*, there is increasing evidence that they may play a major role in signal transduction within the brain. The highest levels of cyclophilins and the FKBPs are not found within the blood or lymph glands, but in the brain parenchyma *(145–147)*. Furthermore, the distribution of [^{3}H]FK-506 binding activity, FKBP12 mRNA, and calcineurin overlap to a considerable degree *(145,147)*. Pharmacological studies using FK-506 have demonstrated effects on a variety of indices of brain function, such as synaptic plasticity, epileptiform discharges, and neuronal development *(148–154)*. These diverse responses are generally assumed to be mediated via an action on calcineurin, which has been shown to regulate the activity of dynamin *(155)*, NMDA receptors *(156)*, L-type Ca^{2+} channels *(157)*, nitric oxide synthase *(142)*, and neuromodulin *(147)*. Alternatively, functional effects of FK-506 in the brain may involve several other proteins, such as the ryanodine receptor *(158,159)*, the inositol 1,4,5-triphosphate receptor (IP3; *(160)*) and the type 1 transforming growth factor β receptor *(161)*, which are physically associated with FKBP12. Indeed, modification of ryanodine and IP3 receptor associated Ca^{2+} channel activity by FK-506 has been demonstrated *(158–160)*.

Whereas the role of calcineurin in the pathophysiology of cerebral ischemia is unclear, a number of intracellular mechanisms that could be relevant have been proposed. These include inhibition of nitric oxide production and suppression of apoptosis.

2.5.2.4. INHIBITION OF NITRIC OXIDE PRODUCTION

Snyder and his associates have proposed that FK-506 and CsA mediate their neuroprotective action in neuronal cell cultures by inhibiting nitric oxide synthase (NOS) activity *(127,142)*. This hypothesis has been elegantly composed around the following observations:

1. NMDA receptor-mediated toxicity is considered to be a major contributing factor to neuronal damage in cerebral ischemia (*see* Chapters 1, 5, and 6).
2. Stimulation of NMDA receptors elevates the concentrations of intracellular calcium, which in turn activates NOS, a calcium–calmodulin dependent enzyme.
3. Phosphorylation of NOS inhibits its catalytic activity.

In support of their hypothesis, Snyder and coworkers *(142)* reported that FK-506 and CsA inhibit NOS activity and block NMDA-induced neurotoxicity in cortical cell cultures. Their hypothesis is also consistent with reports that cyclosporin-induced hypertension may be mediated via inhibition of the NO-dependent relaxation of vascular smooth muscle *(161a,162)*. However, anti-ischemic doses of FK-506 do not inhibit excitotoxic brain damage produced by either intrastriatal or intracortical microinjection of NMDA or quinolinate (Butcher and Sharkey; unpublished data). In this model, in which MK-801 afforded almost total protection, L-NAME at doses that reduced NOS activity by >90% *(163)* was also without effect (Butcher and Sharkey; unpublished data). These findings should obviously be replicated by independent groups before such a promising model is discarded. An alternative mechanism to explain the neuroprotective action of FK-506 is via inhibition of superoxide production, as demonstrated in neutrophils *(120)*. Evidence supporting the role of superoxide in ischemic brain damage has been provided by studies using transgenic mice *(164)*. Superoxide anions have been shown to interact with nitric oxide to form peroxynitrite radicals, which are powerful biological oxidants *(165,166)*. Inhibition of superoxide production by FK-506 or CsA might be expected to reduce peroxynitrite levels, and since peroxynitrite mediated cell death is reported to involve apoptosis *(167,168)*, these drugs may protect against ischemic cell damage by inhibiting programmed cell death. In support of this hypothesis, FK-506 is also reported to inhibit apoptotic cell death induced by the HIV-1 envelope glycoprotein, gp120 *(169)*.

2.5.2.5. SUPPRESSION OF APOPTOSIS

A novel and potentially exciting mechanism to explain the neuroprotective actions of FK-506 and related compounds involves the inhibition of ischemia-induced programmed cell death (apoptosis). Apoptosis describes a type of cell death in which the target cell actively participates in its own death *(170)*. It plays a fundamental role in physiological processes such as embryo morphogenesis, tissue homeostasis, aging, and the development of immunological tolerance. Apoptosis can be distinguished from necrosis on the basis of the following morphological features: progressive condensation of the chromaffin to the inner surface of the nuclear membrane, cell shrinkage with subsequent detachment from neighboring cells, formation of acidophilic globules ("apoptotic bodies"), and fragmentation of DNA. Following agarose-gel electrophoresis, these fragments can be readily seen as a characteristic "ladder" pattern. In contrast, the progressive disappearance of chromatin observed in necrotic cell death is accompanied by random DNA breakdown that appears as a diffuse smear on the agarose gel.

Apoptosis has also been seen in response to a variety of pathological stimuli, including iatrogenic hepatic and renal hyperplasias, carcinogenesis and hepatic ischemia (for reviews, *see* refs. *(171* and *172)*. Whether apoptosis occurs in cerebral ischemia remains somewhat controversial *(173)*. However, a pattern of DNA laddering consistent with apoptosis has been reported following MCA occlusion in rat *(174–178)*, and forebrain ischemia in rat *(1790*, mouse *(180)*, and gerbil *(181,182)*. A role for programmed cell death is further supported by the existence of pro (e.g., p53) and antiapoptotic genes (e.g., bcl-2), which are increased *(183)* or reduced *(184)* following ischemia. In keeping with this view, recombinant knockout of p53 expression in mice has been shown to afford some degree of neuroprotection against MCA occlusion *(185)*, whereas overexpression of bcl-2 reduced the volume of damage associated with mouse MCA occlusion by 50% *(113)*. Furthermore, Shimazaki and coworkers *(184)* demonstrated that ischemic preconditioning, in which a brief exposure to ischemia produces subsequent protection against subsequent ischemic episodes, is associated with increased bcl-2 levels in the CA1 field of the hippocampus.

In 1993, Linnik and coworkers *(177)* demonstrated the presence of DNA laddering 24 h after permanent MCA occlusion in the rat. Using terminal deoxynucleotidyl-mediated dUTP-biotin nick end labeling (TUNEL), Linnik later reported the presence of TUNEL positive cells in the ischemic core as early as 1 h postocclusion. Thereafter, the appearance of TUNEL positive cells expanded in a radial fashion as the infarct developed *(186)*. These findings are in keeping with the hypothesis proposed by Yabuuchi and coworkers *(173)*, which predicts that tissue loss in the ischemic core is necrotic, whereas that found in the penumbra may be due to apoptosis *(173)*. More recently, Li and coworkers reported the appearance of apoptotic bodies and DNA laddering as little as 30 min after transient rat MCA occlusion *(174–176,180)*. Again, the distribution of apoptotic cells appeared to delimit the infarct boundary and expanded as the infarct developed, a pattern similar to that observed following permanent occlusion of the vessel.

Whereas there is no direct evidence to suggest that FK-506 and CsA mediate their anti-ischemic activity in vivo by inhibiting apoptosis, evidence from in vitro studies might lead one to speculate that such a mechanism is possible. Numerous studies have shown that CsA and FK-506 inhibit TCR-mediated apoptosis in T-cell hybridomas in vitro, and thymocyte selection in vivo, via a calcium-dependent mechanism *(124,187–191)*. Furthermore, drug affinity for inhibition of calcineurin activity is reported to correlate with that for preventing apoptosis *(126)*. In a recent study, Shibasaki and McKeon *(10)* provided additional evidence that calcineurin plays a critical role in calcium activated apoptosis. Using a fibroblast cell line (BHK-21) transfected with a constitutively active calcineurin mutant (CnA), they demonstrated that it is increased calcineurin activity, in preference to other calcium-mediated events, that stimulates apoptosis. Moreover, in this cell line calcineurin-induced apoptosis was inhibited by CsA, and by overexpression of bcl-2, thereby linking immunophilins, calcineurin and bcl-2 in the same apoptotic pathway *(10)*.

3. CONCLUSIONS

The study of inflammatory responses to cerebral ischemia is still in its infancy. However, we do know that suppression of the inflammatory response can have a significant impact on pathological and neurological outcome. A variety of anti-inflammatory agents have been shown to be effective in reducing infarct size in animal models of ischemia with reperfusion, possibly by attenuating the reperfusion injury response. Other agents (IL-1ra and immunosuppressants) can, in addition, attenuate ischemic damage in models of permanent vessel occlusion. With the development of new and ever more selective agents that block microglial activation, PMNL adhesion and infiltration, and ischemia-induced apoptosis, we will gain a better understanding of the role of inflammation and the immune system in cerebral ischemia. Doubtless, a better understanding of these events will result in new therapies that will have a major impact on the future clinical treatment of stroke.

REFERENCES

1 Schmid-Schonbein, G. W., Editorial comment, *Stroke,* **26** (1995) 681.

2 Perry, V. H. and Gordon, S., Macrophage and microglia in the nervous system, *TINS,* **6** (1988) 273–277.

3 Barone, F. C., Schmidt, D. B., Hillegrass, L. M., Price, W. J., White, R. F., Feuerstein, G. Z., Clark, R. K., Lee, E. V., Griswold, D. E., Sarau, H. M., and Beckman, J. S., Reperfusion increases neutrophils and leukotriene B4 receptor binding in focal cerebral ischemia, *Stroke,* **23** (1992) 1337–1348.

4 Chen, H., Chopp, M., and Bodzin, G., Neutropaenia reduces the volume of cerebral infarction after middle cerebral artery occlusion in the rat, *Neurosci. Res. Commun.,* **11** (1992) 93–99.

5 Hallenbeck, J. M., Dutka, A. J., Tanishima, T., Kochanek, P. M., Kumaroo, K. K., Obrenovitch, Y. P., and Contreras, T. J., Polymorphonuclear leukocyte accumulation in brain regions with low blood flow during the early postischemic period, *Stroke,* **17** (1986) 246–253.

6 Kochanek, P. M. and Hallenbeck, J. M., Polymorphonuclear leukocytes and monocytes/ macrophages in the pathogenesis of cerebral ischemia and stroke, *Stroke,* **23** (1992) 1367–1379.

7 Pozzilli, C., Lenzi, G. L., Argentino, C., Bozzao, L., Rasura, M., Giuabilei, F., and Fieschi, C., Peripheral white cell blood count in cerebral ischemic infarction, *Acta Neurol. Scand.,* **71** (1985) 396–400.

8 Frei, K., Siepl, C., Groscurth, P., Bodmer, S., Schwerdel, C., and Fontana, A., Antigen presentation and tumour cytoxicity by interferon-gamma-treated microglial cells, *Eur. J. Immunol.,* **17** (1987) 1271–1278.

9 Del Zoppo, G. J., Schmid-Schonbein, G. W., Mori, E., Copeland, B. R., and Chang, C. M., Polymorphonuclear leukocytes occlude capillaries following middle cerebral artery occlusion and reperfusion in baboons, *Stroke,* **22** (1991) 1276–1283.

10 Shibasaki, F. and McKeon, F., Calcineurin functions in Ca^{2+}-activated cell death in mammalian cells, *J. Cell. Biol.,* **131** (1995) 735–743.

11 Ames, A., Wright, L., Kowade, M., Thurston, J. M., and Majors, G., Cerebral ischaemia: the no-reflow phenomenon, *Am. J. Pathol.,* **52** (1968) 437–453.

12 Mori, E., Del Zoppo, G. J., Cambers, J. D., Copeland, B. R., and Arfors, K. E., Inhibition of polymorphonuclear leukocyte adherance suppresses no-reflow after focal cerebral ischemia in baboons, *Stroke,* **23** (1992) 712–718.

13 Garcia, J. H., Liu, K. F., Yoshida, Y., Lian, J., Chen, S., and Del Zoppo, G. J., Influx of leukocytes and platelets in an evolving brain infarct (wistar rat), *Am. J. Pathol.,* **144** (1994) 188–199.

14 Prentice, R. L., Szatrowski, T. P., Kato, H., and Mason, M. W., Leukocyte counts and cerebrovascular disease, *J. Chronic. Dis.,* **35** (1982) 703–714.

15 Fisher, M. and Francis, R., Altered coagulation in cerebral ischemia: platelet, thrombin and plasmin activity, *Arch. Neurol.,* **47** (1990) 1075–1079.

16 Grau, A. J., Graf, T., and Hacke, W., Altered influence of polymorphonuclear leukocytes on coagulation in acute ischaemic stroke, *Thromb. Res.,* **76** (1994) 541–549.

17 Grau, A. J., Sigmund, R., and Hacke, W., Modification of platelet aggregation by leukocytes in acute ischemic stroke, *Stroke,* **25** (1994) 2149–2152.

18 Grau, A., Seitz, R., Immel, A., Steichen-Wiehn, C., and Hacke, W., Increased levels of leukocyte elastase in ischemic stroke, *Stroke,* **24** (1993) 166.

19 Matsuo, Y., Kihara, T., Ikeda, M., Ninomiya, M., Onodera, H., and Kogure, K., Role of neutrophils in radical production during ischaemia and reperfusion of the rat brain: effect of neutrophil depletion on extracellular ascorbyl radical formation, *J. Cereb. Blood Flow Metabol.,* **15** (1995) 941–947.

20 Clark, R. K., Lee, E. V., White, R. F., Jonak, Z. L., Feuerstein, G. Z., and Barone, F. C., Reperfusion following focal stroke hastens inflammation and resolution of ischemic injured tissue, *Brain Res. Bull.,* **35** (1994) 387–392.

21 Garcia, J. H. and Kamijyo, Y. Cerebral infarction: evolution of histopathological changes after occlusion of the middle cerebral artery in primates. *J. Neuropath. Exp. Neurol.,* **33** (1974) 408–421.

22 Jander, S., Kraemer, M., Schroeter, M., Witte, O. W., and Stoll, G., Lymphocyte infiltration and expression if intercellular adhesion molecule-1 in photochemically induced ischemia of the rat cortex, *J. Cereb. Blood Flow Metabol.,* **15** (1995) 42–51.

23 Granger, D. N., Role of xanthine oxidase and granulocytes in ischemia-reperfusion injury, *Am. J. Physiol.,* **25** (1988) H1269–H1275.

24 Harlan, J. M., Leukocyte-endothelial interaction, *Blood,* **65** (1985) 513–525.

25 Lucchesi, B. R., Werns, S. W., and Fantone, J. C., The role of neutrophils and free radicals in ischemic myocardial injury, *J. Mol. Cell. Cardiol.,* **21** (1989) 1241–1261.

26 Mullane, K. M., Westlin, W., and Kraemer, R., Activated neutrophils release mediators that may contribute to myocardial dysfunction associated with ischemia and reperfusion, *Ann. NY Acad. Sci.,* **524** (1988) 103–121.

27 Ferrari, R., Ceconi, C., Curello, S., Cargnoni, A., Alfieri, O., Pardini, A., Marzollo, P., and Visioli, O., Oxygen free radicals and myocardial damage: protective role of thiol containing agents, *Am. J. Med.,* **95** (1991) 95s–105s.

28 Ward, P. A., Mechanisms of endothelial killing by H_2O_2 or products of activated neutrophils. *Am. J. Med.,* **91** (1991) 89s–94s.

29 Bowe, M. P., Zivin, J. A., and Rothlein, R., Monoclonal antibody to the ICAM-1 adhesion site reduces neurological damage in a rabbit cerebral embolism stroke model. *Exp. Neurol.,* **119** (1993) 215–219.

30 Chopp, M., Zhan, R. L., Chen, H., Li, Y., Jiang, N., and Rusche, R., Post-ischaemic administration of anti-ICAM-1antibody reduces ischaemic cell damage after transient middle cerebral artery occlusion in rats, *Stroke,* **25** (1994) 869–876.

31 Clark, W. M., Madden, K. P., Rothlein, R., and Zivin, J. A., Reduction of central nervous system ischaemic injury in rabbits using leukocyte adhesion antibody treatment, *Stroke,* **22** (1991) 877–883.

32 Jiang, N., Moyle, M., Soule, H. R., Rote, W. E., and Chopp, M., Neutrophil inhibitory factor is neuroprotective after focal ischemia in rats. *Ann. Neurol.,* **38** (1995) 935–942.

33 Matsuo, Y., Onodera, H., Shiga, Y., Nakamura, M., Ninomiya, M., Kihara, T., and Kogure, K., Correlation between myeloperoxidase-quantified neutrophil accumulation and ischaemic brain injury in the rat: effects of neutrophil depletion, *Stroke,* **25** (1994) 1469–1475.

34 Matsuo, Y., Onodera, H., Shiga, Y., Shozuhara, H., Ninomiya, M., Kihara, T., Tamatani, T., Miyasaki, M., and Kogure, K., Role of cell adhesion molecules in brain injury after transient middle cerebral artery occlusion in the rat, *Brain Res.,* **656** (1994) 344–352.

35 Jaeschke, H., Farhood, A., and Smith, C. W., Neutrophils contribute to ischemia/reperfusion injury in rat liver in vivo, *FASEB J.,* **4** (1990) 3355–3359.

36 Jolly, S. R., Kane, W. J., and Hook, B. G., Reduction of myocardial infarct size by neutrophil depletion: effect of duration of occlusion. *Am. Heart J.,* **112** (1986) 682–690.

37 Romson, J. L., Hood, B. G., Kunkel, S. L., Abrams, G. D., Schork, M. A., and Lucchesi, B. R., Reduction of the extent of ischemic myocardial injury by neutrophil depletion in the dog, *Circulation,* **67** (1983) 1016–1023.

38 Simpson, P. J., Fantone, J. C., and Mickelson, J. K., Identification of a time window for therapy to reduce experimental canine myocardial injury: suppression of neutrophil activation during 72 hours of reperfusion. *Circ. Res.,* **63** (1988) 1070–1079.

39 Hernandez, L. A., Grishma, M. B., Twohig, B., Arfors, K. E., Harlan, J. M., and Granger, D. N., Role of neutrophils in ischemia-reperfusion-induced microvascular injury, *Am. J. Physiol.,* **253** (1987) H699–H703.

40 Lo, S. K., Garcia-Szabo, R. R., and Malik, A. B., Leukocyte repletion reverses the protective effect of neutropaenia in thrombin-induced increase in lung vascular permeability, *Am. J. Physiol.,* **259** (1990) H149–H155.

41 Guilian, D. and Robertson, C., Inhibition of mononuclear phagocytes reduces ischemic injury in the spinal cord. *Ann. Neurol.,* **27** (1990) 33–42.

42 Means, E. D. and Anderson, D. K., Neuronophagia by leukocytes in experimental spinal cord injury, *J. Neuropath. Exp. Neurol.,* **42** (1983) 707–719.

43 Dutka, A. J., Kochanek, P. M., and Hallenbeck, J. M., Influence of granulocytopaenia on canine cerebral ischemia induced by air embolus, *Stroke,* **20** (1989) 390–395.

44 Helps, S. C. and Gorman, D. F. Air embolism of the brain of rabbits pretreated with mechlorethamine. *Stroke,* **22** (1991) 351–354.

45 Shiga, Y., Onodera, H., Kogure, K., Yamasaki, Y., Yashima, Y., Shouzuhara, H., and Sendo, F., Neutrophil as a mediator of ischemic edema formation in the brain, *Neurosci. Lett.,* **125** (1991) 110–112.

46 Chen, H., Chopp, M., Zhang, R. L., Bodzin, G., Chen, Q., Rusche, J. R., and Todd, R. F., Anti-CD11b monoclonal antibody reduces ischemic cell damage after transient focal cerebral ischemia in rat, *Ann. Neurol.,* **35** (1994) 458–463.

47 Morioka, T., Kalehua, T., and Streit, W. J., Characterization of microglial reaction after middle cerebral artery occlusion in rat brain, *J. Comp. Neurol.,* **327** (1993) 123–132.

48 Del Rio-Hortega, P., Microglia. In W. Penfeld (ed.) *Cytology and Cellular Pathology of the Nervous System*, Paul P. Hocker, New York, 1932, pp., 481–584.

49 Guilian, D., Corpuz, M., Chapman, S., Mansouri, M., and Robertson, C., Reactive mononuclear phagocytes release neurotoxins after ischemic and traumatic injury to the central nervous system, *J. Neurosci. Res.,* **36** (1993) 681–693.

50 Du Bois, M., Bowman, P. D., and Goldstein, G. W., Cell proliferation after ischemic infarction in gerbil brain, *Brain Res.,* **347** (1993) 245–252.

51 Colton, C. A. and Gilbert, D. L., Production of superoxide anions by a CNS macrophage: the microglia, *FEBS Lett.,* **223** (1987) 284–288.

52 Guilian, D. and Vaca, K., Inflammatory glia mediated delayed neuronal damage after ischemia in the central nervous system, *Stroke,* **24** (1993) I-84–I-90.

53 Piani, D., Spranger, M., Frei, K., Schaffner, A., and Fontana, A., Macrophage-induced cytotoxicity of N-methyl-D-aspartate receptor positive neurons involves excitatory amino acids rather than reactive oxygen intermediates and cytokines, *Eur. J. Immunol.,* **22** (1992) 2429–2436.

54 Guilian, D., Vaca, K., and Corpuz, M., Brain glia release factors with opposing actions upon neuronal survival, *J. Neurosci.,* **13** (1993) 29–37.

55 Vaca, K. and Wendt, E., Divergent effects of astroglia and microglia secretions on neuron growth and survival, *Exp. Neurol.,* **118** (1992) 62–72.

56 Chao, C. C., Shuxian, H., Molitor, T. W., Shaskan, E. G., and Peterson, P. K., Activated microglia mediate neuronal cell injury via a nitric oxide mechanism, *J. Immunol.,* **149** (1992) 2736–2741.

57 Shimojo, M., Nakajima, K., Takei, N., Hamanoue, M., and Kohsaka, S., Production of basic fibroblast growth factor in cultured brain microglia, *Neurosci. Lett.,* **123** (1991) 229–231.

58 Guilian, D., Ameboid microglia as effectors of inflammation in the central nervous system, *J. Neurosci. Res.,* **18** (1987) 155–171.

59 Korematsu, K., Goto, S., Nagahiro, S., and Ushio, Y., Microglial response to transient focal cerebral ischemia: an immunocytochemical study on the rat cerebral cortex using anti-phosphotyrosine antibody, *J. Cereb. Blood Flow Metabol.,* **14** (1994) 825–830.

60 Korematsu, K., Goto, S., Nagahiro, S., and Ushio, Y., Changes in the immunoreactivity for synaptophysin ("protein p38) following a transient cerebral ischemia in the rat, *Brain Res.,* **616** (1993) 320–324.

61 Kim, J. S., Chopp, M., and Gautam, S. C., High dose methylprednisolone therapy reduces expression of JE/MCP-1 mRNA and macrophage accumulation in the ischemic rat brain, *J. Neurol. Sci.,* **12** (1995) 28–35.

62 Kato, H., Kogure, K., Araki, T., and Itoyama, Y., Graded expression of immunomolecules on activated microglia in the hippocampus following ischemia in a rat model of ischemic tolerance, *Brain Res.,* **694** (1995) 85–93.

63 Berry, M., Maxwell, W. L., Logan, A., Mathewson, A., McConnell, P., Ashhurst, D. E., and Thomas, G. E., Deposition of scar tissue in the central nervous system, *Acta Neurochir.,* **32** (1983) 31–53.

64 Reier, P. J., Stenasaas, L. J., and Guth, L., Astrocytic scar as an impediment to regeneration in the central nervous system. In C. C. Kao, R. P. Bunge and P. J. Reier (eds.) *Spinal Cord Regeneration*, Raven, New York, 1983, pp. 163–195.

65 Ferrara, N., Ousley, F., and Gospodarowicz, D., Bovine brain astrocytes express basic fibroblast growth factor. a neurotropic and angiogenic mitogen, *Brain Res.,* **462** (1988) 223–232.

66 Guilian, D. and Lachman, L. B. Interleukin-1 stimulation of astroglail proliferation after brain injury, *Science,* **228** (1985) 497–499.

67 Guilian, D., Woodward, J., Young, D. G., Krebs, J. F., and Lachman, L. B., Interleukin-1 injected into mammalian brain stimulates astrogliosis and neovascularization, *J. Neurosci.,* **8** (1988) 2485–2490.

68 Gordon, C. R., Merchant, R. S., Marmarou, A., Rice, C. D., Marsh, T. J., and Young, H. F., Effect of murine recombinant interleukin-1 on brain edema in the rat, *Acta Neurochir.,* **51** (1990) 268–270.

69 Garcia, J. H., Liu, K. F., and Ho, K.-L. Neuronal necrosis after middle cerebral artery occlusion in wistar rats progresses at different time intervals in the caudatoputamen and the cortex. *Stroke,* **26** (1995) 636–643.

70 Rothwell, N. J. and Luheshi, G., Pharmacology of interleukin-1: actions in the brain. *Adv. Pharmacol.*, **25** (1994) 1–20.
71 Kishimoto, T. K. and Rothlein, R., Integrins, ICAMs, and selectins: role and regulation of adhesion molecules in neutrophil recruitment to inflammatory sites, *Adv. Pharmacol.*, **25** (1994) 117–169.
72 Montefort, S. and Holgate, S. T., Adhesion molecules and their role in inflammation, *Resp. Med.*, **85** (1991) 91–99.
73 McLain, C. J., Cohen, D., Ott, L., Dinarello, C. A., and Young, B., Ventricular fluid interleukin-1 activity in patients with head injury, *J. Lab. Clin. Med.*, **15** (1987) 48–54.
74 Buttini, M., Sauter, A., and Boddeke, H. W. G. M., Induction of interleukin-1b mRNA after focal cerebral ischemia in the rat, *Mol. Brain Res.*, **23** (1994) 126–134.
75 Liu, T., McDonnell, P. C., Young, P. R., White, R. F., Siren, A. L., Hallenbeck, J. M., and Feuerstein, G. Z., Interleukin-1β mRNA expression in ischemic rat cortex, *Stroke*, **24** (1993) 1746–1751.
76 Minami, M., Kuraishi, Y., Yabuuchi, K., Yamazaki, A., and Satoh, M., Induction of interleukin-1β mRNA in rat brain after transient forebrain ischaemia, *J. Neurochem.*, **58** (1992) 390–392.
77 Rothwell, N. J. and Relton, J. K., Involvement of cytokines in acute neurodegeneration in the CNS, *Neurosci. Biobehav. Rev.*, **17** (1993) 217–227.
78 Wang, X., Yue, T. L., Barone, F. C., White, R. F., Gagnon, R. C., and Feuerstein, G. Z. Concomitant expression of TNFa and IL-1b mRNAs follows early response gene expression in transient focal ischemia. *Mol. Chem. Neuropath.*, **23** (1994) 103–114.
79 Rothwell, N. J. and Strijbos, P. J. L. M., Cytokines in neurodegeneration and repair, *Int. J. Devl. Neurosci.*, **13** (1995) 179–185.
80 Breder, C. D., Dinarello, C. A., and Saper, C. B., Interleukin-1 immunoreactive innervation of the human hypothalamus, *Science*, **240** (1988) 321–324.
81 Lechan, R. M., Toni, R., and Clark, B. D., Immunoreactive interleukin-1b localisation in the rat forebrain, *Brain Res.*, **514** (1990) 134–140.
82 Ianotti, F., Interleukin-1β in focal cerebral ischaemia in rats. *J. Cereb. Blood Flow Metabol.*, **13** (1993) 125.
83 Yabuuchi, K., Minami, M., Katsumata, S., Yamazaki, A., and Satoh, M., An in situ hybridization study on interleukin 1β mRNA induced by transient forebrain ischemia in the rat, *Mol. Brain Res.*, **26** (1994) 135–142.
84 Guilian, D., Backer, Y. J., Shih, L. N., and Lachman, L. B., Interleukin-1 of the central nervous system is produced by ameboid microglia, *J. Exp. Biol. Med.*, **164** (1986) 594–604.
85 Schindler, R., Clark, B. D., and Dinarello, C. A., Dissociation between interleukin-1b and protein synthesis in human peripheral blood mononuclear cells, *JBC*, **265** (1990) 10,232–10,237.
86 Yamasaki, Y., Matsuura, N., Shozuhara, H., Onodera, H., Itoyama, Y., and Kogure, K., Interleukin-1 as a pathological mediator of ischemic brain damage in rats, *Stroke*, **26** (1995) 676–681.
87 Woodroofe, M. N., Sarna, G. S., Wadhwa, M., Hayes, G. M., Loughlin, A. J., Tinker, A., and Cuzner, M. L., Detection of interleukin-1 and interleukin-6 in adult brain, following mechanical injury, by in vivo microdialysis: evidence of a role for microglia in cytokine production, *J. Neuroimmunol.*, **33** (1991) 227–236.
88 Betz, A. L., Yang, G. Y., and Davidson, B. L., Attenuation of stroke size in rats using an adenoviral vector to induce overexpression of interleukin-1 receptor antagonist in brain, *J. Cereb. Blood Flow Metabol.*, **15** (1995) 547–551.
89 Martin, D., Chinookoswong, N., and Miller, G., The interleukin-1 receptor antagonist (rhIL-1ra) protects against cerebral infarction in a rat model of hypoxia-ischemia, *Exp. Neurol.*, **130** (1994) 362–367.

90 Relton, J. K. and Rothwell, N. J., Interleukin-1 receptor antagonist inhibits ischaemic and excitotoxic neuronal damage in the rat, *Brain Res. Bull.,* **29** (1992) 246.

91 Garcia, J. H., Liu, K. F., and Relton, J. K., Interleukin-1 receptor antagonist decreases the number of necrotic neurons in rats with middle cerebral artery occlusion, *Am. J. Pathol.,* **147** (1995) 1477–1486.

92 Sakr, M. F., McClain, C. J., Gavaler, J. S., Zetti, G. M., Starzl, T. E., and Van Thiel, D. H., FK506 pre-treatment is associated with reduced levels of tumour necrosis factor and interleukin-6 following hepatic ischemia/reperfusion. *J. Hepatol.,* **17** (1993) 301–307.

93 Buttini, M., Appel, K., Gebicke-Haerter, P.-J., and Boddeke, H. W. G. M., Expression of tumour necrosis factor alpha after focal cerebral ischaemia in the rat, *Neurosci.,* **71** (1996) 1–16.

94 Liu, T., Clark, R. K., McDonnell, P. C., Young, P. R., White, R. F., Barone, F. C., and Feuerstein, G. Z., Tumour necrosis factor-a expression in ischemic neurons, *Stroke,* **25** (1994) 1481–1488.

95 Uno, H., Matsuyama, T., Tagaya, M., Hata, R., Akita, H., Nakamura, H., and Sugita, M., Cerebral ischaemi triggers early prouction of tumour necrosis factor-α by microglia, *J. Cereb. Blood Flow Metabol.,* **15** (1995) s118.

96 Laskin, D. L. and Pendino, K. J., Macrophages and inflammatory mediators in tissue injury, *Annu. Rev. Pharmacol. Toxicol.,* **35** (1995) 655–677.

97 Maeyeri, P., Abrahams, C. S., Temesvari, P., Kovacs, J., Vas, T., and Speer, C. P., Recombinant tumour necrosis factor a constricts pial arterioles and increases blood-brain barrier permeability in newborn piglets, *Neurosci. Lett.,* **148** (1992) 137–140.

98 Sharief, M. K. and Thompson, E. J., In vivo relationship of tumour necrosis factor-alpha to blood-brain barrier damage in patients with active multiple sclerosis, *J. Neuroimmunol.,* **38** (1992) 27–34.

99 Estrada, C., Gomez, C., and Martin, C., Effects of TNF-a on the production of vasoactive substances by cerebral endothelial and smooth muscle cells in culture, *J. Cereb. Blood Flow Metabol.,* **15** (1995) 920–928.

100 Pober, J. S. and Cotran, R. S., Cytokines and endothelial cell biology, *Physiol. Rev.,* **70** (1990) 427–451.

101 Beamer, N. B., Coull, B. M., Clark, W. M., Hazel, J. S., and Silberger, J. R., Interleukin-6 and interleukin-1 receptor antagonist in acute stroke, *Ann. Neurol.,* **37** (1995) 800–804.

102 Tarkowski, E., Rosengren, L., Blomstrand, C., Wikkelso, C., Jensen, C., Ekholm, S., and Tarkowski, A., Early intrathecal production of interleukin-6 predicts the size of brain lesions in stroke, *Stroke,* **26** (1995) 1393–1398.

103 Wang, X., Yue, T.-L., Young, P. R., Barone, F. C., and Feuerstein, G. Z., Expression of interleukin-6, c-fos, and zif68mRNAs in rat ischemic cortex, *J. Cereb. Blood Flow Metabol.,* **15** (1995) 166–171.

104 Tilg, H., Trehu, E., Atkins, M. B., Dinarello, C. A., and Mier, J. W., Interleukin-6 as an anti-inflammatory cytokine: induction of circulating IL-1 receptor antagonist and soluble tumor necrosis factor receptor p55, *Blood,* **873** (1994) 113–118.

104a Chiang, C. S., Stalder, A., Samimi, A., and Campbell, I. L., Reactive gliosis as a consequence of interleukin-6 expression in the brain: studies in transgenic mice, *Dev. Neurosci.,* **16** (1994) 212–221.

105 Brett, F. M., Mizisin, A. P., Powell, H. C., and Campbell, I. L., Evolution of neuropathological abnormalities associated with blood-brain barrier breakdown in transgenic mice overexpressing IL-6 in astrocytes, *J. Neuropath. Exp. Neurol.,* **54** (1995) 766–775.

106 Cronstein, B. N. and Weissman, G., Targets for antiinflammatory drugs, *Annu. Rev. Pharmacol. Toxicol.,* **35** (1995) 449–462.

107 Dreyer, W. J., Michael, L. H., West, M. S., Smith, C. W., Rothlein, R., Rossen, R. D., Anderson, D. C., and Entman, M. L., Neutrophil accumulation in ischemic canine myocardium: insights into timecourse, distribution, and mechanism of localization during early reperfusion, *Circulation*, **84** (1991) 400–411.

108 Wang, X., Yue, T. L., Barone, F. C., and Feuerstein, G. Z., Demonstration of increased endothelial-leukocyte adhesion molecule-1 mRNA expression in rat ischemic cortex, *Stroke*, **26** (1995) 1665–1669.

109 Okada, Y., Copeland, B. R., Mori, E., Tung, M. M., Thomas, W. S., and Del Zoppo, G. J., P-selectin and intercellular adhesion molecule-1 expression after focal brain ischemia and reperfusion, *Stroke*, **25** (1994) 202–211.

110 Sobel, R. A., Mitchell, M. E., and Fondren, G., Intracellular adhesion molecule-1 (ICAM-1) in cellular immune reactions in the human brain, *Am. J. Pathol.*, **136** (1990) 1309–1316.

111 Takeshima, R., Kirsch, J. R., Koehler, R. C., Gomoll, A. W., and Traystman, R. J., Monoclonal leukocyte antibody does not decrease the injury of transient focal cerebral ischemia in cats, *Stroke*, **23** (1992) 247–252.

112 Forbes, A. D., Slimp, J. C., Winn, R. K., and Verrier, E. D., Inhibition of neutrophil adhesion does not prevent ischemic spinal cord injury, *Ann. Thorac. Surg.*, **58** (1994) 1064–1068.

113 Martinou, J.-C., Dubois-Dauphin, M., Staple, J. K., Rodriguez, I., Frankowski, H., Missotten, M., Albertini, P., Talabot, D., Catsicas, S., Pietra, C., and Huarte, J., Overexpression of bcl-2 in transgenic mice protects neurons from naturally occurring cell death and experimental ischemia, *Neuron*, **13** (1994) 1017–1030.

114 Kunz, J. and Hall, M. N., Cyclosporin A, FK506 and rapamycin: more than just immunosuppressants, *TIBS*, **18** (1993) 334–338.

115 Liu, J., FK506 and cyclosporin: molecular probes for studying intracellular signal transduction, *TIPS*, **14** (1993) 182–188.

116 Clipstone, N. A. and Crabtree, G. R., Identification of calcineurin as a key signalling enzyme in T-lymphocyte activation, *Nature*, **357** (1992) 695–697.

117 Jain, J., McCaffrey, P. G., Miner, Z., Kerppola, T. K., Lambert, J. N., Verdine, G. L., Curran, T., and Rao, A., The T-cell transcription factor NFAT$_p$ is a substrate for calcineurin and interacts with Fos and Jun, *Nature*, **365** (1993) 352–355.

118 Flannagan, W. M., Corthesy, B., Bram, R. J., and Crabtree, G. R., Nuclear association of a T cell transcription factor blocked by FK506 and cyclosporin A, *Nature*, **352** (1991) 803–807.

119 Liu, J., Farmer, J. D., Lane, W. S., Friedman, J., Weissman, I., and Schreiber, S. L., Calcineurin is a common target of cyclophilin-cyclosporin A and FKBP-FK506 complexes, *Cell*, **66** (1991) 807–815.

120 Nishinaka, Y., Sugiyama, S., Yokota, M., Saito, H., and Ozawa, T., Protective effect of FK506 on ischemia-reperfusion-induced myocardial damage in canine heart, *J. Cardiovasc. Pharmacol.*, **21** (1993) 448–454.

121 Nishiyama, M., Izumi, S., and Okuhara, M., Discovery and development of FK506 (Tacrolimus), a potent immunosuppressant of microbial origin. In V. J. Merluzzi and J. Adams (eds.) *The Search for Anti-Inflammatory Drugs*, Birkhauser, Boston, 1995, pp., 65–104.

122 Harding, M. W., Galat, A., Uehling, D. E., and Schreiber, S. L., A receptor for the immunosuppressant FK506 is a cis-trans peptidyl-prolyl isomerase, *Nature*, **341** (1989) 758–760.

123 Siekerka, J. J., Hung, S. H. Y., Poe, M., Lin, C. S., and Sigal, N. H., A cytosolic binding protein for the immunosuppressant FK506 has peptidyl-prolyl isomerase activity but is distinct from cyclophilin, *Nature*, **341** (1989) 755–757.

124 Fruman, D. A., Klee, C. B., Bierer, B. E., and Burakoff, S. J., Calcineurin phosphatase activity in T lymphocytes is inhibited by FK506 and cyclosporin A, *Proc. Natl. Acad. Sci. USA,* **89** (1992) 3686–3690.

125 Kino, T., Hatanaka, H., Miyata, S., Inamura, N., Yajima, T., Goto, T., Okuhara, M., Aoki, H., and Ochiai, T., FK506, a novel immunosuppressant isolated from a streptomyces. II. Immunosuppressive effects of FK506 in vitro, *J. Antibiot.,* **40** (1987) 1256–1265.

126 Bierer, B. E., Cyclosporin A, FK506, and rapamycin: binding to immunophilins and biological action. In L. E. Samelson (ed.) *Lymphocyte Activation,* Krager, Basel, 1994, pp., 128–155.

127 Snyder, S. H. and Sabatini, D. M., Immunophilins and the nervous system, *Nature Med.,* **1** (1995) 32–37.

128 Dumont, F. J., Melino, M. R., Staruch, M. J., Koprak, S. L., Fischer, P. A., and Sigal, N. H., The immunosuppressive macrolides FK506 and rapamycin act as reciprocal antagonists in murine T cells, *J. Immunol.,* **144** (1990) 1418–1424.

129 Owunwanne, A., Shihab-Eldeen, A., Sadek, D., Junaid, T., Yacob, T., and Abdel-Dayem, H. M., Radionuclide and histopathological assessment of any brain perfusion abnormalities in rats treated with the immunosuppressant drug (CyA), *Eur. J. Nucl. Med.,* **16** (1990) 519.

130 Toung, T. J. K., Bunke, F. J., Grayson, R. F., Kontos, G. J., Fraser, C. D., Baumgartner, W. A., Reitz, B. A., and Traystman, R. J., Effects of cyclosporin on cerebral blood flow and metabolism in dogs, *Transplantation,* **53** (1992) 1082–1088.

131 Sharkey, J. and Butcher, S. P., Calcineurin: a target for anti-ischaemic drug therapy, *J. Cereb. Blood Flow Metabol.,* **15** (1995) s387.

132 Shiga, Y., Onodera, H., Matsuo, Y., and Kogure, K., Cyclosporin-A protects against ischemia-reperfusion injury in the brain, *Brain Res.,* **595** (1992) 145–148.

133 Kondo, Y., Ogawa, N., Asanuma, M., Nishibayashi, S., Iwata, E., and Moori, A., Cyclosporin A prevents ischemia-induced reduction of muscarinic acetylcholine receptors with suppression of microglial activation in gerbil hippocampus, *Neurosci. Res.,* **22** (1995) 123–127.

134 Ogawa, N., Tanaka, K., Kondo, Y., Asanuma, M., Mizukawa, K., and Mori, A., The preventive effect of cyclosporin A, an immunosuppressant, on the late onset reduction of muscarinic acetylcholine receptors in gerbil hippocampus after transient forebrain ischemia, *Neurosci. Lett.,* **152** (1993) 173–176.

135 Sharkey, J., and Butcher, S. P., Immunophilins mediate the neuroprotective effects of FK506 in focal cerebral ischaemia, *Nature,* **371** (1994) 336–339.

136 Kawano, K., Kim, Y. I., Goto, S., Ono, M., and Kobayashi, M., A protective effect of FK506 in ischemically injured rat livers, *Transplantaion,* **52** (1991) 143–145.

137 Sakr, M. F., Zetti, G. M., Farghali, H., Hassanein, T. H., Gavaler, J. S., Starzl, T., and Van Thiel, D. H., Protective effect of FK506 against hepatic ischemia in rats, *Transplant. Proc.,* **23** (1991) 340–341.

138 Wakabayashi, H., Karasawa, Y., Tsubouchi, T., Maeba, T., and Tanaka, S. Hepatic protective effect of FK506 on warm ischemia-reperfusion injury in rats: suppressive effect on ICAM-1 expression. *Jpn. J. Gastreoenterol.,* **26** (1993) 1752.

139 Sakr, M., Zetti, G., McClain, C., Gavaler, J., Nalesnik, M., Todo, S., and Van Thiel, D. H., Protective effect of FK506 pre-treatment against renal ischemia-reperfusion injury in rats. *Transplantation,* **53** (1992) 987–991.

140 Hunter, J., Kubes, P., and Granger, D. N., Effects of cyclosporin A and FK506 on ischemia/reperfusion-induced neutrophil infiltration in cat intestine. *Gastroenterology,* **98** (1990) A176.

141 Wakita, H., Tomimoto, H., Akiguchi, I., and Kimura, J., Protective effect of cyclosporin A on white matter changes in the rat after chronic cerebral hypoperfusion. *Stroke,* **26** (1995) 1415–1422.

142 Dawson, T., Steiner, J. P., Dawson, V. L., Dinerman, J. L., Uhl, G. R., and Snyder, S. H., Immnuosuppressant FK506 enhances phosphorylation of nitric oxide synthase and protects against glutamate neurotoxicity, *Proc. Natl. Acad. Sci. USA,* **90** (1993) 9808–9812.

143 Gold, B. G., Storm-Dickerson, T., and Austin, D. R., The immunosuppressant FK506 increases functional recovery and nerve regeneration following peripheral nerve injury, *Restorative Neurol. Neurosci.,* **6** (1994) 287–296.

144 De Franco, A. L., Signal transduction: immunosuppressants at work, *Nature,* **352** (1991) 754–755.

145 Asami, M., Kuno, T., Mukai, H., and Tanaka, C., Detection of FK506-FKBP-calcineurin complex by simple binding assay, *Biochem. Biophys. Res. Comm.,* **192** (1993) 1388–1394.

146 Dawson, T. M., Steiner, J. P., Lyons, W. E., Fotuhi, M., Blue, M., and Snyder, S. H., The immunophilins, FK506 binding protein and cyclophilin, are discretely localised in the brain: relationship to calcineurin, *Neuroscience,* **62** (1994) 569–580.

147 Steiner, J. P., Dawson, T. M., Fotuhi, M., Glatt, C. E., Snowman, A. M., Cohen, N., and Snyder, S. H., High brain densities of the immunophilin FKBP colocalized with calcineurin, *Nature,* **358** (1992) 584–587.

148 Chang, H. Y., Takel, K., Syder, A. M., Born, T., Rusnak, F., and Jay, D. G., Asymmetric retraction of growth cone filopodia following focal inactivation of calcineurin, Nature, **376** (1995) 686–690.

149 Funauchi, M., Haruta, H., and Tsumoto, T., Effects of the inhibitor for calcium/calmodulin-dependent protein phosphatase, calcineurin, on induction of long-term potentiation in rat visual cortex, *Neurosci. Res.,* **19** (1994) 269–278.

150 Lyons, W. E., George, E. B., Dawson, T. M., Steiner, J. P., and Snyder, S. H., Immunosuppressant FK506 promotes neurite outgrowth in cultures of PC12 cells and sensory ganglia, *Proc. Natl. Acad. Sci. USA,* **91** (1994) 3191–3195.

151 Lyons, W. E., Steiner, J. P., Snyder, S. H., and Dawson, T. M., Neuronal regeneration enhances the expression of the immunophilin FKBP-12, *J. Neurosci.,* **15** (1995) 3985–3994.

152 Moia, L. J. M. P., Lizomar, H., Matsui, H., de Barros, G. A. M., Tomizawa, K., Miyamoto, K., Kuyata, Y., Tokuda, M., Itano, T., and Hatase, O., Immunosuppressant and calcineurin inhibitors, cyclosporin A and FK506, reversibly inhibit eleptogenesis in amygdaloid kindled rat, *Brain Res.,* **648** (1994) 337–341.

153 Mulkey, R., Endo, S., Shenolikar, S., and Malenka, R. C., Involvement of the calcineurin/inhibitor 1 phosphatase cascade in hippocampal long term depression, *Nature,* **369** (1994) 486–488.

154 Torii, N., Kamishita, T., Otsu, Y., and Tsumoto, T., An inhibitor for calcineurin, FK506, blocks induction of long-term depression in rat visual cortex, *Science,* **265** (1994) 674–676.

155 Liu, J.-P., Sim, A. T. R., and Robinson, P. J., Calcineurin inhibition of dynamin-1 GTPase activity coupled to nerve depolarization, *Science,* **265** (1994) 970–973.

156 Lieberman, D. N. and Mody, I., Regulation of NMDA channel function by endogenous Ca^{2+}-dependent phosphatase. *Nature,* **369** (1994) 235–239.

157 Armstrong, D. A., Calcium channel regulation by calcineurin, a Ca^{2+}-activated phosphatase in the mammalian brain, *TINS,* **12** (1989) 117–122.

158 Brillantes, A. B., Ondrias, K., Scott, A., Kobrinsky, E., Ondriasova, E., Moschella, M. C., Jayaraman, T., Landers, M., Ehrlich, B. E., and Marks, A. R., Stabilization of calcium release channel (ryanodine receptor) function by FK506-binding protein, *Cell,* **77** (1994) 513–523.

159 Timerman, A. P., Ogunbumni, E., Freund, E., Wiederrecht, G., Marks, A. R., and Fleischer, S., The calcium release channel of the sarcoplasmic reticulum is modulated by FK506-binding protein, *JBC,* **268** (1993) 22,992–22,999

160 Cameron, A. M., Steiner, J. P., Sabatini, D. M., Kaplin, A. I., Walensky, L. D., and Snyder, S. H., Immunophilin FK506 binding protein associated with inositol 1,4,5-triphosphate receptor modulates calcium influx, *Proc. Natl. Acad. Sci. USA,* **92** (1995) 1784–1788.

161 Wang, T., Donahoe, P. K., and Zervos, A. S. Specific interaction of type 1 receptors of the TGF-β family with the immunophilin FKBP-12. *Science,* **265** (1994) 674–676.

161a Roullet, J.-B., Xue, H., McCarron, D. A., Holcomb, S., and Bennett, W. M., Vascular mechanisms of cyclosporin-induced hypertension in the rat. *J. Clin. Invest.,* **93** (1994) 2244–2250.

162 Sudhir, K., MacGregor, J. S., DeMarco, T., De Groot, C. J. M., Taylor, R. N., Chou, T. M., Yock, P. G., and Chatterjee, K., Cyclosporin impairs release of endothelium-derived relaxing factors in the epicardial and resistance coronary arteries, *Stroke,* **90** (1994) 3018–3023.

163 Bannerman, D. M., Chapman, P. F., Kelly, P. A. T., Butcher, S. P., and Morris, R. G., M. Inhibition of nitric oxide synthase does not prevent the induction of long-term potentiation, *J. Neurosci.,* **14** (1994) 7415–7425.

164 Kamii, H., Kinouchi, H., Sharp, F. R., Koistinaho, J., Epstein, C. J., and Chan, P. H., Prolonged expression of hsp70 mRNA following transient focal cerebral ischemia in transgenic mice overexpressing CuZn-superoxide dismutase, *J. Cereb. Blood Flow Metabol.,* **14** (1994) 478–486.

165 Huie, R. E. and Padmaja, S., The reaction rate of nitric oxide with superoxide, *Free Radic. Res. Commun.,* **18** (1993) 195–199.

166 Lipton, S. A., Choi, Y. B., Pan, Z. H., Lei, S. Z., Chen, H. S. V., Sucher, N. J., Loscalzo, J., and Stamler, J. S., A redox-based mechanism for the neuroptotective and neurodestructive effects of nitric oxide and related nitroso-compounds, *Nature,* **364** (1993) 626–631.

167 Estevez, A. G., Radi, R., Barbeito, L., Shin, J. T., Thompson, J. A., and Beckman, J. S., Peroxynitrite-induced cytotoxicity in PC12 cells: evidence for an apoptotic mechanism differentially modulated by neurotrophic factors, *J. Neurochem.,* **65** (1995) 1543–1550.

168 Lin, K. T., Xue, J. Y., Nomen, M., Spur, B., and Wong, P. Y. K., Peroxynitrite-induce apoptosis in HL-60 cells. *JBC,* **270** (1995) 16,487–16,490.

169 Sekigawa, I., Hishikawa, T., Kanedo, H., Hashimoto, H., Hirose, S., Ingaki, Y., and Yamamoto, N., FK506 can inhibit apoptotic cell death induced by the HIV-1 envelope glycoprotein gp120, *AIDS,* **8** (1994) 1623,1624.

170 Kerr, J. F. R., Wyllie, A. H., and Currie, A. R., Apoptosis: a basic biological phenomenon with wide ranging implications, *Br. J. Cancer,* **26** (1972) 239–257.

171 Columbano, A., Cell death: current difficulties in discriminating apoptosis from necrosis in the context of pathological processes in vivo, *J. Cell Biochem.,* **58** (1995) 181–190.

172 Wright, S. C., Zhing, J., and Larrick, J. W., Inhibition of apoptosis as a mechanism for tumour promotion, *FASEB J.,* **8** (1994) 654–660.

173 Bonfoco, E., Krainc, D., Ankarcrona, M., Nicotera, P., and Lipton, S. A., Apoptosis and necrosis; two distinct events induced, respectively by mild and intense insults with N-methyl-D-aspartate or nitric oxide/superoxide in cortical cell cultures, *Proc. Natl. Acad. Sci. USA,* **92** (1995) 7162–7166.

174 Li, Y., Chopp, M., Jiang, N., Yao, F., and Zaloga, C., Temporal profile of in situ DNA fragmentation after transient middle cerebral artery occlusion in the rat, *J. Cereb. Blood Flow Metabol.,* **15** (1995) 389–397.

175 Li, Y., Chopp, M., Jiang, N., Zhang, Z. G., and Zaloga, C., Induction of DNA fragmentation after 10 to 120 minutes of focal cerebral ischemia in rats, *Stroke,* **26** (1995) 1252–1258.

176 Li, Y., Sharov, V. G., Jiang, N., Zaloga, C., Sabbah, H. N., and Chopp, M., Ultrastructural and light microscopic evidence of apoptosis after middle cerebral artery occlusion in the rat, *Am. J. Pathol.,* **146** (1995) 1045–1051.

177 Linnik, M. D., Zobrist, R. H., and Hatfield, M. D., Evidence supporting a role for programmed cell death in focal cerebral ischemia in rats, *Stroke,* **24** (1993) 2002–2009.

178 Tominaga, T., Kure, S., Narisawa, K., and Yoshimoto, T., Endonuclease activation following focal ischemic injury in the rat brain, *Brain Res.,* **608** (1993) 21–26.

179 Heron, A., Pollard, H., Dessi, F., Moreau, J., Lasbennes, F., Ben-Ari, Y., and Charriaut-Marlangue, C., Regional variability in DNA fragmentation after global ischemia evidenced by combined histological and gel electrophoresis observations in the rat brain, *J. Neurochem.,* **61** (1993) 1973–1976.

180 Li, Y., Chopp, M., Jiang, N., and Zaloga, C., In situ detection of DNA fragmentation after focal cerebral ischemia in mice, *Mol. Brain Res.,* **28** (1995) 164–168.

181 Iwai, T., Hara, A., Niwa, M., Nozaki, M., Uematsu, T., Sakai, N., and Yamada, H., Temporal profile of nuclear DNA fragmentation in situ in gerbil hippocampus following transient forebrain ischemia, *Brain Res.,* **671** (1995) 305–308.

182 Okamoto, M., Matsumoto, M., Ohtsuki, T., Taguchi, A., Mikoshiba, K., Yanagihara, T., and Kamada, T., Internucleosomal DNA cleavage involved in ischemia-induced neuronal death, *Biochem. Biophys. Res. Comm.,* **196** (1993) 1356–1362.

183 Li, Y., Chopp, M., Zhang, Z. G., Zaloga, C., Niewehuis, L., and Gautam, S., p53 immunoreactive protein and p53 mRNA expression after transient middle cerebral artery occlusion, *Stroke,* **25** (1994) 849–856.

184 Shimizaki, K., Ishida, A., and Kawai, N., Increase in bcl-2 oncoprotein and tolerance to ischemia-induced neuronal death in the gerbil hippocampus, *Neurosci. Res.,* **20** (1994) 95–99.

185 Crumrine, R. C., Thomas, A. L., and Morgan, P. F., Attenuation of p53 expression protects against focal ischemic damage in transgenic mice, *J. Cereb. Blood Flow Metabol.,* **14** (1994) 887–891.

186 Linnik, M. D., Miller, J. A., Sprinle-Cavallo, J., Mason, P. J., Thompson, F. Y., Montgomery, L. R., and Schroeder, K. K., Apoptotic fragmentation in the rat cerebral cortex induced by permanent middle cerebral artery occlusion, *Mol. Brain Res.,* **32** (1995) 116–124.

187 Bierer, B. E., Schreiber, S. L., and Burakoff, S. J., The effect of the immunosuppressant FK506 on alternative pathways of T-cell activation, *Eur. J. Immunol.,* **21** (1991) 439–445.

188 Bierer, B. E., Somers, P. K., Wandless, T. J., Burakoff, S. J., and Schreiber, S. L. Probing immunosuppressive action with a nonnatural immunophilin ligand. *Science,* **250** (1990) 556–559.

189 Genistier, L., Dearden-Badet, M. T., Bonnefon-Bernard, N., Lizard, G., and Revillard, J. P., Cyclosporin A and FK506 inhibit activation-induced cell death in the murine WEHI-231 B cell line, *Cell. Immunol.,* **155** (1994) 283–291.

190 Shi, Y., Sahai, B. M., and Green, D. R., Cyclosporin A inhibits activation-induced cell death in T-cell hybridomas and thymocytes, *Nature,* **339** (1989) 625–626.

191 Staruch, M., Sigal, N. H., and Dumont, F. J., Differential effects of the immunosuppressive macrolides FK506 and rapamycin on activation-induced T-cell apoptosis, *Int. J. Immunol.,* **13** (1991) 677–685.

192 Zhang, Z. G., Chopp, M., Tang, W. X., Jiang, N., and Zhang, R. L., Postischemic treatment (2–4 h) with anti-CD11b and anti CD18 monoclonal antibodies are neuroprotective after transient (2 h) focal cerebral ischemia in the rat, *Brain Res.,* **698** (1995) 79–85.

193 Zhang, R. L., Chopp, M., Li, Y., Zaloga, C., Jones, M. L., Miyasaka, M., and Ward, P. A. Anti-ICAM-1 antibody reduces ischemic cell damage after transient middle cerebral artery occlusion in the rat. *Neurology,* **44** (1994) 1747–1751.

11

Animal Models Used in Cerebral Ischemia and Stroke Research

Akira Tamura, Kensuke Kawai, and Kiyoshi Takagi

1. INTRODUCTION

1.1. Prologue

The central nervous system (CNS) is extremely vulnerable to ischemia. In humans, the brain accounts for approx 2% of body weight, but receives about 15% of cardiac output. The amounts of energy metabolites (glucose and glycogen) and oxygen stored in the brain are so small that cessation of blood supply for only a few minutes leads to severe CNS damage. This vulnerability of the CNS mainly derives from the vulnerability of neurons, the major component of the CNS. Therefore, many therapeutic modalities have been developed to protect neurons from ischemic damage. In this sense, cultured neurons (in vitro model) are enough to test the efficacy of a therapy. However, the results of such tests indicate only the level of cytotoxicity of a therapy. The cultured neurons never show neurological deficits. Ischemic insult to the human CNS may lead to a wide variety of signs and symptoms, from death to very slight neurological deficits. Rational therapies for cerebral ischemia should be established on a detailed understanding of the pathomechanisms involved. This is why we need experimental stroke models.

Many different types of cerebral ischemia models have been developed. From the standpoint of degree of complexity, ischemia models can be classified as follows:

1. Cultured neurons without synaptic formation *(1)*.
2. Cultured neurons with synapse formation *(2)*.
3. Cultured neurons with synapse formation and glia *(3)*.
4. Cultured brain slice *(4)*.
5. Rodent.
6. Mid-sized animal.
7. Subhuman primate.
8. Human (clinical observation and neuroimaging studies).

An animal model is a living experimental system that contains whole elements, neurons, glia, vasculature, and cerebrospinal fluid (CSF). We will limit the scope of this chapter to animal models of cerebral ischemia.

From: Clinical Pharmacology of Cerebral Ischemia *Edited by: G. J. Ter Horst and J. Korf Humana Press Inc., Totowa, NJ*

265

1.2. Requirements for Animal Ischemia Models

Reflecting the importance of animal ischemia models as experimental counterparts of stroke in humans, with which newly developed therapeutic modalities are tested, several conditions that the model must fulfill have been proposed for focal ischemia.

1.2.1. Hudgins (5)

1. A high percentage of infarcts with a predictable average size.
2. No surgical manipulation of the brain or exposure of it to the air.
3. In vivo perfusion fixation.

1.2.2. Garcia (6)

1. A single artery can be reproducibly occluded.
2. The vascular occlusion results in predictable changes in blood flow, i.e., focal or regional ischemia.
3. No barbiturates are used at the time of the arterial occlusion.
4. Inducement of the above conditions should always result in a parenchymal lesion closely resembling a human brain infarction.
5. The method of arterial occlusion should be compatible with subsequent reperfusion of the ischemic territory.

These conditions focus mainly on focal ischemia and not global ischemia.

As stated below, pathological and functional evaluations are indispensable in determining the efficacy of a therapeutic modality. Autoradiography is now a very important part of animal ischemia studies. These studies of cerebral ischemia require statistic analysis. Thus, animal ischemia models must fulfill the following conditions:

1. Economical, with many animals with genetic homogeneity easily available.
2. Constant, reliable, and reproducible ischemia can be induced.
3. Blood flow can be reestablished.
4. Neurological and pathological evaluation is easy.
5. Physiological variables are controllable without difficulty.
6. Neuroimaging studies such as autoradiography, MRI, MRS, and PET can be carried out economically and easily.
7. Long-term follow-up is possible.
8. Resistant to infection.

1.3. Classification of Animal Ischemia Models

Experimental cerebral ischemia models can be classified with respect to completeness of blood interruption, duration of ischemia, and spatial extent (global or focal).

From the practical standpoint of view, the following animal models are possible:

1. Transient complete global ischemia
2. Transient incomplete global ischemia
3. Permanent incomplete global ischemia
4. Transient incomplete focal ischemia
5. Permanent incomplete focal ischemia

1.4. Animal Species

Several animal species are candidate hosts for both focal and global ischemia models; subhuman primates *(5,7–10)*, cats *(11–13)*, dogs *(14,15)*, rabbits *(16–18)*, rats *(19–26)*, gerbils *(27,28)*, and mice *(29)*.

The focal ischemia model in subhuman primates closely resembles human stroke due in part to the similarities between subhuman primates and humans in the behavior, motor and sensory integration, amount of neocortex, and construction of cerebrovasculature *(30)*. Although subhuman primates are extremely valuable for the study of ischemic stroke *(6)*, it is currently quite difficult to use them as hosts for experimental models for economic and ethical reasons.

Among the animal species listed above, rats are now most widely used. This is because pure strains are available, detailed anatomical studies have been established *(31–36)*, they are inexpensive *(24,37)*, they are small (easy to handle and good for autoradiographic study), their intracranial circulation is similar to that of humans *(38,39)*, infarction can be induced consistently and reliably in them, and a large amount of neurochemical data on rats is available *(40)*.

In fact, the rat meets all the conditions for an animal focal ischemia model. However, when it comes to introducing a newly developed therapy whose efficacy has been confirmed in a rat focal ischemia model, other focal ischemia models in mid-sized animals, particularly the cat, still hold validity.

1.5. Factors Affecting the Results of Ischemic Experiments

Through meticulous studies of the mechanisms of ischemic brain damage, many factors affecting the extent of ischemic brain damage have been identified.

1.5.1. Experimental Conditions Including Physiological Variables

1.5.1.1. TEMPERATURE

Busto et al. were the first to describe the effects of small differences in brain temperature on ischemic cerebral damage *(41)*; since then, the importance of control of brain temperature in experiment of global ischemia has been emphasized *(42,43)*. The importance of temperature control was also demonstrated in focal ischemia *(44–48)*. Because postischemic hypothermia may have a protective effect on ischemic cerebral damage *(49–51)*, temperature management of experimental animals after surgery may also affect the infarct volume. Minamisawa et al. *(42)* used a closed chamber equipped with a heating fan and a heated water bath in order to strictly control the brain temperature during and after ischemia. Recently, Ginsberg summarized the effects of brain temperature on cerebral ischemia *(52)*.

The site at which (intracerebral, subdural, in the temporal muscle, or rectal) brain temperature is monitored must be carefully selected. Although the temperature of the temporal muscle correlates well with brain temperature in the rat middle cerebral artery (MCA) occlusion model with craniectomy *(48)*, only a slight change in the direction of the light beam of a heat lamp may easily alter the brain temperature. It is very important to keep experimental conditions constant.

1.5.1.2. GLUCOSE LEVEL

Blood glucose level is also an important physiological parameter in in vivo ischemia models. Many experimental studies *(10,53–58)* and one clinical study *(59)* have revealed the aggravating effect of hyperglycemia on ischemic brain damage. However, in some studies an inhibiting effect of hyperglycemia on ischemic brain damage, particularly in focal ischemia models was observed *(18,60,61)*. Regardless of the reported effects of hyperglycemia, the blood glucose level should be maintained at a constant level. Overnight fasting is a useful method in rodents to maintain the blood glucose level in a narrow range.

1.5.1.3. BLOOD PRESSURE

Theoretically, blood pressure during ischemia has no effect on the results of the experiment in complete global ischemia without hypotension. The situation is quite different in global ischemia with hypotension *(20,62,63)* and focal ischemia. In global ischemia with hypotension, cessation of blood flow is not achieved by only slight elevation of blood pressure during the ischemia. Hypotension during focal ischemia markedly aggravates ischemic brain damage *(64,65)*. Therefore, blood pressure should be carefully controlled during experiment.

1.5.1.4. OTHERS

Anesthetic agents *(66)*, pH *(67)*, hematocrit, and blood gas concentrations are also important factors influencing the results. A long duration of surgery and/or anesthesia generally means complicated surgery and may result in increased ischemic damage. For example, we make it a rule not to include in our studies animals who underwent MCA occlusion surgery that lasted longer than 30 min.

1.5.2. Factors Relating to the Animal Itself

1.5.2.1. AGE

The body weights of rats and gerbils, both widely used in ischemia research, are generally 200–400 g and 60–80 g, respectively. Precise brain atlases of these animals in these weight ranges are available *(33,68)*. However, it should be kept in mind that the rat and gerbil in these weight ranges are young adults. Whereas, stroke usually occurs in humans older than 50 yr with underlying cardiovascular or hematologic predisposing conditions, the rodents usually used in experiments are correspondingly younger and healthier. It is said that the ages of 50–80 yr in humans correspond to the ages of 2–3 years in rats.

Ischemic lesion has been suggested to be worse in aged rats than in younger animals *(56,69–71)*.

1.5.2.2. SEX

While animals of either sex can be used in cerebral ischemia research, in the vast majority of ischemia studies, male animals are employed. The preference for male animals can partly be explained as an effort to avoid the possible effects of periodic hormonal changes on ischemic injury. Furthermore, in several studies the ischemic damage induced was found to be more extensive in male than in female rodents. Payan and Conard *(72)* reported that young gerbils prior to sexual maturity and sexually mature female gerbils were more resistant to ischemia than sexually mature male gerbils following bilateral carotid ligation, as evaluated with mortality. Similarly, metabolic distress following bilateral carotid artery occlusion in spontaneously hypertensive rats (SHR) was reported to be more severe in the male than in the female animals *(73)*. Although the sex difference in susceptibility to ischemia in these studies was attributed to differences in the collateral circulation *(74)* or the influence of the gonads on the blood flow *(72)* and blood pressure *(73)*, Hall et al. recently suggested the involvement of antioxidant effects of estrogen *(75)*.

1.5.2.3. OTHERS

In the focal ischemia model in the rat, the size of the infarction strongly depends on the animal strain *(56,76)*. Although the reason is unclear, the severity of ischemic lesion also depends on the animal supplier *(25,56)*. Currently, SHR is frequently used in experimental stroke studies. The high blood pressure of this strain may be easily decreased by changing the type of food pellets fed to them.

To sum it up, it is fundamentally important to maintain all conditions as strictly as possible during a set of experiments. Changes in conditions should be avoided, particularly in studies involving small animals such as the rat.

1.6. MONITORING

1.6.1. What Can Be Monitored? To Monitor or Not to Monitor?

The monitoring systems currently used vary widely. For example, physiological variables such as hematocrit, blood glucose level, blood gas concentrations, and information on the brain itself such as EEG, regional cerebral blood flow (rCBF), data on extracellular fluid collected by microdialysis, and cortical vascular changes through cranial window are monitored.

In a small, anesthetized animal, physiological variables are easy to alter even with slight changes in atmospheric temperature. Therefore, physiological variables should be monitored when the effectiveness of a therapeutic modality in one experimental group is to be compared with that in one or more others. However, the use of needles and/or probes to monitor rCBF or obtain extracellular information may result in damage to the brain tissue and frequent blood sampling may induce changes in the systemic hemodynamics. Because of this dilemma, monitored variables should be as limited as little as possible and no methodological process should be altered during one set of experiments.

1.6.2. What Should Be Monitored?

All variables affecting ischemic damage should be monitored. Monitoring of the physiological parameters during and after ischemia is indispensable in evaluating the efficacy of some therapeutic strategies.

Blood pressure, blood glucose levels, blood gas concentrations, pH, body temperature, and brain temperature should be monitored at least. CBF and other variables are to be monitored as necessary.

2. GLOBAL ISCHEMIA MODELS

2.1. Total Body Ischemia

2.1.1. Decapitation and Cardiac Arrest Without Resuscitation

Decapitation of experimental animals is the easiest way to achieve global cerebral ischemia. Obviously the major disadvantage of this model is that studies involving postischemic recirculation is impossible. Therefore, adoption of this model is limited to studies on morphological and biochemical changes during very early phases of ischemia. Similarly categorized is a model of cardiac arrest without resuscitation. Xie et al. *(77)* examined involvement of ion channels in anoxic depolarization in a very early phase of global ischemia in a rat model of cardiac arrest induced by iv injection of magnesium chloride.

2.1.2. Cardiac Arrest and Resuscitation

Complete global cerebral ischemia is achieved by cardiac arrest. Systemic cessation of blood circulation excludes the possibly remaining collateral circulation into the brain, which can be a cause of variability in vascular occlusion models. When animals can be resuscitated successfully, this model can represent global cerebral ischemia very similar to that seen in clinical settings and can be a good model for clinical pathologies such as postresuscitation encephalopathy. However, injury to the systemic organs, especially to the myocardium, and disturbance in the systemic physiological parameters such as the arterial blood gas and blood electrolyte concentrations are inevitable and their effects on ischemic cerebral injury are unknown. The fact that it is usually difficult to obtain repeatedly the same degree of transient ischemia even if resuscitation is started at the same time may reduce the validity of this model for drug studies where groups of animals are compared.

Cardiac arrest and resuscitation have been attempted in various experimental animals including monkeys *(10)*, dogs *(14)*, cats *(13)*, and rats *(78–80)*. Cardiac arrest (or ventricular fibrillation) can be induced by injection of potassium chloride *(10,78,79)*, electrical shock *(13,14)*, or mechanical obstruction of the ascending aorta *(80)*. Resuscitation consists of artificial ventilation, closed chest cardiac massage, and in some cases electrical defibrillation. As the duration of cardiac arrest increases, the rate of successful resuscitation decreases. De Garavilla et al. *(78)* reported that in their rat model of cardiac arrest as the duration of ischemia was increased from 8 to 18 min, the rate of survival immediately following resuscitation decreased from 100 to 25%, and survival at 48 h after ischemia decreased from 60 to 0%. Moreover, as the duration of cardiac arrest increases, the degree of recirculation instability increases. When resuscitation was started 3, 5, 7, or 9 min after the start of cardiac arrest in the rat, the duration from start of arrest to the time when the mean arterial blood pressure recovered to 50 mmHg was 4.5 ± 0.5, 8.3 ± 0.7, 10.7 ± 0.9, or 13.7 ± 1.2 min, respectively *(81)*. Thus, whereas longer ischemia can create severe brain damage, the greater variability in the severity of ischemic damage cannot be avoided. Furthermore, attempts of resuscitation after long cardiac arrest must involve intensive care including use of various drugs, further complicating studies of postischemic brain damage, although this is often the case in clinical settings. Therefore, the duration of experimental cardiac arrest, thus far, has been shorter than 10–15 min if chronic brain injury is the objective of the study. Leonov et al. *(82)* studied the effect of hypothermia in a dog model of 17 min cardiac arrest in which they used cardiopulmonary bypass in order to stabilize the circulation in early resuscitation phases.

Whereas cardiac arrest models in larger animals are more desirable for pharmacological studies to evaluate the efficacy of various drugs against cerebral ischemia owing to their resemblance to cardiac arrest seen in clinical settings, use of larger animals has been increasingly limited due to economic and ethical problems, and the necessity for complicated experimental preparation. If experiments are designed for specific aspects of cerebral ischemia, a rodent model of cardiac arrest is appropriate for economic reasons and because of the possibility of mass production of ischemic animals. One of the authors studies histological brain damage and development of susceptibility to audiogenic seizures using a rat model of 10 min cardiac arrest, which allows long-term observation *(80,81)*. An L-shaped stainless steel hook inserted into the mediastinum was used to induce cardiac arrest without thoracic surgery. Artificial ventilation and closed chest massage were enough to resuscitate the animals and no use of drugs or intensive care was necessary. In spite of some variability in ischemic duration, early damage was uniformly observed in the thalamus when resuscitation was started at longer than 5 min, and delayed damage in the hippocampus was uniformly observed when resuscitation was started at longer than 7 min.

Temperature control is a complicated matter in cardiac arrest models. Whereas the difference between the brain (or skull) temperature and rectal temperature is much smaller compared with that in vascular occlusion models, a decrease in temperature during cardiac arrest is inevitable where modification with circulating blood or breathing air is impossible. Theoretically, the use of a temperature-control chamber as reported by Minamisawa et al. *(43)* allows maintenance of a constant temperature. Leonov et al. *(82,83)* controlled temperature by changing the temperature of the cardiopulmonary bypass circuit. We found that in small animals such as rats it is possible to control the temperature using only a heating lamp and heating pad or ice pad.

Cardiac arrest and resuscitation of experimental animals has advantages in the terms of the completeness of the global cerebral ischemia that can be induced and resemblance to the clinical settings. However, the difficulties involved in setting the exact ischemic duration limit the adoption of this model to specific experiments. One experimental aspect in this context seems to be effects of systemically circulating mediators for other organs subjected to transient ischemia on ischemic cerebral injury.

2.1.3. Profound Systemic Hypotension

Profound systemic hypotension induced by pharmacological agents results in incomplete but almost global cerebral ischemia. Brierley et al. *(84)* examined changes in various physiological parameters and histopathological changes in the brain using this model in the rhesus monkey. The main advantage of a profound systemic hypotension model seems to be the ease of resuscitation of the animal. However, without rCBF study, possible variabilities in severity and regional distribution of cerebral ischemia may limit adoption of such models to experimental studies focusing on phenomenology in the postischemic brain. Yatsu et al. *(85)* overcame these drawbacks by combining systemic hypotension (mean arterial blood pressure 30–35 mmHg) and hypoxia (4% oxygen) and, using a standard of five min of isoelectric electroencephalography, observed a protective effect of a rapid-acting barbiturate on ischemic damage.

2.2. Global Cerebral Ischemia

2.2.1. Increased Intracranial Pressure

Transient global cerebral ischemia can also be achieved by increasing the intracranial pressure. Since cerebral perfusion pressure is the difference between the arterial blood pressure and the intracranial pressure, cisternal injection of artificial CSF, which causes an increase in the intracranial pressure to greater than the systolic arterial blood pressure, results in a CBF of almost zero. This model has been established in various experimental animals including dogs *(86)*, rabbits *(16)*, and rats *(87)*. One merit of this model is the ease of control of the brain temperature during ischemia by changing the temperature of the fluid injected *(88)*. Although in this model, global cerebral ischemia can easily be achieved, the primary pathological process is not ischemia but intracranial hypertension. Treatment with Cushing's phenomenon, i.e., systemic hypertension, resulting from the intracranial hypertension is necessary. Ross and Duhaime *(88)* reported that they obtained ischemic brain lesions similar to those in other models of rat global cerebral ischemia.

2.2.2. Asphyxia or Cervical Compression

Occlusion of the cervical blood vessels by tourniquet when combined with systemic hypotension was reported to result in ischemic damage in the brain. The mean arterial pressure was reduced to 50 Torr by trimetaphan camsylate and a high-pressure neck cuff was inflated to 30 psi in the rhesus monkey *(89)* and the cat *(90)*. Complete arrest of the brain circulation was verified by brain scan after intra-arterial injection of radio-isotopes and isoelectric electroencephalography within 15 s after the start of ischemia. However from an anatomical point of view, it seems difficult to completely occlude the vertebral and anterior spinal arteries. In a strict sense, this model should be considered not a global cerebral ischemia but a forebrain ischemia one. Furthermore, the effect of compression of the various cervical nerves and ganglia on ischemic cerebral injury, and especially postischemic cerebral circulation, is unknown.

2.2.3. Combination of Surgical Occlusion of the Major Arteries

Attempts to obtain global cerebral ischemia have been made in large animals with combination of surgical occlusion of the major arteries.

Hossmann and Zimmermann *(63)* produced global cerebral ischemia of 60 min in adult monkeys *(Macaca mulatta)* by intrathoracal clamping of the brachiocephalic and left subclavian arteries close to their origin at the aortic arch with ligation of the internal mammary arteries and lowering of the arterial blood pressure to 80 mmHg. Complete arrest of the cerebral circulation was confirmed with injection of Evans Blue. They observed collateral blood flow through the ascending spinal arteries when the arterial blood pressure was above 150 mmHg.

In the dog, extensive collateral circulation makes complete interruption of cerebral circulation quite difficult. Osgood et al. *(91)* investigated the efficacy of mammary-carotid anastomosis in a dog model of global cerebral ischemia by ligation and transection of both common carotid, vertebral, costocervical, mammary, omocervical, and axillary arteries via left thoracotomy. The circulatory arrest was permanent and its completeness was not confirmed in their model. Snyder et al. *(92)* occluded the aorta and venae cavae for 15 min by right thoracotomy confirming cardiac outflow to the coronary and pulmonary circulation and examined postischemic intracranial pressures, cerebral blood flow, and glucose uptake for 2 h. Miller et al. *(93)* added clamping of the descending aorta to transient occlusion of the brachiocephalic and left subclavian arteries in order to exclude the collateral circulation to the brain arising from the descending aorta. In this case, catheterization from the femoral artery into the aortic arch was necessary to provide a pressure vent for the left ventricle to prevent blood congestion in the heart and lungs. In contrast, Kayama et al. *(94)* attempted to create global cerebral ischemia with regulated perfusion by occlusion of the major cerebral arteries comprising the circle of Willis and cannulation into the unilateral MCA via craniotomy under a surgical microscope.

Using cats, Hossmann and Sato *(62)* reported a full recovery of the pyramidal responses and evoked electroencephalographic activity after global cerebral ischemia of 1 h. It is also known that postischemic cats can survive for as long as one year *(95)*. They ligated the internal mammillary artery, temporarily clamped the innominate and subclavian arteries, and lowed the systemic blood pressure to 80 mmHg. The total ischemia of the brain and spinal cord to the upper thoracic level was confirmed by the intravenous injection of Evans Blue. Zaren et al. *(96)* produced global cerebral ischemia in the cat by ligation of the bilateral subclavian and vertebral arteries and temporary occlusion of the common carotid arteries.

A rabbit model of global cerebral ischemia induced using this approach was also reported *(97,98)*. The basilar artery was transected via an anterior cervical approach and several days later both common carotid arteries were temporarily clipped and a cervical pressure cuff was inflated.

As described, extended intrathoracic or intracranial surgery is necessary when induction of global cerebral ischemia is attempted by combining multiple occlusions of the major arteries. Whereas myocardial injury can be avoided, one must keep in mind the effects of extended surgical intervention.

2.3. Forebrain Ischemia

Models of forebrain ischemia in rodents were first reported in the late 1970s and early 1980s. Since then, these models have been extensively used for studies on various aspects of cerebral ischemia including pharmacological studies. Since the cerebral ischemia in these models is not completely global, but restricted to the forebrain, these models do not accurately represent cerebral ischemia encountered in clinical settings in which bilateral forebrain ischemia rarely occurs. However, the remaining blood flow in the hindbrain excludes the respiratory and systemic circulatory effects unlike in total body ischemia models and complete global cerebral ischemia models. These rodent models of forebrain ischemia have contributed in the last 15 yr to accumulation of vast amounts of knowledge on the underlying mechanisms of and therapeutic approaches against ischemic neuronal injury such as the roles of excitatory amino acids and protective effects of glutamate antagonists *(99–101)*, changes in and roles of intracellular signal transduction pathways *(102–104)*, changes in levels of expression of nucleic and mitochondrial genes, especially ones encoding stress proteins *(105–107)*, changes in rates of protein synthesis *(108–110)*, involvement of free radical injury and nitric oxide *(111,112)*, protective effects of neurotrophic factors *(113)*, and involvement of apoptosis *(114)*.

2.3.1. Bilateral Common Carotid Occlusion in Mongolian Gerbils

The absence of anastomosis between the vertebral and internal carotid circulation in gerbils enables forebrain ischemia to be induced by merely occluding both common carotid arteries *(27)*. In young gerbils, communication exists between the two circulations *(69)*, and gerbils weighing more than 60 g are generally used. Originally unilateral carotid artery occlusion was used as a model of focal ischemia since, in some gerbils, the anterior communicating artery is aplastic *(115)*. Although the ischemia induced by bilateral carotid artery occlusion is not completely global, ischemic damage to the forebrain structures including the caudoputamen and the hippocampus is reproducible, and this model has been widely used since the report of delayed neuronal death in the hippocampus by Kirino *(116)*. Bilateral carotid artery occlusion for 10–15 min resulted in selective neuronal loss in the hippocampal CA1 region, the striatum and layers III, V, and VI of the neocortex. In contrast, 5-min ischemia resulted in delayed neuronal death in the hippocampus, in which no morphological changes in the CA1 neurons were detected by light microscopy 1 d after ischemia but neuronal loss progressed a few days after ischemia. There is some interanimal variability in the severity of ischemia induced by bilateral occlusion of the common carotid arteries, which can usually be compensated for by increasing the number of animals used in studies with statistic management. Gerbils are known to have a high percentage of postischemic seizures especially when the duration of ischemia is longer than 15 min. Monitoring and control of the body or brain temperature is essential in gerbils *(117)*.

Occlusion of bilateral common carotid arteries involves surgical dissection of the anterior aspects of the neck and clipping of the arteries under anesthesia using a surgical microscope. Slightly extending the neck makes finding and manipulating the arteries easy. It is important to dissect the arteries free from the accompanying jugular veins and vagus nerves and to confirm cessation of blood flow in the distal portion of the arteries during clipping. In order to occlude both common carotid arteries in awake gerbils, a simple occlusion/release device has been used *(118)*. Tomida et al. *(119)* designed another type of simple occlusion/release device to produce repeated ischemia and reported the cumulative effect of repeated ischemic insults when the interval of the repeated ischemia was <1 h. In contrast, it was reported that sublethal 2-min ischemia a few days prior to subsequent lethal 5-min ischemia reduced the severity of ischemic damage of the hippocampal CA1 neurons by the second ischemia in the gerbil forebrain ischemia model *(120,121)*.

2.3.2. Four-Vessel Occlusion in the Rat (Pulsinelli's Model)

This is also a widely used model since reported by Pulsinelli and Brierley *(19)*. On the first experimental day, occlusion devices *(122)* are placed around both common carotid arteries following surgical dissection of the anterior aspects of the neck, and then both vertebral arteries are electrocoagulated through the alar foramina at the first cervical level following occipital-nuchal dissection *(123)*. Twenty-four hours after the surgical preparation, both common carotid arteries are occluded in the awake rat. In general, during occlusion, 77% of the rats become unresponsive and lose the righting reflex; these rats are selected for the experiment. Fifteen percent of the rats become lethargic, and 8% die of respiratory failure within a few minutes; these should be excluded from the experiment. When occlusion must be performed under anesthesia and selection of animals based on the degree of responsiveness and absence or presence of the righting reflex is impossible, flattening of the electroencephalogram and mydriasis should be confirmed. Some rats develop postischemic seizures and they also should be excluded. In the original work, 10 , 20, and 30 min of occlusion resulted in seizure in 0, 8, and 40% of the rats, respectively. During successful ischemia, blood flow to most of the forebrain including the frontal and parietal cortices, striatum and hippocampus is reduced to 3% or less of control values, whereas blood flow to the diencephalon, cerebellum, and brainstem ranges from 10–30% of control values *(124)*.

Obviously the relative ease of the procedure and use of the rat, which has become more and more important and popular in neuroscience research, are major advantages of this model. A drawback may be the inconsistency of ischemia owing to the difficulty in confirming complete electrocauterization of the vertebral arteries and the existence of collateral pathways mainly from the anterior spinal artery *(125,126)*. Although the drawbacks can be overcome by adhering to animal selection criteria *(123,124)*, care must be taken because selection of animals may lead to a significantly decreased efficiency of experimental studies in some laboratories *(125)*. It is known that the rate of successful occlusion and mortality vary among strains, animals from different suppliers, and even seasons *(123)*. Use of Sprague-Dawley rats in this model is not recommended. Kameyama et al. *(126)* modified the model in order to improve the inconsistency of ischemia by electrocauterizing the basilar artery instead of the vertebral arteries via anterior cervical dissection. However, maybe owing to the necessity of complicated surgical preparation, their model does not appear to be widely used. Another modification was reported by Boehme et al. *(127)* in which graded ischemia can be produced by computer-controlled compression of the unilateral common carotid artery following permanent occlusion of the other carotid and bilateral subclavian arteries.

2.3.3. Bilateral Common Carotid Artery Occlusion with Hypotension (Smith's Model)

Another model of global ischemia in the rat that has been widely accepted and used for studies of various aspects of cerebral ischemia is bilateral common carotid artery occlusion with systemic hypotension. This model was first described by a group at the University of Lund in the early 1970s *(128)*. They reported no significant changes in the energy state in the tissue following ligation of the carotid arteries only, but a severely affected energy state when carotid artery ligation was combined with systemic hypotension. Although at that time the carotid artery occlusion was permanent and their studies were limited to early recirculation phases, later they presented extensive studies using transient occlusion of the carotid arteries *(20,129)*. Histological examination 7 d following ischemia >4 min revealed consistent damage in CA1 pyramidal and CA4 hilar neurons in the hippocampus, the thalamic reticular nucleus, and the neocortical layers III-V. When ischemia was longer, damage was also seen in the caudoputamen and the pars reticulata of the substantia nigra. They concluded the rank order of vulnerability as CA4 > subiculum, CA1 > neocortex > CA3 > caudoputamen. However, since the degree of ischemic sensitivity to temperature changes differs among structures in the brain, the rank order of vulnerability is meaningful only at a given brain temperature *(42)*. They also examined the deleterious effect of hyperglycemia on ischemic brain injury using this model *(130)*. Marked exacerbation of ischemic damage by hyperglycemia was found to occur in the pars reticulata of the substantia nigra, which they associated with postischemic seizure activity *(131)*.

Transient global cerebral ischemia was also induced in the cat by combining bilateral carotid artery occlusion and systemic hypotension *(132)*. Changes in the levels of energy metabolites in the cerebral cortex have been investigated using this model.

2.3.4. Bilateral Common Carotid Artery Occlusion in Spontaneously Hypertensive Rats

Whereas bilateral common carotid artery occlusion by itself does not consistently produce ischemic changes in the brain of normal rats, bilateral common carotid artery occlusion in the SHR is known to produce ischemic changes in the brain *(133,134)*. The carotid artery occlusion used in these studies *(133,134)* was permanent and the histopathological nature of the ischemic changes was infarction in the forebrain. In this context, this model may be categorized as a focal ischemia model but not a global ischemia model. However, this model is very important, since hypertension is undoubtedly one of the major risk factors for cerebral infarction, and the occurrence of severe vascular changes secondary to hypertension may be operative in cerebrovascular diseases.

Bilateral carotid artery occlusion in SHR resulted in massive cerebral infarction in the carotid artery territory and an approximately 70% mortality within 24 h in one study. Vulnerability of the forebrain in SHR cannot be attributed to hypoplasty of the posterior communicating arteries or other major cerebral arteries. Rather, the reactive dilatative capacity in the pial arteries is reduced in SHR *(135)*. The reduced collateral flow through the leptomeningeal arteries may explain the susceptibility to ischemia of the forebrain, especially cerebral cortex, in SHR. In another study, bilateral carotid artery occlusion reduced the cortical blood flow to less than 10 mL/100 g/min whereas 20–40% of the blood flow remained in the thalamus even after bilateral carotid artery occlusion *(136)*.

Spontaneously hypertensive rats are much more expensive than more widely used rats such as Sprague-Dawley and Wistar rats. Development of hypertension significantly varies depending on age and sex. Therefore, experimental studies must be carefully planned. Selection of appropriate artificial food pellets is also important for maintaining hypertension. Care must be taken in regard to salt and protein content of food. Since SHR is more vulnerable to respiratory infection than conventionally used rats, the breeding environment should be carefully monitored. It is also known that the female SHR is very sensitive to sensory stimulation such as light, sound, and pain. The blood pressure and blood glucose level tend to increase easily in response to such stimuli and deeper anesthesia than applied to conventionally used rats may be required *(137)*.

3. FOCAL ISCHEMIA MODEL

3.1. The Importance of Focal Ischemia Models

Cerebral ischemia is the most commonly encountered type of stroke in human. Among the many causes of cerebral ischemia, occlusion of a single trunk artery, particularly the internal carotid or MCA, is the most frequent. Despite clinical application of many therapeutic strategies, we have not yet seen dramatic improvement of the outcome.

Treatment of ischemic stroke should be based on good understanding of the pathophysiology of focal ischemia. To investigate the pathophysiological mechanisms underlying the development of ischemic brain damage, animal ischemia models are indispensable as experimental counterparts of human focal cerebral ischemia. In this sense, animal models are ideal when they resemble the human disease.

3.2. What Has Been Elucidated Through Experimental Focal Ischemia?

Since all neuronal activities of vertebrates are supported by continuous blood supply, determination of rCBF is one of the essential parts in experimental focal ischemia.

Hence it is only since rCBF data have become available *(138–140)* that systematic studies of focal ischemia have become possible.

The following concepts have been established with regard to rCBF. These concepts were not developed without use of focal ischemia models.

3.2.1. Ischemic Threshold and Penumbra

In the MCA occlusion model of focal ischemia in subhuman primates, rCBF was measured using a hydrogen clearance technique *(141,142)* and it was disclosed that the reduction of rCBF was gradual. Using this model, Symon and coworkers demonstrated that evoked response magnitude sharply declined in the area of rCBFs of 14–16 mL/ 100 g/min, and they have termed this phenomenon flow threshold for failure of neuronal electrical function *(143)*. It was also disclosed that the extracellular potassium concentration suddenly increases when rCBF decreases to below 10 mL/100 g/min *(144)*. The condition of the ischemic brain with rCBF between the two thresholds has been termed "penumbra" *(145–147)*.

The existence of an ischemic threshold was also confirmed for pathological changes (infarction) *(148,149)*, and this threshold was found to be lower than that for electrical activity. However, this threshold is time dependent *(150)*.

Although the original definition of "ischemic penumbra" was based on the level of electrical activity, the meaning of this concept has been expanded *(151)*.

Current understandings of ischemic threshold and penumbra were summarized well by Hossmann *(152)*.

3.2.2. Therapeutic Window

The ischemic penumbra disappears as the duration of ischemia increases. After a certain time (1.5–4 h), ischemic damage becomes as severe as permanent ischemia *(153–155)*. Any treatments including reperfusion are thought to be effective when they are started only before this period. The term therapeutic window *(153)* has been used as an analogy for this period open to therapeutic strategies.

3.3. Classification of Focal Ischemia Models

Focal ischemia is different from global ischemia in that the ischemic region is focal, the intensity of ischemia is basically incomplete and gradual (ischemic core and penumbra), and the ischemia can be permanent or transient. Reflecting these differences, there are several possible means of classification with regard to the criteria concerned. Some criteria for classification of focal ischemia are:

1. Duration of ischemia (permanent or transient).
2. Occluded vessel (single arterial [carotid artery {CA} or MCA] or multiarterial [MCA + unilateral CA, MCA + bilateral CA]).
3. Site of occlusion (proximal or distal).
4. Occlusion method (electrocoagulation, ligation, clip, thrombosis with solid occluder, photochemical thrombosis, autologous embolus). Thrombosis with multiple solid occluders including an autologous blood clot seems to be most similar to human stroke, so it was once used frequently by some investigators. However, it is no longer used as frequently because of the difficulty of controling the lodging site and consequent difficulties in statistic analysis.
5. Position of occluder (intra-arterial or extra-arterial).
6. Animal species (subhuman primate, cat, dog, rabbit, rodent [gerbil, rat]).

Many combinations of these categories are possible. However, from the practical point of view, experimental focal ischemia models currently used can be classified as follows.

3.3.1. Rodent Focal Ischemia Models

3.3.1.1. MCA Occlusion in the Rat

The focal ischemia models in this category are diverse and the most frequently used *(37)*. Therefore, these models will be described in detail later. In a radical argument, it can be said that current neuroscience is established on the rat.

3.3.1.2. MCA Occlusion in the Mouse

Mice are very small and not easy to handle with frequent sampling of blood to maintain physiological variables constant. However, transgenic mice or knockout mice of some particular gene that is considered to play an important role in the development of ischemic injury have recently been developed *(112,156)*. Transgenic rats are not currently available. It is important to investigate the effect of ischemic insult on these genetically controlled animals. Hence this model is now an important model and deserves careful description *(29)*.

3.3.1.3. Unilateral Carotid Artery Occlusion in the Gerbil

Unilateral carotid arterial occlusion leads to severe ischemia. In some animals, the severity of ischemia is almost the same as that of global ischemia *(157)*. The method is quite simple but the results are inconsistent. The major drawbacks of gerbil models are variability in the incidence and severity of the ischemia *(158–160)*. This variability is attributed to the variety of cerebral vascular patterns in this animal *(158,159)*. In one study, half of the gerbils did not show any neurological deficits, one-tenth showed only transient neurological symptoms, and two-fifths showed varying neurological signs after unilateral carotid artery occlusion. Gerbils with severe neurological symptoms are susceptible to seizures *(157)*. To overcome these problems, Matsumoto et al. have proposed a neurological scoring system to predict the ischemic severity (157). This means that this model is quite uneconomical because at most only half of the animals can be used.

3.3.2. MCA Occlusion via Transorbital Approach
in Subhuman Primates and Mid-Sized Animals (Cat, Rabbit)

For large and mid-sized animals, the transorbital approach is an established method of inducing focal ischemia. In this category, the cat MCA occlusion model is still frequently used.

Among mid-sized animals used in the laboratory, the cat is almost the only species in which focal ischemia can be constantly and reproducibly induced by the single arterial occlusion (MCA occlusion). The transorbital approach to the MCA was developed by O'Brien in 1973 *(11)*. Selective transorbital MCA occlusion has been considered by many investigators as the ideal procedure for inducing cerebral ischemia.

Upon enucleation of the orbital contents and after performing a small craniectomy by enlarging the orbital fissure under an operative microscope, the bifurcation of the internal carotid artery (ICA) can be identified immediately inside the dura *(11,12)*. The amount of brain exposed is negligible, the volume of CSF lost is limited, and the intracranial fluid dynamics can be quickly re-established by careful obliteration of the cranial defect *(11)*. The origin of the MCA can be identified without difficulty.

However, the transorbital approach has some disadvantages *(161)*. Manipulation of the MCA can damage the autonomic nerve fibers that lie on its wall, potentiating the ischemic insult because the neurogenic factors of the autoregulatory mechanism are lost.

Modifications to this approach that have been proposed include performing MCA occlusion under awake condition *(12,162)* and not enucleating the orbit *(163)*. Because of the high mortality when the proximal portion of the MCA is occluded *(163)*, some modifications are necessary, for example occlusion of only the distal portion of the MCA *(163)*.

3.3.3. Others

Although it is difficult to obtain consistent infarction in rabbits by single arterial occlusion, this method is still used, with some modification *(18,61,164)*. Other animal models—MCA and/or posterior cerebral artery occlusion in gerbils *(28)*, dog focal ischemia *(15,165)* are not widely used today.

3.4. Rat Ischemia Models

For the reasons above mentioned , rat middle cerebral arterial occlusion models are now the most widely used to study focal ischemia.

Technical aspects of these models can be found in the original papers. The purpose of this section is to describe the wide variety of methods used to study focal ischemia.

To mimic the human embolic state, rat MCA occlusion models are modified in several ways. There are three different types of MCA occlusion:

1. Mechanical or electrical arterial occlusion with craniotomy and dural opening.
2. Intraluminal arterial occlusion without craniotomy.
3. Photochemical occlusion with craniotomy without dural opening.

3.4.1. Proximal MCA Occlusion (Tamura Model)

This is the oldest model of proximal MCA occlusion developed by Tamura et al. *(22)*. Prior to the development of this model, Robinson reported the results of a series of experiments performed using Sprague-Dawley rat MCA occlusion model *(21,166,167)*. Robinson's model was a distal MCA occlusion model in which the infarction was limited to a small cerebral cortical area *(21)*. Even such a small cortical infarct can produce a dramatic interhemispheric difference in behavior.

According to the original method *(22)*, the operation was carried out in the lateral position under controlled ventilation. The coronoid process of the mandible and zygoma were removed and a burr hole was opened lateral to the foramen ovale. When the dura was opened, the MCA was identified through the burr hole. This is a modification of the retroorbital approach already established in the cat and monkey. The purpose of this approach is to minimize the brain damage during exposure of the MCA.

The advantages of this method are that:

1. Arterial occlusion is confirmed directly through operative microscope.
2. Both permanent and temporary MCA occlusion are possible.
3. This approach is possible in a rat fixed on a stereotaxic frame and almost any kind of monitoring system is applicable.

4. A low mortality is achieved with minor modification in which the zygomatic arch is preserved *(168)*.

The disadvantages of this method are that:

1. Exposure of the brain to air during the craniectomy may alter intracranial pressure and blood-brain barrier (BBB) permeability *(5,169)*.
2. Clipping or cauterizing of the proximal MCA may cause damage to autonomic nerves around the MCA and autoregulation of CBF may be lost *(170)*.
3. The eyeball on the operative side is damaged (this may be avoided with surgical modification *[171]*).
4. This method requires surgical skillfulness under an operative microscope.

Several fundamental studies concerning mainly the occlusion point and neurological aspects have been carried out using this model *(35,172)*. Occlusion of only the origin of the MCA did not produce consistent infarcts *(172)*.

Recently, instability of the thrombus induced by electrocoagulation was reported *(173)*. However, from a neurosurgical standpoint of view, this phenomenon is encountered in clinical neurosurgical cases not infrequently and we routinely sever the cauterized artery to ensure arterial blood flow cessation. Bederson et al. also reported that a coagulated MCA should be severed to ensure homogeneity of the size of infarction *(172)*.

3.4.2. Intravascular Thread Model (Koizumi Model)

Koizumi et al. developed a model of MCA occlusion without craniectomy *(23)*, for which several modifications have been proposed *(174–176)*. The procedure originally proposed *(23)* is as follows: expose the carotid bifurcation and common carotid, internal carotid, and external carotid arteries with a straight neck skin incision under spontaneous breathing; insert a nylon thread from the common carotid artery toward the internal carotid artery with common and external carotid artery ligated; and ligate internal carotid artery at its origin to fix the thread.

It has been reported that two technical factors may seriously affect the size of the resulting infarct: the shape of the thread *(177)*, and the supplier of the nylon monofilament *(178)*.

The advantages of this model are that:

1. MCA can be occluded and recanalized without craniectomy.
2. The simplicity of the technique.

For these reasons, this model has achieved wide popularity *(155,174–179)*.

The major drawback of this model is the high mortality when it is used as a permanent ischemia model. According to the original work of Koizumi et al. *(23)*, the mortality in the permanent ischemia group was 100%. All animals died within 32 h of ischemia. Death occurred most frequently between 12–16 h after ischemic insult. Therefore, this is actually a model for transient MCA occlusion model. The infarct volume in this model is the largest among those in 5 MCAO models. In this model, the anterior cerebral artery (ACA) is also blocked *(23)*, which means that collateral circulation from the ACA to the MCA is markedly diminished. This may be the reason for the large infarct volume. The clinical counterpart of this model is MCA or ICA occlusion in a patient with poor crossflow through anterior communication artery.

The external carotid artery is sacrificed, which may result in a reduction in the blood flow to the scalp on the ischemia side and hence a change in brain temperature.

3.4.3. Distal MCA Occlusion with Bilateral CCAs Occlusion (Chen Model)

This model was developed as a means of producing consistent cerebral infarctions with relatively noninvasive surgery *(24)*. The right distal MCA and right CCA are ligated and the left CCA is clipped temporarily (60 min). This procedure can be used to produce consistent cortical infarction with an overall mortality of 7% *(24)*, and this model can be used as a transient focal ischemia model.

In this model, a small contralateral infarction is encountered occasionally *(24)*.

3.4.4. Tandem Occlusion Model (Brint Model)

Distal MCA (immediately superior to rhinal fissure) is permanently occluded, with ipsilateral permanent occlusion of the CCA (25). This method can be used to induce consistent cortical infarction in SHR. Although the Brint model is usually called the tandem occlusion model, their original work demonstrated that a single distal MCA occlusion in SHR can produce large consistent cortical infarcts *(25)*.

From the standpoint of statistic analysis of the infarct volume, Brint evaluated several rat strains. They recommended use of SHR in drug studies with a limited number of animals because of its small value of coefficient variance of the infarct volume.

3.4.5. Photothrombosis Model (Watson Model)

All the methods mentioned (*see* Sections 3.4.1.–3.4.4.) involve mechanical occlusion of the MCA. This photothrombosis model is quite different from others. Since Watson et al. first reported the photothrombotic stroke in rats *(180)*, they have continuously improved their method *(26,181–184)*. After exposing the artery to be occluded (dura is intact), a photosensitizing dye, Rose Bengal, is administered intravenously with simultaneous laser irradiation (argon laser-activated dye laser operating at 562 nm). Actually, this method can be used to occlude any vessels that can be exposed. Recently, Watson et al. summarized their work in a publication *(36)*. Although the stability of the clot made by this method was confirmed in rabbit cornea *(185)*, it is not clear whether the thrombus permanently occludes the artery.

3.4.6. Miscellaneous

Although there are several other different focal ischemia models (autologous blood clot *[186,187]*, graded focal ischemia *[188]*), they are not widely used currently.

3.5. Evaluation System

To investigate the effects of therapeutic strategies on stroke, it is essential to establish a method of determining the degree of ischemic damage. First, the mortality under given conditions should be stated clearly. In a drug study, it is especially important to clarify the mortality for each study group. Morphological methods have a long history and most of the techniques involved are well established. Because the rat MCA occlusion model is currently most frequently used to test the effects of newly developed therapies, we will only describe the recently introduced evaluation systems for the rat here.

3.5.1. Pathological Evaluation

3.5.1.1. TTC STAIN AND VOLUME CALCULATION

Bederson et al. reported that staining using 2,3,5-triphenyltetrazolium chloride (TTC) is convenient for reliably assessing the infarction 24 h after MCA occlusion *(189)*. This method has gained popularity because of its simplicity and rapidity. However, limitations of TTC staining have also been reported *(190)*. TTC staining is useful but one should limit its use to samples obtained 24 h after onset of ischemia. After 36 h of ischemia, infiltration of macrophages with intact mitochondria, neovasculalization, and astrocytic reactions occur *(5)*. These events obscure the margin of ischemia.

Osborne et al. *(64)* proposed a pathological evaluation system involving use of data for multiple brain slices plotted on a line graph to estimate the three-dimensional extent of ischemic brain damage. Many investigators use this system for the pathological evaluation of drug effects. In recent reports, the ischemic lesion is frequently expressed in terms of absolute volume. Uncontrollable shrinkage of fixed tissues *(191)* must be borne in mind. The shrinkage rate of the brain upon fixation in paraffin is approx 30%, but the rate is inconsistent.

3.5.1.2. SWANSON'S METHOD

Brain edema induced by focal ischemia results in enlargement of the infarct area and overestimation of infarct size. To overcome these effects, Swanson et al. proposed using a semiautomatic calculation system in which the infarct areas are obtained by reducing the areas of normal brain tissue of ischemic side with vital staining from the areas of nonischemic hemisphere *(192)*.

3.5.1.3. LONG-TERM PATHOLOGY

Aside from the secondary changes of ischemic damage (193, 194), long-term evaluation of ischemic damage may be necessary, since a recent study demonstrated that postischemic hypothermia only delays the process of cell death and does not change the final (2 mo after ischemia) outcome *(195)*.

When the MCA-occluded rat survives for a long time, infarcted area becomes liquefied. At this chronic phase, it is difficult to measure infarct volume. Measuring the hemispheric weight separately is a time-saving and convenient way to assess the long-term ischemic damage *(196)*. The hemispheric weight at chronic phase reflects the total effect of ischemia, including primary effects and secondary changes in the thalamus and substantia nigra. This is a very simple and quite accurate method to evaluate the effects of ischemia and/or treatment. Usually, the infarct volume is calculated from the infarct areas of given slices and the distances of the neighboring slices. This is a very useful way to evaluate the ischemic effects. However, it is based on the hypothesis that the mass of infarction can be represented as a barrel or a truncated cone model and that the distance between neighboring sections is constant.

3.5.2. Functional Evaluation

The ultimate aim of all therapies is functional recovery. The extent of pathological lesion is strongly correlated with the functional outcome in nontreated animals *(172,196)*. However, when a drug is administered, the extent of functional recovery may not correlate with that of pathological damage *(197,198)*. Recently, it was suggested that a laboratory environment is important for functional recovery to occur *(199)*.

Exposure to a rich environment can result in improved function without pathological improvement. Functional recovery is certainly based on morphological changes such as reorganization of synaptic networks *(200)*. However, such changes are not detectable with methods used in conventional pathological studies. Hence functional evaluation is inevitable for the assessment of the therapy. Fortunately, it is possible to detect functional changes even in a rat with a very small cortical infarction *(167)*. Functional differences between the right and left hemispheres can also be investigated in such a small animal *(21,167,201)*.

3.5.2.1. MOTOR FUNCTION

Several methods have been proposed for evaluating motor function in rats and they well correlate the extent of the ischemic damage *(35,172,196,202)*. These are useful in short-term observations. However, motor function recovery is quite fast in the rat focal ischemia model *(203)*. Contrary to the motor function, cognitive impairment persists considerably long period (203).

3.5.2.2. COGNITIVE (BEHAVIORAL) FUNCTION

Before evaluating the cognitive function, total activity must be evaluated. If the activity is diminished, there is no reason to do a cognitive test.

To test cognitive function in the rat, water mazes *(204)*, passive avoidance tests, active avoidance tests, the Y-maze, and the eight-arm radial maze are frequently employed.

4. CONCLUDING REMARKS

Classification of ischemia, factors affecting the result of ischemic insult, experimental models for each category, and evaluation systems for the rat focal ischemia model were described.

Recent advances in medical technology are astonishing; it is possible to obtain some metabolic information in human stroke using PET or MRI. In a sense, we are in a new era of clinical ischemic study. However, we cannot obtain complete information of ischemia without sampling brain tissue itself.

In recent years, old concepts such as the no-reflow phenomenon *(98)* and neuronal plasticity *(200)* have been reevaluated *(205,206)*, and new aspects of ischemic brain damage such as lymphocytic infiltration *(207)* have been disclosed. These new and reevaluated findings may open new therapeutic windows. With increasing accumulation of information regarding mechanisms of cerebral ischemia, new therapeutic strategies aimed at specific targets are being developed *(208–210)*. Animal models are indispensable in investigating the efficacy of these therapeutic modalities.

REFERENCES

1 Choi, D. W., Glutamate neurotoxicity in cortical cell culture is calcium dependent, *Neurosci. Lett.,* **58** (1985) 293–297.

2 Rothman, S. M., Synaptic activity mediates death of hypoxic neurons, *Science,* **220** (1983) 536–537.

3 Vibulsreth, S., Hefti, F., Ginsberg, M. D., Dietrich, W. D., and Busto, R., Astrocytes protect cultured neurons from degeneration induced by anoxia, *Brain Res.,* **422** (1987) 303–311.

4 Seren, M. S., Aldino, C., Zanoni, R., Leon, A., and Nicoletti, F., Stimulation of inositol phospholipid hydrolysis by excitatory amino acids is enhanced in brain slices from vulnerable regions after transient global ischemia, *J. Neurochem.,* **53** (1989) 1700–1705.

5 Hudgins, W. R. and Garcia, J. H., Transorbital approach to the middle cerebral artery of the squirrel monkey: a technique for experimental cerebral infarction applicable to ultrastructural studies, *Stroke,* **1** (1970) 107–111.

6 Garcia, J. H., Experimental ischemic stroke: a review, *Stroke,* **15** (1984) 5–14.

7 Harvey, J. and Rasmussen, T., Occlusion of the middle cerebral artery. An experimental study, *Arch. Neurol. Psychiatry,* **66** (1951) 20–29.

8 Symon, L., Experimental model of stroke in baboon, *Adv. Neurol.,* **10** (1975) 211–213.

9 Bremer, A. M., Watanabe, O., and Bourke, R. S. Artificial embolization of the middle cerebral artery in primates: description of an experimental model with extracranial technique, *Stroke,* **6** (1975) 387–390.

10 Myers, R. E. and Yamaguchi, S., Nervous system effects of cardiac arrest in monkeys: preservation of vision, *Arch. Neurol.,* **34** (1977) 65–74.

11 O'Brien, M. D. and Waltz, A. G., Transorbital approach for occluding the middle cerebral artery without craniectomy, *Stroke,* **4** (1973) 201–206.

12 Hayakawa, T. and Waltz, A. G., Immediate effects of cerebral ischemia: evolution and resolution of neurological deficits after experimental occlusion of one middle artery in conscious cats, *Stroke,* **6** (1975) 321–327.

13 Todd, M. M., Dunlop, B. J., Shapiro, H. M., Chadwick, H. C., and Powell, H. C., Ventricular fibrillation in the cat: a model for global cerebral ischemia, *Stroke,* **12** (1981) 808–815.

14 Safar, P., Stezoski, W., and Nemoto, E., Amelioration of brain damage after 12 minutes' cardiac arrest in dogs, *Arch. Neurol.,* **33** (1976) 91–95.

15 Suzuki, J., Yoshimoto, T., Tnanka, S., and Sakamoto, T., Production of various models of cerebral infarction in the dog by of occlusion of intracranial trunk arteries, *Stroke,* **11** (1980) 337–41.

16 Marshall, L. F., Durity, F., Lounsbury, R., Graham, G. I., Welsh, F., and Langfitt, T. W., Experimental cerebral oligemia and ischemia produced by intracranial hypertension: Part 1: pathophysiology, electroencephalography, cerebral blood flow, blood-brain barrier, and neurological function, *J. Neurosurg.,* **43** (1975) 308–317.

17 Yamamoto, K., Yoshimine, T., and Yanagihara, T., Cerebral ischemia in rabbit: a new experimental model with immunohistochemical investigation, *J. Cereb. Blood Flow Metab.,* **5** (1985) 529–536.

18 Zasslow, M. A., Pearl, R. G., Shuer, L. M., Steinberg, G. K., Lieberson, R. E., and Larson, C. P., Hyperglycemia decreases acute neuronal ischemic changes after middle cerebral artery occlusion in cats, *Stroke,* **20** (1989) 519–523.

19 Pulsinelli, W. A. and Brierley, J. B., A new model of bilateral hemispheric ischemia in the unanesthetized rat, *Stroke,* **10** (1979) 267–272.

20 Smith M-L., Bendek G., Dahlgren N., Rosén I., Wieloch T., K. SB. Models for studying long-term recovery following forebrain ischemia in the rat. 2. A2-vessel occlusion model. *Acta Neurol. Scand.,* **69** (1984) 385–401.

21 Robinson, R. G. and Coyle, J. T., The differential effect of right versus left hemispheric cerebral infarction on catecholamines and behavior in the rat, *Brain Res.,* **188** (1980) 63–78.

22 Tamura, A., Graham, D. I., McCulloch, J., and Teasdale, G. M., Focal cerebral ischemia in the rat: 1. Description of technique and early neuropathological consequences following middle cerebral artery occlusion, *J. Cereb. Blood Flow Metab.,* **1** (1981) 53–60.

23 Koizumi, J., Yoshida, Y., Nakazawa, T., and Ohneda, G., Experimental studies of ischemic brain edema, I: a new experimental model of cerebral embolism in rats in which recirculation can be introduced in the ischemic area, *Jpn. J. Stroke,* **8** (1986) 1–8.

24 Chen, S. T., Hsu, C. Y., Hogan, E. L., Maricq, H., and Balentine, J. D., A model of focal ischemic stroke in the rat: Reproducible extensive cortical infarction, *Stroke,* **17** (1986) 738–743.

25 Brint, S., Jacewicz, M., Kiessling, M., Tanabe, J., and Pulsinelli, W., Focal brain ischemia in the rat: Methods for reproducible neocortical infarction using tandem occlusion of the distal middle cerebral and ipsilateral common carotid arteries, *J. Cereb. Blood Flow Metab.*, **8** (1988) 474–485.

26 Nakayama, H., Dietrich, W. D., Watson, B. D., Busto, R., and Ginsberg, M. D., Photothrombotic occlusion of rat middle cerebral artery: histopathological and hemodynamic sequelae of acute recanalization, *J. Cereb. Blood Flow Metab.*, **8** (1988) 357–366.

27 Levine, S. and Payan, H., Effect of ischemia and other procedures on the brain and retina of the gerbil *(Meriones unguiculatus)*, *Exp. Neurol.*, **16** (1966) 255–262.

28 Yoshimine, T. and Yanagihara, T., Regional cerebral ischemia by occlusion of the posterior cerebral artery and the middle cerebral artery in gerbils, *J. Neurosurg.*, **58** (1983) 362–367.

29 Barone, F. C., Knudsen, D. J., Nelson, A. H., Feuerstein, G. Z., and Willette, R. N., Mouse strain differences in susceptibility to cerebral ischemia are related to cerebral vascular anatomy, *J. Cereb. Blood Flow Metab.*, **13** (1993) 683–692.

30 Moossy, J., Morphological validation of ischemic stroke models. In Price, T. R. and Nelson, E., (eds.) *Eleventh Princeton Conference*, Raven Press, New York, 1979, pp. 3–10.

31 Zeman, W. and Innes, J. R. M., Craigle's Neuroanatomy of the rat. Academic Press, London, 1963.

32 Rieke, G. K., Bowers, D. E. J., and Penn, P., Vascular supply pattern to rat caudatoputamen and globus pallidus: Scanning electronmicroscopic study of vascular endocasts of stroke-prone vessels, *Stroke*, **12(6)** (1981) 840–847.

33 Paxinos, G. and Watson, C., *The Rat Brain in Sterotaxic Coodinates.* 2nd ed., Academic, Sydney, 1986.

34 Rubino, G. J. and Young, W., Ischemic cortical lesions after permanent occlusion of individual cerebral artery branches in rats, *Stroke,* **19** (1988) 870–877.

35 Menzies, S. A., Hoff, J. T., and Betz, A. L., Middle cerebral artery occlusion in rats: A neurological and pathological evaluation of a reproducible model, *Neurosurgery,* **31** (1992) 100–107.

36 Watson, B. D., Dietrich, W. D., Prado, R., Nakayomia, H., Kanemitsu, H., Futrell, N. N., Yao, H., Markgraf, C. G., and Wester, P., Concepts and techniques of experimental stroke induced by cerebrovascular photothrombosis. *Central Nervous System Trauma: Research Techniques*, CRC, Boca Raton, FL, 1995, pp. 169–194.

37 Ginsberg, M. D. and Busto, R., Rodent models of cerebral ischemia, *Stroke,* **20** (1989) 1627–1642.

38 Coyle, P., Arterial patterns of the rat rhinencephalon and related structures, *Exp. Neurol.,* **49** (1975) 671–690.

39 Yamori, Y., Horie, R., Handa, H., Sato, M., and Fukase M., Pathogenetic similarity of strokes in stroke-prone spontaneously hypertensive rats and humans, *Stroke,* **7** (1976) 46–53.

40 Iversen, L. L., Iversen, S. D., and Snyder, S. H., *Handbook of Phychopharmacology: Chemical Pathways in the Brain.* Plenum, London, 1978, vol. 9.

41 Busto, R., Dietrich, W. D., Globus, M. Y.-T., Valdés, I., Scheinberg, P., and Ginsberg, M. D., Small differences in intraischemic brain temperature critically determine the extent of ischemic neuronal injury, *J. Cereb. Blood Flow Metab.,* **7** (1987) 729–738.

42 Minamisawa, H. and Smith, M.-L., and Siesjö, B. K., The effect of mild hyperthermia and hypothermia on brain damage following 5, 10, and 15 minutes of forebrain ischemia, *Ann. Neurol.,* **28** (1990) 26–33.

43 Minamisawa, H., Mellergård P., Smith, M.-L., et al., Preservation of brain temperature during ischemia in rats, *Stroke,* **21** (1990) 758–764.

44 Onesti, S. T., Baker, C. J., Sun, P. P., and Solomon, R. A., Transient hypothermia reduces focal ischemic brain injury in the rat, *Neurosurgery,* **29** (1991) 369–373.

45 Baker, C. J., Onesti, S. T., and Solomon, R. A., Reduction by delayed hypothermia of cerebral infarction following middle cerebral artery occlusion in the rat: a time-course study, *J. Neurosurg.,* **77** (1992) 438–444.

46 Kader, A., Brisman, M. H., Maraire, N., Huh, J.-T., and Solomon, R. A., The effect of mild hypothermia on permanent focal ischemia in the rat, *Neurosurgery,* **31** (1992) 1056–1061.

47 Morikawa, E., Ginsberg, M. D., Dietrich, W. D., Duncan, R. C., Kraydieh, S., Globus, M. Y. T., and Busto, R., The significance of brain temperature in focal cerebral ischemia: histopathological consequences of middle cerebral artery occlusion in the rat, *J. Cereb. Blood Flow Metab.,* **12** (1992) 380–389.

48 Takagi, K., Ginsberg, M. D., Globus, M. Y.-T., Martinez, E., and Busto, R., Effect of hyperthermia on glutamate release in ischemic penumbra after middle cerebral artery occlusion in rats, *Am. J. Physiol.,* **266** (1994) H1770–H1776.

49 Carroll, M. and Beek, O., Protection against hippocampal CA1 cell loss by post-ischemic hypothermia is dependent on delay of initiation and duration, *Metab. Brain Dis.,* **7** (1992) 45–50.

50 Zhang, Z. G., Chopp, M., and Chen, H., Duration dependent post-ischemic hypothemia alleviates cortical damage after transient middle cerebral artery occlusion in the rat, *J. Neurol. Sci.,* **117** (1993) 240–244.

51 Karibe, H., Chen, J., Zarow, G. J., Graham, S. H., and Weinstein, P. R., Delayed induction of mild Hypothermia to reduce infarct volume after temporary middle cerebral artery occlusion in rats, *J. Neurosurg.,* **80** (1994) 112–119.

52 Ginsberg, M. D., Sternau, L. L., Globus, M. Y.-T., Dietlich, W. D., and Busto, R., Therapeutic modulation of brain temperature: Relevance to ischemic brain injury, *Cerebrovasc. Brain Metab. Rev.,* **4** (1992) 189–225.

53 Ginsberg, M. D., Welsh, F. A., and Budd, W. W., Deleterious effects of glucose pretreatment on recovery from diffuse cerebral ischemia in cat. I. Local cerebral blood flow and glucose utilization, *Stroke,* **11** (1980) 347–354.

54 Pulsinelli, W. A., Waldman, S., Rawlinson, D., and Plum, F., Moderate hyperglycemia augments ischemic brain damage: a neuropathologic study in the rat, *Neurology,* **32** (1982) 1239–1246.

55 Nedergaard, M. and Diemer, N. H., Focal ischemia of the rat brain, with special reference to the influence of plasma glucose consentration, *Acta Neuropathol. (Berl.),* **73** (1987) 131–137.

56 Duverger, D. and MacKenzie, E. T., The quantification of cerbral infarction following focal ischemia in the rat: Influence of strain, arterial pressure, blood glucose concentration, and age, *J. Cereb. Blood Flow Metab.,* **8** (1988) 449–461.

57 de Courten-Myers, G. M., Myers, R. E., and Schoolfield, L., Hyperglycemia enlarges infarct size in cerebrovascular occlusion in cats, *Stroke,* **19** (1988) 623–630.

58 Prado, R., Ginsberg, M. D., Dietrich, W. D., Watson, B. D., and Busto, R., Hyperglycemia increases infarct size in collaterally perfused but not end-arterial vascular territories, *J. Cereb. Blood Flow Metab.,* **8** (1988) 186–192.

59 Kushner, M., Nencini, P., Reivich, M., Rango, M., Jamieson, D., Fazekas, F., Zimmerman, R., Chawluk, J., Alavi, Al, and Alres, W., Relation of hyperglycemia early in ischemic brain infarction to anatomy, metabolism, and clinical outcome, *Ann. Neurol.,* **28** (1990) 129–135.

60 Ginsberg, M. D., Prado, R., Dietrich, W. D., Busto, R., and Watson, B. D., Hyperglycemia reduces the extent of cerebral infarction in rats, *Stroke,* **18** (1987) 570–574.

61 Kraft, S. A., Larson, C. P., Shuer, L. M., Steinberg, G. K., Benson, G. V., and Pearl, R. G., Effect of hyperglycemia on neuronal changes in a rabbit model of focal cerebral ischemia, *Stroke,* **21** (1990) 447–450.

62 Hossmann, K.-A. and Sato, K., Recovery of neuronal function after prolonged cerebral ischemia, *Science,* **168** (1970) 375–376.

63 Hossmann, K.-A. and Zimmermann, V., Resuscitation of the monkey brain after 1 h complete ischemia. I. Physiological and morphological observations, *Brain Res.,* **81** (1974) 59–74.

64 Osborne, K. A., Shigeno, T., Balarsky, A. M., Ford, I., McCullock, J., Teasdale, G. M., and Graham, D. I., Quantitative assessment of early brain damage in a rat model of focal cerebral ischaemia, *J. Neurol. Neurosurg. Psychiatry,* **50** (1987) 402–410.

65 Zhu, C. Z. and Auer, R. N. Graded hypotension and MCA occlusion duration: effect in transient focal ischemia. *J. Cereb. Blood Flow Metab.,* **15** (1995) 980–988.

66 Warner, D. S., Zhou, J. G., Ramani, R., and Todd, M. M., Reversible focal ischemia in the rat: effects of halothane, isoflurane, and methohexital anesthesia, *J. Cereb. Blood Flow Metab.,* **11** (1991) 794–802.

67 Kraig, R. P., Petito, C. K., Plum, F., and Pulsinelli, W. A., Hydrogen ions kill brain at concentrations reached in ischemia, *J. Cereb. Blood Flow Metab.,* **7** (1987) 379–386.

68 Loskota, W. J., Lomax, P., Verity M. A. A stereotaxic atlas of the mongolian gerbil brain *(Meriones unguiculatus),* Ann Arbor Science, Ann Arbor, 1974.

69 Matsuyama, T., Matsumoto, M., Fujisawa, A., Handa, N., Tanaka, K., Yoneda, S., Kimuya, K., and Abe, H., Why are infant gerbils more resistant than adults to cerebral infarction after carotid ligation?, *J. Cereb. Blood Flow Metab.,* **3** (1983) 381–385.

70 Yao, H., Sadoshima, S., Ooboshi, H., Sato, Y., Uchimura, H., and Fujishima, M., Age-related vulnerability to cerebral ischemia in spontaneously hypertensive rats, *Stroke,* **22** (1991) 1414 1418.

71 Davis, M., Mendelow, A. D., Perry, R. H., Chambers, I. R., James, O. F. W., Experimental stroke and neuroprotection in the aging rat brain, *Stroke,* **26** (1995) 1072–1078.

72 Payan, H. M. and Conrad, J. R. Carotid ligation in gerbils. Influence of age, sex, and gonads, *Stroke,* **8** (1977) 194–196.

73 Nakatomi, Y., Fujishima, M., Tamaki, K., Ishitsuka, T., Ogata, J., and Omae, T., Influence of sex on cerebral ischemia following bilateral carotid occlusion in spontaneously hypertensive rats: a metabolic study, *Stroke,* **10** (1979) 196–9.

74 Berry, K., Wisniewxki, J. M., Svarzbein, L., and Baez, S., On the relationship of brain vasculature to production of neurological deficit and morphological chagnes following acute unilateral common carotid artery ligation in gerbils, *J. Neurol. Sci.,* **25** (1975) 75–92.

75 Hall, E. D., Pazara, K. E., and Linseman, K. L., Sex differences in postischemic neuronal necrosis in gerbils, *J. Cereb. Blood Flow Metab.,* **11** (1991) 292–298.

76 Coyle, P. and Jokelainen, P. T., Differential outcome to middle cerebral artery occlusion in spontaneously hypertensive stroke-prone rats (SHRSP) and Wistar Kyoto (WKY) rats, *Stroke,* **14** (1983) 605–611.

77 Xie, Y., Zacharias, E., Hoff, P., and Tegtmeier, F., Ion channel involvement in anoxic depolarization induced by cardiac arrest in rat brain, *J. Cereb. Blood Flow Metab.,* **15** (1995) 587–594.

78 De Garavilla, L., Babbs, C., and Tacker, W., An experimental circulatory arrest model in the rat to evaluate calcium antagonists in cerebral resuscitation, *Am. J. Emerg. Med.,* **2** (1984) 321–326.

79 Blomqvist, P. and Wieloch, T., Ischemic brain damage in rats following cardiac arrest using a long-term recovery model, *J. Cereb. Blood Flow Metab.,* **5** (1985) 420–431.

80 Kawai, K., Nitecka, L., Ruetzler, C. A., Nagashima, G., Joo, F., Mies, G., Nowak, T. S., Jr., Saito, N., Lohr, J. M., and Klatzo, I., Global cerebral ischemia associated with cardiac arrest in the rat.I. Dynamics of early neuronal changes, *J. Cereb. Blood Flow Metab.,* **12** (1992) 238–249.

81 Kawai, K., Penix, L., Kawahara, N., Reutzler, C. A., and Klatzo, I., Development of susceptibility to audiogenic seizures following cardiac arrest cerebral ischemia in rats, *J. Cereb. Blood Flow Metab.,* **15** (1995) 248–258.

82 Leonov, Y., Sterz, F., Safar, P., and Radovsky, A., Moderate hypothermia after cardiac arrest of 17 minutes in dogs on cerebral and cardiac outcome, *Stroke,* **21** (1990) 1600–1606.

83 Leonov, Y., Sterz, F., Safar, P., et al., Mild cerebral hypothermia during and after cardiac arrest improves neurologic outcome in dogs, *J. Cereb. Blood Flow Metab.,* **10** (1990) 57–70.

84 Brierley, J., Brown, A., Excell, B., and Meldrum, B., Brain damage in the rhesus monkey resulting from profound arterial hypotension. I. Its nature, distribution and general physiological correlates, *Brain Res.,* **13** (1969) 68–100.

85 Yatsu, F. M., Diamond, I., Graziano, C., and Lindquist, P., Experimental brain ischemia: protection from irreversible damage with a rapid-acting barbiturate (methohexital), *Stroke,* **3** (1972) 726–732.

86 Hallenbeck, J. M. and Bradley, M. E., Experimental model for systematic study of impaired microvascular reperfusion, *Stroke,* **8** (1977) 238–243.

87 Ljunggren, B., Schutz, H., and Siesjö, B. K., Changes in energy state and acid-base parameters of the rat brain during complete compression ischemia, *Brain Res.,* **73** (1974) 277–289.

88 Ross, D. and Duhaime, A.-C., Degeneration of neurons in the thalamic reticular nucleus following transient ischemia due to raised intracranial pressure: excitotoxic degeneration mediated via non-NMDA receptors?, *Brain Res.,* **501** (1989) 129–143.

89 Nemoto, E. M., Bleyaert, A. L., Stezoski, S. W., Moossy, J., Rao, G. R., and Safar, P., Global brain ischemia: a reproducible monkey model, *Stroke,* **8** (1977) 558–564.

90 Nemoto, E. M., Hossmann, K.-A., and Cooper, H. K, Post-ischemic hypermetabolism in cat brain, *Stroke,* **12** (1981) 666–676.

91 Osgood, C. P., Dujovny, M., and Wisotzkey, H., Acute canine cerebral ischemia: a preliminary model to evaluate microvascular mammary-carotid anastomosis, *Stroke,* **5** (1974) 477–482.

92 Snyder, J. V., Nemoto, E. M., Carroll, R. G., and Safar, P., Global ischemia in dogs: intracranial pressures, brain blood flow and metabolism, *Stroke,* **6** (1975) 21–27.

93 Miller, C. L., Lampard, D. G., Alexander, K., and Brown, W. A., Local cerebral blood flow following trancient cerebral ischemia. I: onset of impaired reperfusion within the first hour following global ischemia, *Stroke,* **11** (1980) 534–541.

94 Kayama, T., Mizoi, K., and Suzuki, J., A canine model of a completely ischemic brain regulated with the perfusion method, *Surg. Neurol.,* **16** (1981) 167–172.

95 Hossmann, K.-A., Schmidt-Kastner, R., and Ophoff, B. G., Recovery of integrative central nervous function after one hour global cerebro-circulatory arrest in normothermic cat, *J. Neurol. Sci.,* **77** (1987) 305–320.

96 Zaren, H. A., Weinstein, J. D., and Langfitt, T. W., Experimental ischemic brain swelling, *J. Neurosurg.,* **32** (1970) 227–235.

97 Kowada, M., Ames, A. I., Majno, G., and Wright, R. L., Cerebral ischemia. I. An improved experimental method for study; cardiovascular effects and demonstration of an early vascular lesion in the rabbit, *J. Neurosurg.,* **28(2)** (1968) 150–157.

98 Ames, A. I., Wright, L., Kowada, M., Thurston, J. M., Majno, G., Cerebral ischemia; II. The non-reflow phenomenon, *Am. J. Pathol.,* **52** (1968) 437–453.

99 Benveniste, H., The excitotoxin hypothesis in relation to cerebral ischemia, *Cerebrovasc. Brain Metab. Rev.,* **3** (1991) 213–245.

100 Gill, R., Foster, A. C., and Woodruff, G. N., Systemic administration of MK-801 protects against ischaemia-induced hippocampal neurodegeneration in the gerbil, *J. Neurosci.,* **7** (1987) 3343–3349.

101　Sheardown, M. J., Nielsen, E. O., Hansen, A. J., Jacobsen, P., and Honore, T., 2,3-dihydroxy-6-nitro-7-sulfamoyl-benzo(F)quinoxaline: a neuroprotectant for cerebral ischemia, *Science,* **247** (1990) 571–574.

102　Hara, H., Onodera, H., Yoshidomi, M., Matsuda, Y., and Kogure, K., Staurosporine, a novel protein kinase C inhibitor, prevents postischemic neuronal damage in the gerbil and rat, *J. Cereb. Blood Flow Metab.,* **10** (1990) 646–653.

103　Lee, K. S., Frank, S., Vanderklish, P., Arai, A., and Lynch, G., Inhibition of proteolysis protects hippocampal neurons from ischemia, *Proc. Natl. Acad. Sci. USA,* **88** (1991) 7233–7237.

104　Wieloch, T., Cardell, M., Bingren, H., Zivin, J., and Saitoh, T., Changes in the activity of protein kinase C and the differential subcellular redistribution of its isozymes in the rat striatum during and following transient forebrain ischemia, *J. Neurochem.,* **56** (1991) 1227–1235.

105　Nowak, T. S. J., Localization of 70kD stress protein mRNA induction in gerbil brain after ischemia, *J. Cereb. Blood Flow Metab.,* **11** (1991) 432–439.

106　Simon, R. P., Cho, H., Gwinn, R., and Lowenstein, D. H., The temporal profile of 72-kD heat-shock protein expression following global ischemia, *J. Neurosci.,* **11** (1991) 881–889.

107　Abe, K., Kawagoe, J., Aoki, M., and Kogure, K., Changes of mitochondrial DNA and heat shock protein gene expression in gerbil hippocampus after transient forebrain ischemia, *J. Cereb. Blood Flow Metab.,* **13** (1993) 773–780.

108　Widmann, R., Kuroiwa, T., Bonnekoh, P., and Hossmann, K.-A., [^{14}C]Leucine incorporation into brain proteins in gerbils after transient ischemia: relationship to selective vulnerability of hippocampus, *J. Neurochem.,* **56** (1991) 789–796.

109　Krause, G. S. and Tiffany, B. R., Suppression of protein synthesis in the reperfused brain, *Stroke,* **24** (1993) 747–756.

110　Hu, B. R. and Wieloch, T., Stress-induced inhibition of protein synthesis initiation: modulation of initiation factor 2 and guanine nucleotide exchange factor activities following transient cerebral ischemia in the rat, *J. Neurosci.,* **13** (1993) 1830–1838.

111　Siesjö, B. K., Agardh C. D., and Bengtsson, F., Free radicals and brain damage, *Brain Metab. Rev.,* **1** (1989) 165–211.

112　Huang, Z., Huang, PL., Panahian, N., Dalkara, T., Fishman, M. C., and Moskowitz, M. A., Effects of cerebral ischemia in mice deficient in neuronal nitric oxide synthase, *Science,* **265** (1994) 1883–1885.

113　Shigeno, T., Mima, T., Takakura, K., Graham, D. I., Kato, G., Hashimoto, Y., and Furukawa, S., Amelioration of delayed neuronal death in the hippocampus by nerve growth factor, *J. Neurosci.,* **11** (1991) 2914–2919.

114　Nitatori, T., Sato, N., Waguri, S., Karasawa, Y., Araki, H., Shibanai, K., Kominami, E., and Uchiyama, Y., Delayed neuronal death in the CA1 pyramidal cell layer of the gerbil hippocampus following transient ischemia is apoptosis, *J. Neurosci.,* **15** (1995) 1001–1011.

115　Ito, U., Spatz, M., Walker, J. T., and Klatzo, I., Experimental cerebral ischemia in mongolian gerbils: I. Light microscopic observations, *Acta Neuropathol.,* **32** (1975) 209–223.

116　Kirino, T., Delayed neuronal death in the gerbil hippocampus following ischemia, *Brain Res.,* **239** (1982) 57–69.

117　Kuroiwa, T., Bonnekoh, P., and Hossmann, K.-A., Prevention of postischemic hyperthermia prevents ischemic injury of CA1 neurons in gerbils, *J. Cereb. Blood Flow Metab.,* **10** (1990) 550 556.

118　Tone, O., Miller, J. C., Bell, J. M., and Papoport, S. I., Regional cerebral palmitate incorporation following transient bilateral carotid occlusion in awake gerbils, *Stroke,* **18** (1987) 1120–1127.

119 Tomida, S., Nowak, T. S. J., Vass, K., Lohr, J. M., and Klatzo, I., Experimental model for repetitive ischemic attacks in the gerbil: the cumulative effect of repeated ischemic insults, *J. Cereb. Blood Flow Metab.,* **7** (1987) 773–782.

120 Kitagawa, K., Matsumoto, M., Tagaya, M., Hata, R., Ueda, H., Niinobe, M., Handa, N., Fukunaga, R., Kimura, K., Mikoshiba, K., and Kamada, T., 'Ischemic tolerance' phenomenon found in the brain, *Brain Res.,* **528** (1990) 21–24.

121 Kirino, T., Tsujita, Y., and Tamura A., Induced tolerance to ischemia in gerbil hippocampal neurons, *J. Cereb. Blood Flow Metab.,* **11** (1991) 299–307.

122 Brown, A. W., Brierley, J. B., Evidence for early anoxic-ischaemic cell damage in the rat brain, *Experientia,* **22** (1966) 546–547.

123 Pulsinelli, W. A. and Buchan, A. M., The four-vessel occlusion rat model: method for complete occlusion of vertebral arteires and control of collateral circulation, *Stroke,* **19** (1988) 913–914.

124 Pulsinelli, W. D., Levy, D. E., and Duffy, T. E., Cerebral blood flow in the four-vessel occlusion rat model, *Stroke,* **14** (1983) 832–833 (Letter to Editor).

125 Furlow, T. W., Cerebral ischemia produced by four-vessel occlusion in the rat: A quantitative evaluation of cerebral bleed flow, *Stroke,* **13** (1982) 852–855.

126 Kameyama, M., Suzuki, J., Shirane, R., and Ogawa, A., A new model of bilateral hemispheric ischemia in the rat - three vessel occlusion model, *Stroke,* **16** (1985) 489–493.

127 Boehme, R. J., Conger, K. A., and Anderson, M. L., Computer-regulated constant pressure ischemia in the rat: the animal model, *J. Cereb. Blood Flow Metab.,* **8** (1988) 236–243.

128 Eklöf, B. and Siesjö, B. K., The effect of bilateral carotid artery ligation upon the blood flow and energy state of the rat brain, *Acta Physiol. Scand.,* **86** (1972) 155–165.

129 Smith, M.-L., Auer, R. N., and Siesjö, B. K., The density and distribution of ischemic brain injury in the rat following 2–10 min of forebrain ischemia, *Acta Neuropathol. (Berl.),* **64** (1984) 319–332.

130 Warner, D. S., Smith, M.-L., and Siesjö, B. K., Ischemia in normo- and hyperglycemic rats: effects on brain water and electrolytes, *Stroke,* **18** (1987) 464–471.

131 Smith, M.-L., Kalimo, H., Warner, D. S., and Siesjö, B. K., Morphological lesions in the brain preceding the development of postischemic seizures, *Acta Neuropathol.,* **76** (1988) 253–264.

132 Welch, F., O'Conner, M., Marcy, V., Spatacco, A., and Johns, R., Factors limiting regeneration of ATP following temporary ischemia in cat brain, *Stroke,* **13** (1982) 234–242.

133 Fujishima, M., Ogata, J., Sugi, T., and Omae, T., Mortality and cerebral metabolism after bilateral carotid artery ligation in normotensive and spontaneously hypertensive rats, *J. Neurol. Neurosurg. Psychiat.,* **39** (1976) 212–217.

134 Ogata, J., Fujishima, M., Morotomi, Y., and Omae, T., Cerebral infarction following bilateral carotid artery ligation in normotensive and spontaneously hypertensive rats: a pathological study, *Stroke,* **7** (1976) 54–60.

135 Coyle, P., Spatial relations of dorsal anastomoses and lesion border after cerebral artery occlusion, *Stroke,* **18** (1987) 1133–1140.

136 Choki, J. I., Yamaguchi, T., Takeya, Y., Morotomi, Y., and Omae, T., Effect of carotid artery ligation on regional cerebral blood flow in normotensive and spontaneously hypertensive rats, *Stroke,* **8** (1977) 374–379.

137 Sadoshima, S., Spontaneously hypertensive rats., In Sano, K., Tamura, A., Hayakawa T., and Kirino T. (eds.) *Handbook of Experimental Stroke Research* (in Japanese). IPC., Tokyo, 1990, pp. 115–121.

138 Kety, S. S. and Schmidt, C. F., The determination of cerebral blood flow in man by the use of nitrous oxide in low concentration, *Am. J. Physiol.,* **143** (1945) 53–66.

139 Aukland, K., Brower, B. F., and Berliner, R. W., Measurement of local blood flow with hydrogen gas, *Circ. Res.,* **16** (1964) 164–187.

140 Sakurada, O., Kennedy, C., Jehle, J., Brown, J. D., Carbin, G. L., and Sokoloff, L., Measurement of local cerebral blood flow with iodo[^{14}C]antipyrine, *Am. J. Physiol.*, **234** (1978) H59–H66.

141 Pasztor, E., Symon, L., Dorsch, N. W. C., and Branston, N. M., The hydrogen clearance method in assesment of blood flow in cortex, white matter and deep nuclei of baboons, *Stroke*, **4** (1973) 556–567.

142 Symon, L., Pasztor, E., and Branston, N. M., The distribution and density of reduced cerebral blood flow following acute middle cerebral artery occlusion: an experimental study by the technique of hydrogen clearance in baboons, *Stroke*, **5** (1974) 355–364.

143 Branston, N. M., Symon, L., Crockard, H. A., and Paszor, E., Relationship between the cortical evoked potential and local cortical blood flow following acute middle cerebral artery occlusion in the baboon, *Exp. Neurol.*, **45** (1974) 195–208.

144 Astrup, J., Symon, L., Branston, N. M., and Lassen, N. A., Cortical evoked potential and extracellular K$^+$ and H$^+$ at critical levels of brain ischaemia, *Stroke*, **8** (1977) 51–57.

145 Symon, L., The relationship between CBF., evoked potentials and the clinical features in cerebral ischemia. *Acta Neurol. Scand.*, **62(Suppl 78)** (1980) 175–190.

146 Astrup, J., Siesjö, B. K., and Symon, L., Thresholds in cerebral ischemia—The ischemic penumbra, *Stroke*, **12** (1981) 723–725.

147 Astrup, J., Energy-requiring cell functions in the ischemic brain. Their critical supply and possible inhibition in protective therapy, *J. Neurosurg.*, **56** (1982) 482–497.

148 Morawetz, R. B., DeGirolami, U., Ojemann, R. G., Marcoux, F. W., and Crowell, R. M., Cerebral blood flow determined by hydrogen clearance during middle cerebral artery occlusion in unanesthetized monkeys, *Stroke*, **9** (1978) 143–149.

149 Tamura, A., Asano, T., and Sano, K., Correlation between rCBF and histological changes following temporary middle cerebral artery occlusion, *Stroke*, **11** (1980) 487–493.

150 Jones, T. H., Morawetz, R. B., Crowell, R. M., Marcoux, F. W., Fitzgibbon, S. I., DeGirolani, U., and Ojemann, R. G., Thresholds of focal cerebral ischemia in awake monkeys, *J. Neurosurg.*, **54** (1981) 773–782.

151 Hakim, A. M., The cerebral ischemic penumbra, *Can. J. Neurol. Sci.*, **14** (1987) 557–559.

152 Hossmann, K. A., Viability thresholds and the penumbra of focal ischemia, *Ann. Neurol.*, **36** (1994) 557–565.

153 Kaplan, B., Brint, S., Tanabe, J., Jacewicz, M., Wang, X.-J., and Pulsinelli, W., Temporal thresholds for neocortical infarction in rats subjected to reversible focal cerebral ischemia, *Stroke*, **22** (1991) 1032–1039.

154 Buchan, A. M., Xue, D., Slivka, A., A new model of temporary focal neocortical ischemia in the rat. *Stroke*, **23** (1992) 273–279.

155 Memezawa, H., Smith, M.-L., B.K. S., Penumbral tissues salvaged by reperfusion following middle cerebral artery occlusion in rats, *Stroke*, **23** (1992) 552–559.

156 Kinouchi, H., Epstein, C. J., Mizui, T., Carlson, E., Chen, S. F., and Chan, P. H., Attenuation of focal cerebral ischemic injury in transgenic mice overexpressing CuZn superoxide dismutase, *Proc. Natl. Acad. Sci. USA*, **88** (1991) 11,158–11,162.

157 Matsumoto, M., Hatakeyama, T., Akai, F., Brengman, J. M., and Yanagihara, T., Prediction of stroke before and after unilateral occlusion of the carotid artery in gerbils, *Stroke*, **19** (1988) 480–487.

158 Horizoe, H., Fukuda, T., and Tamura, A., (JPN) Relationship of cerebral vasculature to infarction following unilateral common carotid artery ligation in the mongolian gerbils (Japanese with English abstract), *Brain Nerve (Tokyo)*, **33** (1981) 825–831.

159 Horizoe, H. and Tamura, A., Relationship of cerebral vasculature to ischemia following unilateral common carotid artery ligation in mongoian gerbils, *J. Med. Soc. Toho. Univ.* **33** (1986) 277–288.

160 Tamura, A., Horizoe, H., and Fukuda, T., Relationship of cerebral vasculature to infarcted areas following unilateral common carotid artery ligation in the mongolian gerbils. *J. Cereb. Blood Flow Metab.*, **1(Suppl. 1)** (1981) S194–195.

161 Molinari, G. F. and Laurent, J. P., A classification of experimental methods of brain ischemia, *Stroke,* **7** (1976) 14–17.

162 Little, J. R., Implanted device for middle cerebral artery occlusion in conscious, *Stroke,* **8** (1977) 258–260.

163 Tamura, A., Asano, T., Sano, K., Tsumagari, T., and Nakajima, A., Protection from cerebral ischemia by a new imidazole derivative (Y-9179) and pentobarbital. A comparative study in chronic middle cerebral occlusion in cats, *Stroke,* **10** (1979) 126–134.

164 Steinberg, G. K., George, C. P., DeLaPaz, R., Shibata, D. K., and Gross, T., Dextrometrophan protects against cerebral injury following transient focal ischemia in rabbits, *Stroke,* **19** (1988) 1112–1118.

165 Okada, Y., Shima, T., Yokoyama, N., and Uozumi, T., Comparison of middle cerebral artery trunk occlusion by silicone embolization and by trapping, *J. Neurosurg.,* **58** (1983) 492–499.

166 Robinson, R. G., Shoemaker, W. J., Schlumpf, M., Valk, T., and Bloom, F. E., Effect of experimental cerebral infarction in rat on catecholamines and behaviour, *Nature,* **255** (1975) 332–334.

167 Robinson, R. G., Differential behavioral and biochemical effects of right and left hemispheric cerebral infarction in the rat, *Science,* **205** (1979) 707–710.

168 Yamamoto, M., Tamura, A., Kirono, T., Shimsh, M., and Sano, K., Behavioral changes after focal cerebral ischemia by left middle cerebral artery occlusion in rats, *Brain Res.,* **452** (1988) 323–328.

169 Coyle, P., Middle cerebral artery occlusion in the young rat, *Stroke,* **13** (1982) 855–859.

170 Kano, M., Moskowitz, M. A., and Yokota, M., Parasympathetic denervation of rat pial vessels significantly increases infarction volume following middle cerebral artery occlusion, *J. Cereb. Blood Flow Metab.,* **11** (1991) 628–637.

171 Hirakawa, M., Tamura, A., Nagashima, H., Nakayama, H., and Sano, K., Disturbance of retention of memory after focal cerebral ischemia in rats, *Stroke,* **25** (1994) 2471–2475.

172 Bederson, J. B., Pitts, L. H., Tsuji, M., Nishimura, M. C., Davis, R. L., and Bartowski, H., Rat middle cerebral artery occlusion: Evaluation of the model and development of a neurologic examination, *Stroke,* **17** (1986) 472–476.

173 El-Sabban, F., Reid, K. H., Zhang Y. P., and Edmonds, H. L., Stability of thrombosis induced by electrocoagulation of rat middle cerebral artery, *Stroke,* **25** (1994) 2241–2245.

174 Nagasawa, H. and Kogure, K., Correlation between cerebral blood flow and histologic examination in new rat model of middle cerebral artery occlusion, *Stroke,* **20** (1989) 1037–1043.

175 Zea Longa, E., Weinstein, P. R., Carlson, S., and Cummins, R., Reversible middle cerebral artery occlusion without craniectomy in rats. *Stroke,* **20** (1989) 84–91.

176 Memezawa, H., Minamisawa, H., Smith, M. L., and Siesjö, B. K., Ischemic penumbra in a model of reversible middle cerebral artery occlusion in the rat, *Exp. Brain Res.,* **89** (1992) 67–78.

177 Laing, R. J., Jakubowski, J., and Laing, R. W., Middle cerbral artery occlusion without craniectomy in rats: Whch method works best?, *Stroke,* **24** (1993) 294–298.

178 Kuge, Y., Minematsu, K., Yamaguchi, T., and Miyake, Y., Nylon monofilament for intraluminal middle cerebral artery occlusion in rats, *Stroke,* **26** (1995) 1655–1657.

179 Chen, H., Chopp, M., Vandelinde, A. M. Q., Dereski, M. O., Garcia, J. H., and Welch, K. M. A., The effects of post-ischemic hypothermia on the neuronal injury and brain metabolism after forebrain ischemia in the rat, *J. Neurol. Sci.,* **107** (1992) 191–198.

180 Watson, B. D., Dietrich, W. D., Busto, R., and Ginsberg, M. D., Phothochemically induced thrombotic stroke in rat brain, *Ann. Neurol.,* **14** (1983) 126.

181 Watson, B. D., Dietrich, W., Busto, R., Wachtel, M. S., and Ginsberg, M. D., Induction of reproducible brain infarction by photochemically initiated thorombosis, *Ann. Neurol.,* **17** (1985) 497–504.

182 Markgraf, C. G., Kraydieh, S., Prado, R., Watson, B. D., Dietrich, W. D., and Ginsberg, M. D., Comparative histopathologic consequences of photothrombotic occlusion of the distal middle cerebral artery in Sprague-Dawley and Wister rats, *Stroke,* **24** (1993) 286–293.

183 Yao, H., Markgraf, C. G., Dietrich, W. D., Prado, R., Watson, B. D., and Ginsberg, M. D., Glutamate antagonist MK-801 attenuates incomplete but not complete infarction in thrombotic distal middle cerebral artery occlusion in Wistar rats, *Brain Res.,* **642** (1994) 117–122.

184 Wester, P., Watson, B. D., Prado, R., and Dietrich, W. D., A photothrombotic 'ring' model of rat stroke-in-evolution displaying putative penumbral inversion, *Stroke,* **26** (1995) 444–450.

185 Huang, A. J.-W., Watson, B. D., Hernandez, E., and Tseng, S. C.-G., Induction of conjictival transdifferentiation on vascularized corneas by photothrombotic occluion of corneal neovascularization, *Ophthalmology,* **95** (1988) 228–235.

186 Kudo, M., Aoyama, A., Ichimori, S., and Fukunaga, N., An animal model of cerebral infarction. Homologous blood clot emboli in rats, *Stroke,* **13** (1982) 505–508.

187 Meden, P., Overgaard, K., Pedersen, H., and Boysen, G., The influence of body temperature on infarct volume and thrombolytic therapy in a rat embolic stroke model, *Brain Res.,* **647** (1994) 131–138.

188 Busto, R. and Ginsberg, M. D., Graded focal cerebral ischemia in the rat by unilateral carotid artery occlusion and elevated intracranial pressure: Hemodynamic and biochemical characterization, *Stroke,* **16** (1985) 466–476.

189 Bederson, J. B., Pitts, L. H., Germano, S. M., Nishimura, M. C., Davis, R. L., and Bartkowski, H. M., Evaluation of 2,3,5-triphenyltetrazolium chloride as a stain for detection and quantification of experimental cerebral infarction in rats, *Stroke,* **17** (1986) 1304–1308.

190 Liszczak, T. M., Hedley-Whyte, E. T., Adams, J. F., Han, D. H., Kolluri, U. S., Vacanti, F. X., Heros, R. C., and Zorvas, N. T., Limitation of tetrazolium salts in delineating infarcted brain, *Acta Neuropathol. (Berl.),* **65** (1984) 150–157.

191 Stowell, R. E., Effect on tissue volume of various methods of fixation, dehydration, and embedding, *Stain Technol.,* **16** (1941) 67–83.

192 Swanson, R. A., Morton, M. T., Tsao-Wu, G., Savalos, R. A., Davidson, C., and Sharp, F. R., A semiautomated method for measuring brain infarct volume, *J. Cereb. Blood Flow Metab.,* **10** (1990) 290–293.

193 Fujie, W., Kirino, T., Tomukai, N., Iwasawa, T., and Tamura, A., Progressive shrinkage of the thalamus following middle cerebral artery occlusion in rats, *Stroke,* **21** (1990) 1485–1488.

194 Tamura, A., Kirino, T., Sano, K., Takagi, K., and Oka, H., Atrophy of the ipsilateral substantia nigra folloing middle cerebral artery occlusion in the rat, *Brain Res.,* **510** (1990) 154–157.

195 Dietrich, W. D., Busto, R., Alonso, O., Globus, M. Y.-T., and Ginsberg, M. D., Intraischemic but not postischemic brain hypothermia protects chronically following global forcbrain ischcmia in rats, *J. Cereb. Blood Flow Metab.,* **13** (1993) 541–549.

196 Grabowski, M., Nordbirg, C., Brundin, P., and Johansson, B. B., Middle cerebral artery occlusion in the hypertensive and normotensive rat: a study of histopathology and behaviour, *J. Hypertens.* **6** (1988) 405–411.

197 Schallert, T., Hernandez T. D., and Barth T. M., Recovery of function after brain damage: severe and chronic disruption by diazepam, *Brain Res., 379* (1986) 140–146.

198 De Ryck, M., Van Reempts, J., Borgers, M., Wauquier, A., and Janssen, P. A., Photochemical stroke model: flunarizine prevents sensorimotor after neocortical infarcts in rats, *Stroke, 20* (1989) 1383–1390.

199 Ohlsson, A.-L. and Johansson, B. B., Environment influences functional outcome of cerebral infarction in rats, *Stroke, 26* (1995) 644–649.

200 Raisman, G., Neuronal plasticity in the septal nuclei of the adult rat, *Brain Res., 14* (1969) 25–48.

201 Tamura, A., Tomukai, N., Iwasawa T., Takagi, K., Olada, M., Fiujiwara, N., and Sano, K., Hemispheric differences in spatial memory after focal cerebral ischemia in rats, *J. Cereb. Blood Flow Metab., 15(Suppl. 1)* (1995) S312.

202 Ungerstedt, U. and Arbuthnott, G. W., Quantitative recording of rotational behaviour in rats after 6-hydroxy-dopamine lesions of the nigro-striatal dopamine system, *Brain Res., 24* (1970) 485–493.

203 Markgraf, C. G., Green, E. J., Hurwitz, B. E., Morikawa, E., Dietrich, W. D., McCabe, P. M., Ginsberg, M. D., and Schneiderman, N., Sensorimotor and cognitive consequences of middle cerebral artery occlusion in rats, *Brain Res., 575* (1992) 238–246.

204 Morris, R. G. M., Garrud, P., Rawlins, J. N. P., and O'Keefe, J., Place navigation impaired in rats with hippocampal lesions, *Nature, 297* (1982) 681–683.

205 del Zoppo, G. J., Schmid-Schönbein, G. W., Mori, E., Copeland, B. R., and Chang, C.-M., Polymorphonuclear leukocytes occlude capillaries following middle cerebral artery occlusion and reperfusion in baboons, *Stroke, 22* (1991) 1276–1283.

206 Stroemer, R. P., Kent, T. A., and Hulsebosch, C. E., Neocortical neural sprouting, synaptogenesis, and behavioral recovery after neocortical infaction in rats, *Stroke, 26* (1995) 2135–2144.

207 Jander, S., Kraemer, M., Schroeter, M., Witte, O. W., Stoll, G., Lymphocytic infiltration and expression of intercellular adhesion molecule-1 in photochemically induced ischemia of the rat cortex, *J. Cereb. Blood Flow Metab., 15* (1995) 42–51.

208 K. SB., Pathophysiology and treatment of focal cerebral ischemia. Part I: Pathophysiology, *J. Neurosurg., 77* (1992) 169–184.

209 K. SB., Pathophysiology and treatment of focal cerebral ischemia. Part II: mechanisms of damage and treatment, *J. Neurosurg., 77* (1992) 337–354.

210 Ginsberg, M. D., Emerging strategies for the treatment of ischemic brain injury. In Waxman SG. (ed.) *Molecular and Cellular Approaches to the Treatment of Neurological Disease*, Raven, New York, 1993, pp. 207–237.